THE THEORY OF MACHINES

The Principles of Mechanism; Elementary Mechanics of Machines

MAXWELL PRESS

THE THEORY OF MACHINES

The Principles of Mechanism; Elementary Mechanics of Machines

Robert W. Angus B.A.SC.,

Member of the American Society of Mechanical Engineers,
Professor of Mechanical Engineering, University of Toronto,
Toronto, Canada

MAXWELL PRESS

Chennai Trichy New Delhi

MAXWELL PRESS

This edition has been published in india by arrangement with Carson Books, UK

ISBN 978-93-5528-061-9

Maxwell Press
No. 44, Nallathambi Street,
Triplicane, Chennai 600 005

Printed and bound in India

MJP 1293

Publisher : C. Janarthanan

Publisher's Note

The legacy of a country is in its varied cultural heritage, historical literature, developments in the field of economy and science. The top nations in the world are competing in the field of science, economy and literature. This vast legacy has to be conserved and documented so that it can be bestowed to the future generation. The knowledge of this legacy is slowly getting perished in the present generation due to lack of documentation.

Keeping this in mind, the concern with retrospective acquiring of rare books has been accented recently by the burgeoning reprint industry. Maxwell Press is gratified to retrieve the rare collections with a view to bring back those books that were landmarks in their time.

In this effort, a series of rare books would be republished under the banner, "Maxwell Press". The books in the reprint series have been carefully selected for their contemporary usefulness as well as their historical importance within the intellectual. We reconstruct the book with slight enhancements made for better presentation, without affecting the contents of the original edition.

Most of the works selected for republishing covers a huge range of subjects, from history to anthropology. We believe this reprint edition will be a service to the numerous researchers and practitioners active in this fascinating field. We allow readers to experience the wonder of peering into a scholarly work of the highest order and seminal significance.

MAXWELL PRESS

PREFACE

The present treatise dealing with the Principles of Mechanism and Mechanics of Machinery is the result of a number of years' experience in teaching the subjects and in practising engineering, and endeavors to deal with problems of fairly common occurrence. It is intended to cover the needs of the beginner in the study of the science of machinery, and also to take up a number of the advanced problems in mechanics.

As the engineer uses the drafting board very freely in the solution of his problems, the author has devised graphical solutions throughout, and only in a very few instances has he used formulæ involving anything more than elementary trigonometry and algebra. The two or three cases involving the calculus may be omitted without detracting much from the usefulness of the book.

The reader must remember that the book does not deal with machine design, and as the drawings have been made for the special purpose of illustrating the principles under discussion, the mechanical details have frequently been omitted, and in certain cases the proportions somewhat modified so as to make the constructions employed clearer.

The phorograph or motion diagram has been introduced in Chapter IV, and appeared in the first edition for the first time in print. It has been very freely used throughout, so that most of the solutions are new, and experience has shown that results are more easily obtained in this way than by the usual methods.

As the second part of the book is much more difficult than the first, it is recommended that in teaching the subject most of the first part be given to students in the sophomore year, all of the second part and possibly some of the first part being assigned in the junior year.

The thanks of the author are due to Mr. J. H. Parkin for his careful work on governor problems, some of which are incorporated, and for assistance in proofreading; also to the various firms and others who furnished cuts and information, most of which is acknowledged in the body of the book.

The present edition has been entirely rewritten and enlarged and all of the previous examples carefully checked and corrected where necessary. The cuts have been re-drawn and many new ones added; further, the Chapter on Balancing is new. Questions at the end of each chapter have been added.

R. W. A.

UNIVERSITY OF TORONTO,
February, 1917.

CONTENTS

PART II

MECHANICS OF MACHINERY

CHAPTER XII

CHAPTER XIII

CHAPTER XIV

CHAPTER XV

CHAPTER XVI

SYMBOLS USED

The following are some of the symbols used in this book, with the meanings usually attached to them.

w = weight in pounds.

g = acceleration of gravity = 32.16 ft. per second per second.

m = mass = $\frac{w}{g}$

v = velocity in feet per second.

n = revolutions per minute.

ω = radians per second = $\frac{2\,\pi n}{60}$

π = 3.1416.

α = angular acceleration in radians per second per second.

θ = crank angle from inner dead center.

I = moment of inertia about the center of gravity.

k = radius of gyration in feet = $\sqrt{I/m}$

J = reduced inertia referred to primary link.

T = torque in foot-pounds.

P, P', P'' represent the point P and its images on the velocity and acceleration diagrams respectively.

PART I

THE PRINCIPLES OF MECHANISM

THE THEORY OF MACHINES

CHAPTER I

THE NATURE OF THE MACHINE

1. General.—In discussing a subject it is important to know its distinguishing characteristics, and the features which it has in common with other, and in many cases, more fundamental matters. This is particularly necessary in the case of the machine, for the problems connected with the mechanics of machinery do not differ in many ways from similar problems in the mechanics of free bodies, both being governed by the same general laws, and yet there are certain special conditions existing in machinery which modify to some extent the forces acting, and these conditions must be studied and classified so that their effect may be understood.

Again, machinery has recently come into very frequent use, and is of such a great variety and number of forms, that it deserves special study and consideration, and with this in mind it will be well to deal with the subject specifically, applying the known laws to the solution of such problems as may arise.

2. Nature of the Machine.—In order that the special nature of the machine may be best understood, it will be most convenient to examine in detail one or two well-known machines and in this way to see what particular properties they possess. One of the most common and best known machines is the reciprocating engine, (whether driven by steam or gas is unimportant) which consists of the following essential, independent parts: (*a*) The part which is rigidly fixed to a foundation or the framework of a ship, and which carries the cylinder, the crosshead guides, if these are used, and at least one bearing for the crankshaft, these all forming parts of the one rigid piece, which is for brevity called the **frame**, and which is always **fixed** in position. (*b*) The piston, piston rod and crosshead, which are also parts of one rigid piece, either made up of several parts screwed together as in large steam and gas engines, or of a single casting as in automobile engines, where the piston rod is entirely omitted and the crosshead is combined with the piston. It will be convenient to refer to this part as the piston, and it is to be noticed that the piston always moves relatively to the frame with a motion of translation,[1] and further always contains **the wristpin,** a round pin to facilitate connection with other parts. The piston then moves relatively to the frame and is so constructed as to **pair** with other parts of the machine such as the frame and connecting rod now to be described. (*c*) The **connecting rod** is the third part, and its motion is peculiar in that one end of it describes a circle while the other end, which is **paired** with the wristpin, moves in a straight line, which latter motion is governed by the piston. All points on the rod move in parallel planes, however, and it is said to have plane motion, as has also the piston. The purpose of the rod is to transmit the motion of the piston, in a modified form, to the remaining part of the machine, and for this purpose one end of it is bored out to fit the wristpin while the other end is bored out to fit a pin on the crank, which two pins are thus kept a fixed distance apart and their axes are always kept parallel to one another. (*d*) The fourth and last essential part is the crank and crankshaft, or, as it may be briefly called, the **crank.** This part also pairs with two of the other parts already named, the frame and the connecting rod, the crankshaft fitting into the bearing arranged for it on the frame and the **crankpin,** which travels in a circle about the crankshaft, fitting into the bored hole in tho connecting rod available for it. The **stroke** of the piston depends upon the radius of the crank or the diameter of the crankpin circle, and is equal to the latter diameter in all cases where the direction of motion of the piston passes through the center of the crankshaft. The flywheel forms part of the crank and crankshaft.

In many engines there are additional parts to those mentioned, steam engines having a valve and valve gear, as also do many internal-combustion engines, and yet a number of engines have no more than the four parts mentioned, so that these appear to be the only essential ones.

3. Lathe.—Another well-known machine may be mentioned, namely, the lathe. All lathes contain a fixed part or frame or

[1] By a motion of translation is meant that all points on the part considered move in parallel straight lines in the same direction and sense and through the same distance.

bed which holds the fixed or tail center, and which also contains bored bearings for the live center and gearshafts. Then there is the live center which rotates in the bearings in the frame and which drives the work, being itself generally operated by means of a belt from a countershaft. In addition to these parts there is the carriage which holds the tool post and has a sliding motion along the frame, the gears, the lead screw, belts and other parts, all of which have their known functions to perform, the details of which need not be dwelt upon.

4. Parts of the Machine.—These two machines are typical of a very large number and from them the definition of the machine may be developed. Each of these machines contains more **than one part,** and in thinking of any other machine it will be seen that it contains at least two parts: thus a crowbar is not a machine, neither is a shaft nor a pulley; if they were, it would be difficult to conceive of anything which was not a machine. The so-called "simple machines," the lever, the wheel and axle, and the wedge cause confusion along this line because the complete machine is not inferred from the name: thus the bar of iron cannot be called a lever, it serves such a purpose only when along with it is a fulcrum; the wheel and axle acts as a machine only when it is mounted in a frame with proper bearings; and so with the wedge. Thus a machine consists of a **combination of parts.**

5. Again, these parts must offer some resistance to change of shape to be of any value in this connection. Usually the parts of a machine are **rigid,** but very frequently belts and ropes are used, and it is well known that these serve their proper purpose only when they are in tension, because only when they are used in this way do they produce motion since they offer **resistance to change of shape.** No one ever puts a belt in a machine in a place where it is in compression. Springs are often used as in valve gears and governors, but they offer resistance wherever used. Thus the parts of a machine must be **resistant.**

6. Relative Motion.—Now under the preceding limitations a ship or building or any other **structure** could readily be included, and yet they are not called machines, in fact nothing is a machine in which the parts are incapable of motion with regard to one another. In the engine, if the frame is stationary, all the other parts are capable of moving, and when the machine is serving its true purpose they do move; in a bicycle, the wheels, chain, pedals, etc., all move relatively to one another, and in all machines the parts must have relative motion. It is to be borne in mind that all the parts do not necessarily move, and as a matter of fact there are very few machines in which one part, which is referred to briefly as the frame, is not stationary, but all parts must move relatively to one another. If one stood on the frame of an engine the motion of the connecting rod would be quite evident if slow enough; and if, on the other hand, one clung to the connecting rod of a very slow-moving engine the frame would appear to move, that is, the frame has a motion relative to the connecting rod, and *vice versa.*

7. In a bicycle all parts move when it is going along a road, but still the different parts have relative motion, some parts moving faster than others, and in this and in many other similar cases, the frame is the part on which the rider is and which has no motion relative to him. In case of a car skidding down a hill, all parts have exactly the same motion, none of the parts having relative motion, the whole acting as a solid body.

8. Constrained Motion.—Now considering the nature of the motion, this also distinguishes the machine. When a body moves in space its direction, sense and velocity depend entirely upon the forces acting on it for the time being, the path of a rifle ball depends upon the force of the wind, the attraction of gravity, etc., and it is impossible to make two of them travel over exactly the same path, because the forces acting continually vary; a thrown ball may go in an approximately straight line until struck by the batter when its course suddenly changes, so also with a ship, that is, in general, the path of a free body varies with the external forces acting upon it. In the case of the machine, however, the matter is entirely different, for the path of each part is predetermined by the designer, and he arranges the whole machine so that each part shall act in conjunction with the others to produce in each a perfectly defined path.

Thus, in a steam engine the piston moves in a straight line back and forth without turning at all, the crankpin describes a true circle, each point on it remaining in a fixed plane, normal to the axis of the crankshaft during the rotation, while also the motion of the connecting rod, although not so simple is perfectly definite. In judging the quality of the workmanship in an engine one watches to see how exact each of these motions is and how nearly it approaches to what was intended; for example, if a point on the crank does not describe a true circle in a fixed plane, or the crosshead does not move in a perfectly straight line the engine is not regarded as a good one.

The same general principle applies to a lathe; the carriage must slide along the frame in an exact straight line and the spindle must have a true rotary motion, etc., and the lathe in which these conditions are most exactly fulfilled brings the highest price.

These motions are fixed by the designer and the parts are arranged so as to constrain them absolutely, irrespective of the external forces acting; if one presses on the side of the crosshead its motion is unchanged, and if sufficient pressure is produced to change the motion the machine breaks and is useless. The carriage of the lathe can move only along the frame whether the tool which it carries is idle or subjected to considerable force due to the cutting of metal; should the carriage be pushed aside so that it would not slide on the frame, the lathe would be stopped and no work done with it till it was again properly adjusted. These illustrations might be multiplied indefinitely, but the reader will think out many others for himself.

This is, then, a distinct feature of the machine, that the relative motions of all parts are completely fixed and do not depend in any way upon the action of external forces. Or perhaps it is better to say that whatever external forces are applied, the relative paths of the parts are unaltered.

9. Purpose of the Machine.—There remains one other matter relative to the machine, and that is its purpose. Machines are always designed for the special purpose of doing work. In a steam engine energy is supplied to the cylinder by the steam from the boiler, the object of the engine is to convert this energy into some useful form of work, such as driving a dynamo or pumping water. Power is delivered to the spindle of a lathe through a belt, and the lathe in turn uses this energy in doing work on a bar by cutting a thread. Energy is supplied to the crank on a windlass, and this energy, in turn, is taken up by the work done in lifting a block of stone. Every machine is thus designed for the express purpose of doing work.

10. Definition of the Machine.—All these points may now be summed up in the form of a definition: **A machine consists of resistant parts, which have a definitely known motion relative to each other, and are so arranged that a given form of available energy may be made to do a desired form of work.**

11. Imperfect Machines.—Many machines approach a great state of perfection, as for example the cases quoted of the steam engine and the lathe, where all parts are carefully made and the motions are all as close to those desired as one could make them. But there are many others, which although commonly and correctly classed as machines, do not come strictly under the definition. Take the case of the block and tackle which will be assumed as attached to the ceiling and lifting a weight. In the ideal case the pulling chain would always remain in a given position and the weight should travel straight up in a vertical line, and in so far as this takes place the machine may be considered as serving its purpose, but if the weight swings, then motion is lost and the machine departs from the ideal conditions. Such imperfections are not uncommon in machines; the endlong motion of a rotor of an electrical machine, the "flapping" of a loose belt or chain, etc., are familiar to all persons who have seen machinery running; and even the unskilled observer knows that conditions of this kind are not good and are to be avoided where possible, and the more these incorrect motions are avoided, the more perfect is the machine and the more nearly does it comply with the conditions for which it was designed.

DIVISIONS OF THE SUBJECT

12. Divisions of the Subject.—It is convenient to divide the study of the machine into four parts:

1. A study of the motions occurring in the machine without regard to the forces acting externally; this study deals with the **kinematics of machinery.**

2. A study of the external forces and their effects on the parts of the machine assuming them all to be moving at uniform velocity or to be in equilibrium; the balancing forces may then be found by the ordinary methods of statics and the problems are those of **static equilibrium.**

3. The study of **mechanics of machinery** takes into account the mass and acceleration of each of the parts as well as the external forces.

4. The determination of the proper sizes and shapes to be given the various parts so that they may be enabled to carry the loads and transmit the forces imposed upon them from without, as well as from their own mass. This is **machine design,**

a subject of such importance and breadth as to demand an entirely separate treatment, and so only the first three divisions are dealt with in the present treatise.

KINDS OF MOTION

13. Plane Motion.—It will be best to begin on the first division of the subject, and to discuss the methods adopted for obtaining definite forms of motion in machines. In a study of the steam engine, which has already been discussed at some length, it is observed that in each moving part the path of any point always lies in one plane, for example, the path of a point on the crankpin lies on a plane normal to the crankshaft, as does also the path of any point on the connecting rod, and also the path of any point on the crosshead. Since this is the case, the parts of an engine mentioned are said to have **plane motion**, by which statement is simply meant that the path of any point on these parts always lies in one and the same plane. In a completed steam engine with slide valve, all parts have plane motion but the governor balls, in a lathe all parts usually have plane motion, the same is true of an electric motor and, in fact, the vast majority of the motions with which one has to deal in machines are plane motions.

14. Spheric Motion.—There are, however, cases where different motions occur, for example, there are parts of machines where a point always remains at a fixed distance from another fixed point, or where the motion is such that any point will always lie on the surface of a sphere of which the fixed point is the center, as in the universal and ball and socket joints. Such motion is called **spheric motion** and is not nearly so common as the plane motion.

15. Screw Motion.—A third class of motions occurs where a body has a motion of rotation about an axis and also a motion of translation along the axis at the same time, the motion of translation bearing a fixed ratio to the motion of rotation. This motion is called **helical** or **screw motion** and occurs quite frequently.

In the ordinary monkey wrench the movable jaw has a plane motion relative to the part held in the hand, the plane motion being one of translation or sliding, all points on the screw have plane motion relative to the part held, the motion being one of rotation about the axis of the screw, and the screw has a helical motion relative to the movable jaw, and *vice versa.*

PLANE CONSTRAINED MOTION

It has been noticed already that plane motion is frequently constrained by causing a body to rotate about a given axis or by causing the body to move along a straight line in a motion of translation, the first form of motion may be called **turning motion,** the latter form **sliding motion.**

16. Turning Motion.—This may be constrained in many ways and Fig. 1 shows several methods, where a shaft runs in a fixed bearing, this shaft carrying a pulley as shown in the upper left

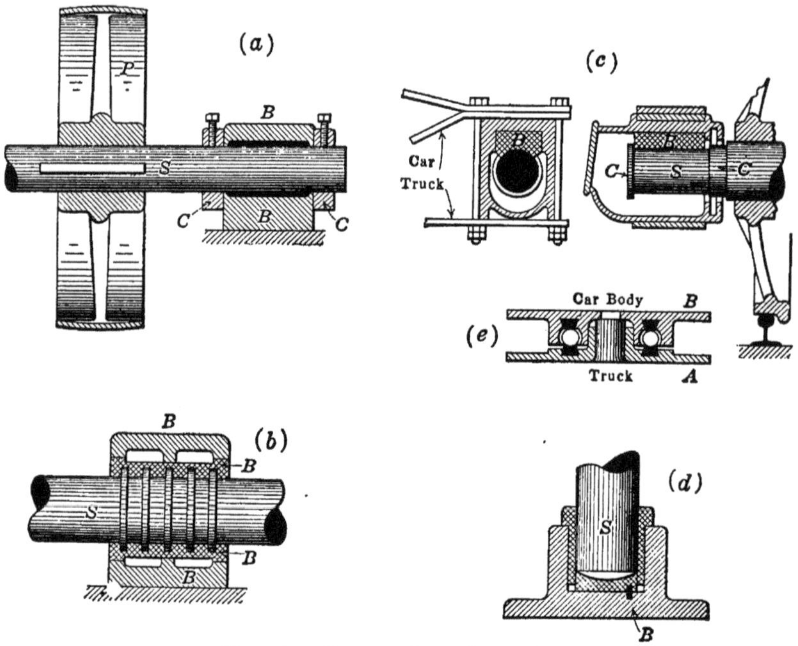

FIG. 1.—Forms of turning pairs.

figure, while the lower left figure shows a thrust bearing for the propeller shaft of a boat. In the figure (*a*), there is a pulley *P* keyed to a straight shaft S which passes through a bearing *B*, and if the construction were left in this form it would permit plane turning motion in the pulley and shaft, but would not constrain it, as the shaft might move axially through *B*. If, however, two collars *C* are secured to the shaft by screws as shown, then these collars effectually prevent the axial motion and make only pure turning possible. On the propeller shaft at (*b*) the collars *C* are forged on the shaft, a considerable number being used on account of the great force tending to push the shaft axially. Thus in both cases the relative turning motion is necessitated by the two bodies, the shaft with its collars forming one and the bearing the other, and these together are called a **turning pair** for obvious reasons, the pair consisting of two **elements.**

It is evident that the turning pair may be arranged by other constructions such as those shown on the right in Fig. 1, the form used depending upon circumstances. The diagram (*c*) shows in outline the method used in railroad cars, the bearing coming in contact with the shaft only for a small part of the circumference of the latter, the two being held in contact purely because of the connection to the car which rests on top of *B*, and the collars *C* are here of slightly different form. At (*d*) is a vertical bearing which, in a somewhat better form is often used in turbines, but here again it is only possible to insure turning motion provided the weight is on the vertical shaft and presses it into *B*. In this case there is only one part corresponding to the collar *C*, which is the part of *B* below the shaft. At (*e*) is a ball bearing used to support a car on top of a truck, the weight of the car holding the balls in action.

17. Chain and Force Closure.—In the cases (*a*) and (*b*), turning motion will take place by construction, and is said to be secured by **chain closure,** which will be referred to later, while in the cases (*c*), (*d*) and (*e*) the motion is only constrained so long as the external forces act in such a way as to press the two elements of the pair together, plane motion being secured by **force closure.** In cases, such as those described, where force closure is permissible, it forms the cheaper construction, as a general rule.

18. Sliding Motion.—The **sliding pair** also consists of two elements, and if a section of these elements is taken normal to the direction of sliding the elements must be non-circular. As in the previous case the sliding pair in practice has very many forms, a few of which are shown in Fig. 2, (*a*), (*b*), (*c*) and (*d*) being forms in common use for the crossheads of steam engines, (*b*) and (*c*) being rather cheaper in general than the others. At (*e*), (*f*) and (*g*) are shown forms which are used in automobile change gears and other similar places where there is little sliding; (*e*) consists of a gear with a long keyway cut in it while the other element has a parallel key, or "feather," fastened to it, so that the outer element may slide along the shaft but cannot rotate upon it. The construction of the forms (*e*) and (*f*) is evident. The reader will see very many forms of this pair in machines and should study them carefully.

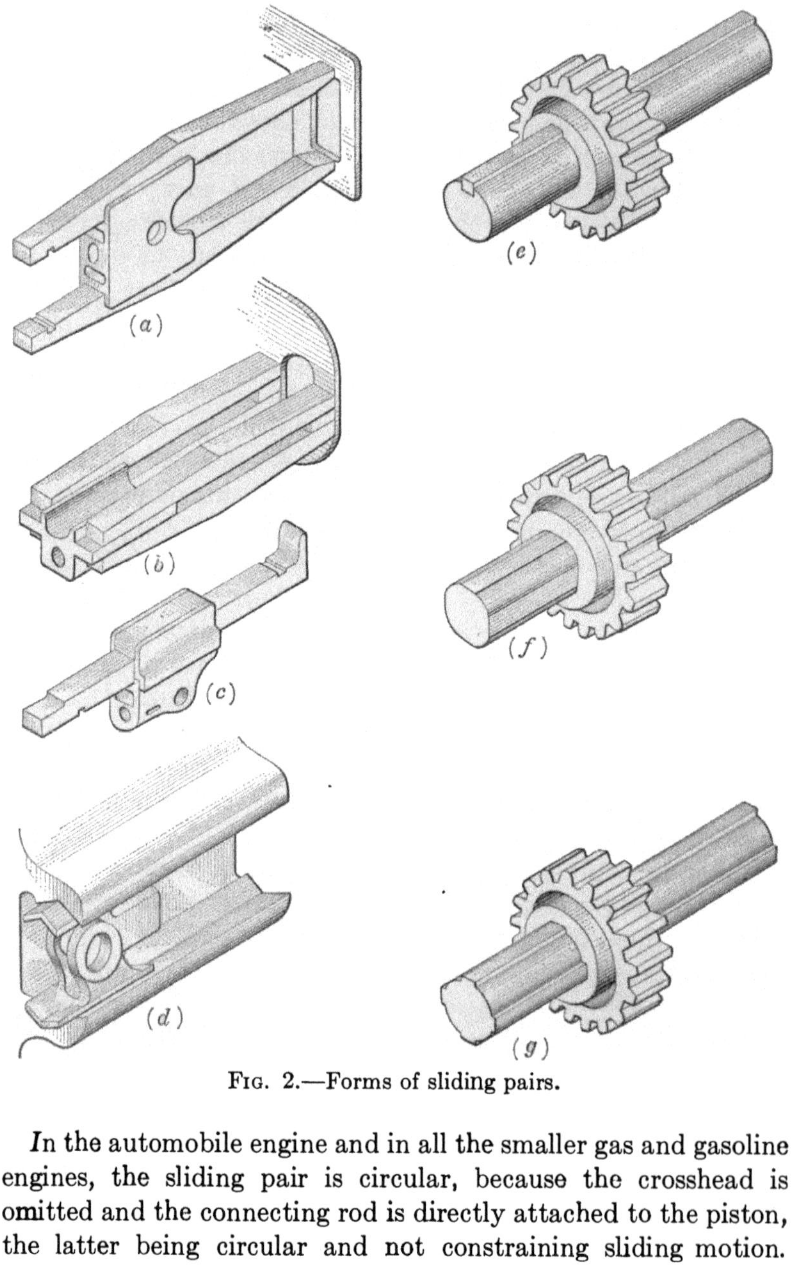

Fig. 2.—Forms of sliding pairs.

In the automobile engine and in all the smaller gas and gasoline engines, the sliding pair is circular, because the crosshead is omitted and the connecting rod is directly attached to the piston, the latter being circular and not constraining sliding motion.

In this case the sliding motion is constrained through the connecting rod, which on account of the pairing at its two ends will not permit the piston to rotate. The real sliding pair, of course, consists of the cylinder and piston, both of which are circular, and constrainment is by force closure.

In the case of sliding pairs also it is possible to have chain closure where constraint is due to the construction, as in the cases illustrated in Fig. 2; in these cases the motion being one of sliding irrespective of the directions of the acting forces. Fre-

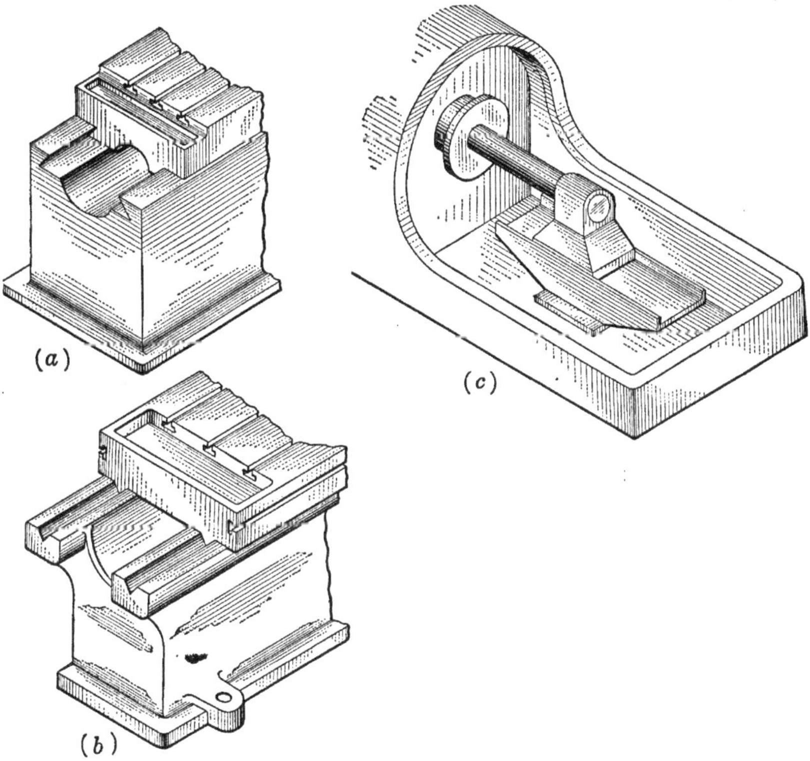

FIG. 3.—Sliding pairs.

quently, however, force closure is used as in the case (b) shown at Fig. 3 which represents a planer table, the weight of which alone keeps it in place. Occasionally through an accident the planer table may be pushed out of place by a pressure on the side, but, of course, the planer is not again used until the table is replaced, for the reason that the design is such that the table is only to have plane motion, a condition only possible if the table rests in the grooves in the frame. In Fig. 3 (a) the same table is constrained by chain closure and the tail sliding piece of the piston rod in Fig. 3 (*c*) by force closure as is evident.

19. Lower and Higher Pairs.—The two principal forms of plane constrained motion are thus turning and sliding, these motions being controlled by turning and sliding pairs respectively, and each pair consisting of two elements. Where contact between the two elements of a pair is over **a surface** the pair is called a **lower pair,** and where the contact is only **along a line or at a point,** the pair is called a **higher pair.** To illustrate this the ordinary bearing may be taken as a very common example of lower pairing, whereas a roller bearing has line contact and a ball bearing point contact and are examples of higher pairing, these illustrations are so familiar as to require no drawings. The

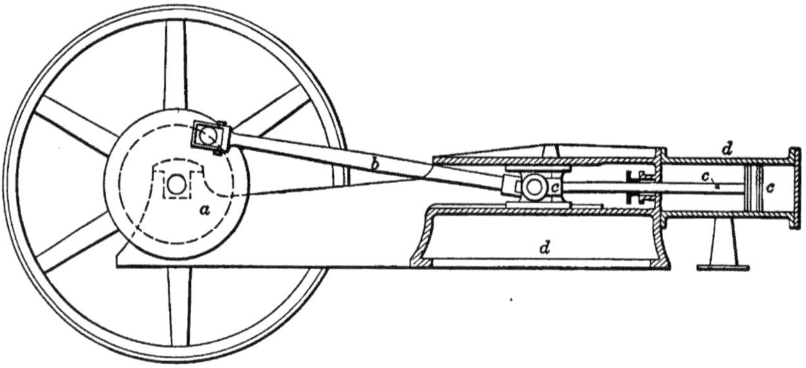

Fig. 4.

contact between spur gear teeth is along a line and therefore an example of higher pairing.

In general, the lower pairs last longer than the higher, because of the greater surface exposed for wear, but the conditions of the problem settle the type of pairing. Thus, lower pairing is used on the main shafts of large engines and turbines, but for automobiles and bicycles the roller and ball bearings are common.

MACHINES, MECHANISMS, ETC.

20. Formation of Machines.—Returning now to the steam engine, Fig. 4, its formation may be further studied. The valve gear and governor will be omitted at present and the remaining parts discussed, these consist of the crank, crankshaft and flywheel, the connecting rod, the piston, piston rod and crosshead, and finally the frame and cylinder. Taking the connecting rod *b* it is seen to contain two turning elements, one at either end, and the real function of the metal in the rod is to keep these two elements parallel and at a fixed distance apart. The crank and crankshaft *a* contains two turning elements, one of which is paired with one of the elements on the connecting rod *b*, and forms the crankpin, and the other is paired with a corresponding element on the frame *d*, forming the main bearing. It is true that the main bearing may be made in two parts, both of which are made on the frame, as in center-crank engines, or one of which may be placed as an **outboard bearing,** but it will readily be understood that this division of the bearing is only a

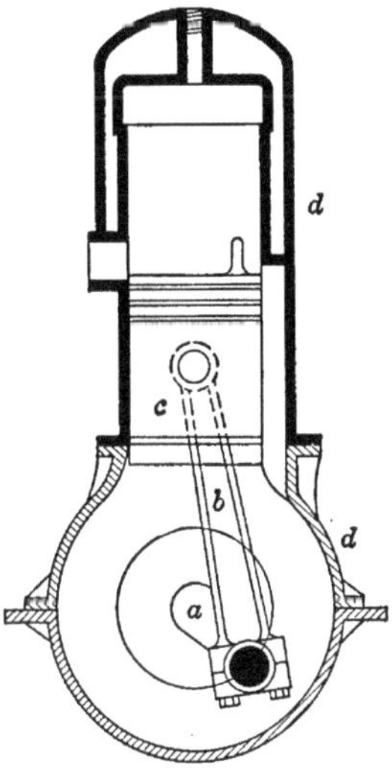

FIG. 5.—Two-cycle gasoline engine.

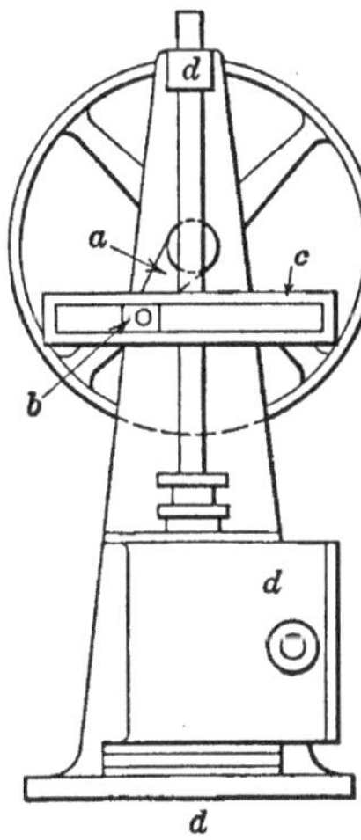

FIG. 6.

matter of practical convenience, for it is quite conceivable that the bearing might be made in one piece, and if this piece were long enough it would serve the purpose perfectly. Thus the crank consists essentially of two turning elements properly connected.

Again, the frame *d* contains the outer element of a turning pair, of which the inner element is the crankshaft, and it also contains a sliding element which is usually again divided into two parts for the purpose of convenience in construction, the parts being the crosshead guides and the cylinder. But the two parts are not absolutely essential, for in the single-acting gasoline engine, the guides are omitted and the sliding element is entirely in the cylinder. Of course, the shape of the element depends upon the purpose to which it is put; thus in the case last referred to it is round.

Then, there is the crosshead *c*, with the turning element pairing with the connecting rod and the sliding element pairing with the sliding element on the frame. The sliding element is usually in two parts to suit those of the frame, but it may be only in one if so desired and conditions permit of it (see Fig. 2).

Thus, the steam engine consists of four parts, each part containing two elements of a pair, in some cases the elements being for sliding, and in others for turning.

Again, on examining the small gasoline engine illustrated in Fig. 5, it will be seen that the same method is adopted here as in the steam engine, but the crosshead, piston and piston rod are all combined in the single piston *c*. Further, in the Scotch yoke, Fig. 6, a scheme in use for pumps of small sizes as well as on fire engines of some makes and for other purposes, there is the crank *a* with two turning elements, the piston and crosshead *c* with two sliding elements, and the block *b*, and the frame *d*, each with one turning element and one sliding element.

21. Links and Chains.—The same will be found true in all machines having plane motion; each part contains at least two elements, each of which is paired with corresponding elements on the adjacent parts. For convenience each of these parts of the machine is called a **link,** and the series of links so connected as to give a complete machine is called a **kinematic chain,** or simply a chain. It must be very carefully borne in mind that if a kinematic chain is to form part of a machine or a whole machine, then all the links must be so connected as to have definite relative motions, this being an essential condition of the machine.

In Fig. 7 three cases are shown in which each link has two turning elements. Case (*a*) could not form part of a machine because the three links could have no relative motion whatever, as is evident by inspection, while at (*b*) it would be quite impossible to move any link without the others having corresponding changes of position, and for a given change in the relative positions of two of the links a definite change is produced in the others. Looking next at case (*c*), it is observed at once that both *DC* and *OD* could be secured to the ground and yet *AB*, *BC*, and *OA* moved, that is a definite change in *AB* produces no necessary change in *OD* or in *CD*, or one link may move without all the others undergoing motion or relative change of position. Such an arrangement could not form part of a machine because the relative motions of the parts are not fixed but variable according to conditions. At (*d*) is a chain which can be used, because if any one link move relatively to any other, all the links move relatively, or if one link, say *OD*, is fastened to the ground and *OA* moved, then must all the other links move.

22. Mechanisms.—When a chain is used as a machine, usually one of the links acts as the frame and is fixed to a foundation or other stationary body. In studying the motions of various links it is not necessary to know the exact shape of the links at all,

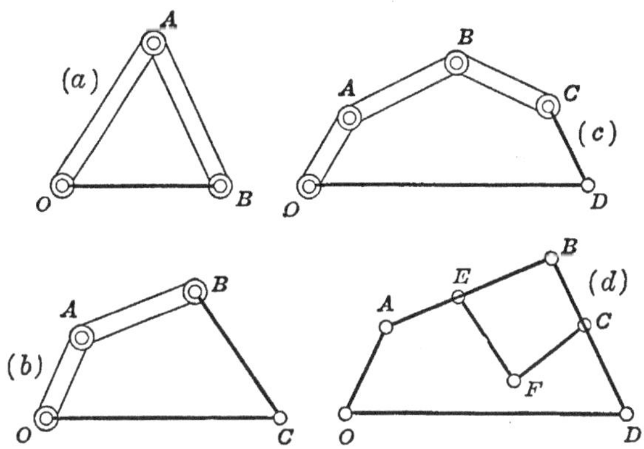

FIG. 7.

for the motion is completely known if the location and form of the pairs of elements is known. Thus, the actual link may be replaced by a straight bar which connects the elements of the link together, and it will always be assumed that this bar never changes its shape or length during motion. Thus, the chain will be represented by straight lines and a chain so represented having the relative motions of all links completely constrained and having one link fixed will be called a **mechanism.**

23. Simple and Compound Chains.—If the links of a chain have only two elements each, the chain is said to be **simple,** but if any link has three or more elements, as *AB* or *BD*, in Fig. 7 (*d*), the chain is **compound.**

24. Inversion of the Chain.—Since in forming a mechanism one link of the chain is fixed, it would appear that since any of the links may be fixed in a given chain, it may be possible to change the nature of the resulting mechanism by fixing various links successively. Take as an example the mechanism shown at (1) Fig. 8, *d* being the fixed link; here *a* would describe a circle, *c* would swing about *C* and *b* would have a pendulum motion, but with a moving pivot *B*. If *b* is fixed instead of *d*, *a* still rotates, *c* swings about *B* and *d* now has the motion *b* originally had, or the mechanism is unchanged.

If *a* is fixed then the whole mechanism may rotate, *b* and *d* rotating about *A* and *O* respectively as shown, and *c* also rotating, the form of the mechanism being thus changed to one in which all the links rotate. If, on the other hand, *c* is fixed, then none of the links can rotate, but *b* and *d* simply oscillate about *B* and *C* respectively. The reader will do well to make a cardboard model to illustrate this point.

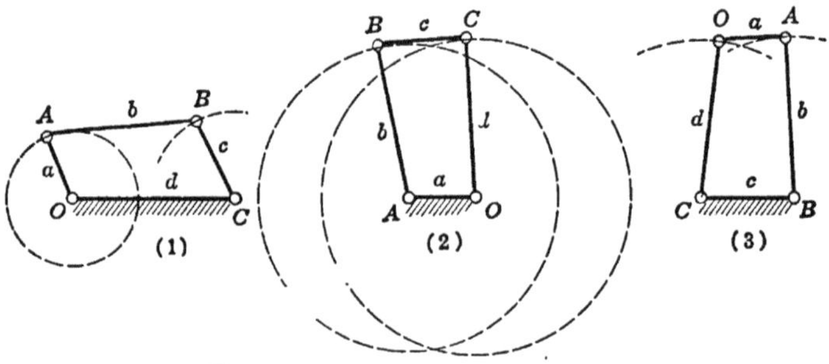

FIG. 8.—Inversion of the chain.

The process by which the nature of the mechanism is altered by changing the fixed link is called inversion **of the chain,** and in general, there are as many mechanisms as there are links in the chain of which it is composed, although in the above illustration there are only three for the four links.

25. Slider-crank Chain.—This inversion of the chain is very well illustrated in case of the chain used in the steam engine, which will be referred to in future as the **slider-crank chain.** The mechanism is shown in Fig. 9 with the crank *a*, connecting rod *b* and piston *c*, the latter having one sliding and one turning element and representing the reciprocating masses, *i.e.*, piston, piston rod and crosshead. The frame *d* is represented by a straight line and although it is common, yet the line of motion of *c* does not always pass through *O*; however, as shown at (1), it represents the usual construction for the ordinary engine. If now, instead of fixing *d*, *b* is fastened to the foundation, *b* being the longer of the two links containing the two turning elements, then *a* still rotates, *c* merely swings about *Q* and *d* has a swinging and sliding motion, and if *c* is a cylinder and a piston is attached

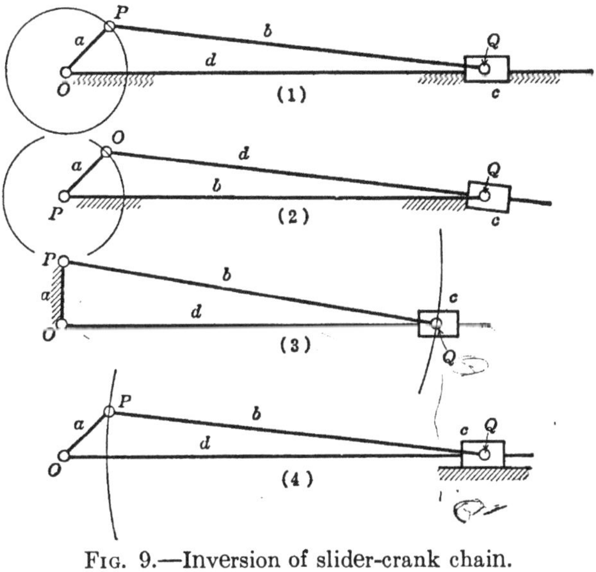

FIG. 9.—Inversion of slider-crank chain.

to *d* the result is the oscillating engine as shown at (2) Fig. 9, and drawn in some detail in Fig. 10.

If instead of fixing the long rod *b* with the two turning elements. the shorter rod *a* is fixed as shown at (3), then *b* and *d* revolve

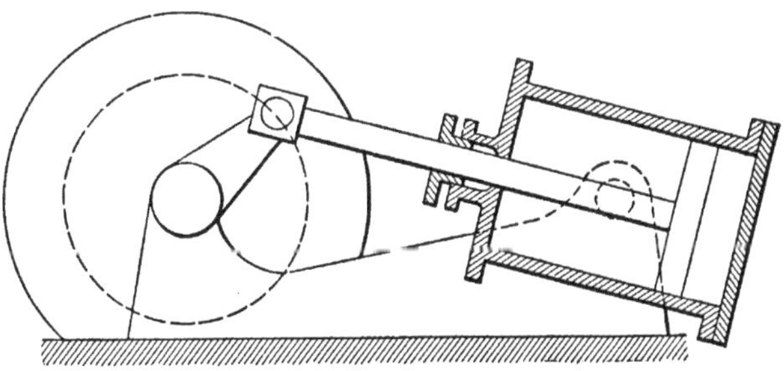

FIG. 10.—Oscillating engine.

about *P* and *O* respectively, and *c* also revolves sliding up and down on *d*. If *b* is driven by means of a belt and pulley at constant speed, then the angular velocity of *d* is variable and the device may be used as a quick-return motion; in fact, it is employed in the Whitworth quick-return motion. The practical form is also shown, Fig. 11, and the relation between the mechanism and the actual machine will be readily discovered with the help of the same letters.

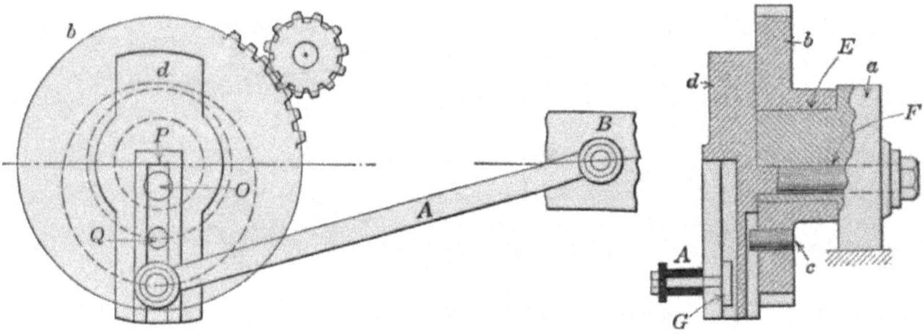

FIG. 11.—Whitworth quick-return motion.

*I*n the Whitworth quick-return motion, Fig. 11, the pinion is driven by belt and meshes with the gear *b*. The gear rotates on a large bearing *E* attached to the frame *a* of the machine, and through the bearing *E* is a pin F, to one side of the center of *E*,

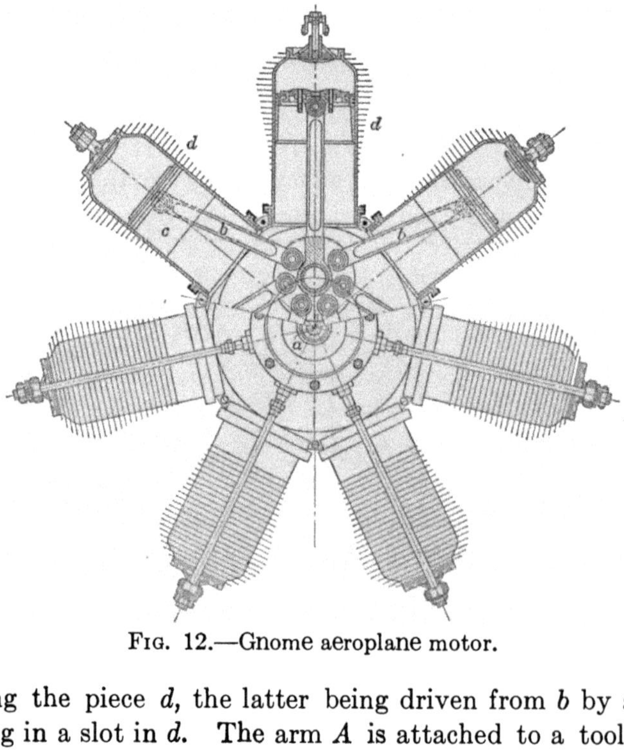

FIG. 12.—Gnome aeroplane motor.

carrying the piece *d*, the latter being driven from *b* by a pin *c* working in a slot in *d*. The arm *A* is attached to a tool holder at *B*.

The Gnome motor used on aeroplanes is also an example of this same inversion. It is shown in Fig. 12 and the cylinder shown at the top with its rod and piston form the same mechanism as the Whitworth quick-return motion, *a* being the link between the shaft and lower connecting-rod centers. Study the mechanism used with the other cylinders.

The fourth inversion found by fixing *c* is rarely used though it is found occasionally. It is shown at (4) Fig. 9.

There are thus four inversions of this chain and it might be further changed slightly by placing *Q* to one side of the link *d*

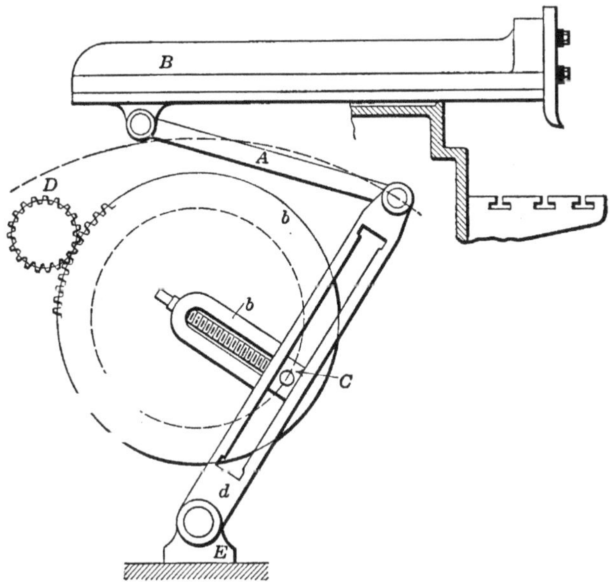

FIG. 13.—Shaper mechanism.

so that the line of motion of *Q*, Fig. 9 (1), passses above *O*, giving the scheme used in operating the sleeves in some forms of gasoline engines, etc.

A somewhat different modification of the slider-crank chain is shown at Fig. 13 a device also used as a quick-return motion in shapers and other machines. On comparing it with the Whitworth motion shown at Fig. 11, and the engine shown at Fig. 10, it is seen that the mechanism of Fig. 9 may be somewhat altered by varying the proportions of the links. The mechanism illustrated at Fig. 13 should be clear without further explanation. *D* is the driving pinion working in with the large gear *b*, the tool is attached to *B* which is driven from *c* by the link *A*. It is readily seen that *B* moves faster in one direction than the other. Further, an arrangement is made for varying the stroke of *B* at pleasure by moving the center of *c* closer to, or further from, that of *b*.

26. Double Slider-crank Chain.—A further illustration of a chain which goes through many inversions in practice is given in Fig. 14 and contains two links, *b* and *d*, with one sliding and one turning element each, also one link *a* with two turning elements and one *c* with two sliding elements. When the link *d* is fixed, *c* has a reciprocating motion and such a setting is frequently used for small pumps driven by belt through the crank *a* (Fig. 14), *c* being the plunger. A detail of this has already been given in Fig. 6.

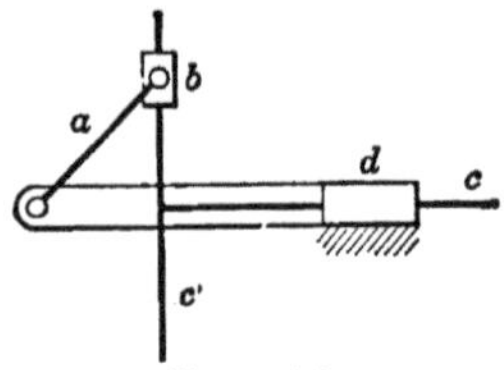

Fig. 14.

With *a* fixed the device becomes Oldham's coupling which is used to connect two parallel shafts nearly in line, Fig. 15. In the figure *b* and *d* are two shafts which are parallel and rotate about fixed axes. Keyed to each shaft is a half coupling with a slot running across the center of its face and between these half couplings is a peice *c* with two keys at right angles to each other, one on each side, fitting in grooves in *b* and *d*. As *b* and *d*

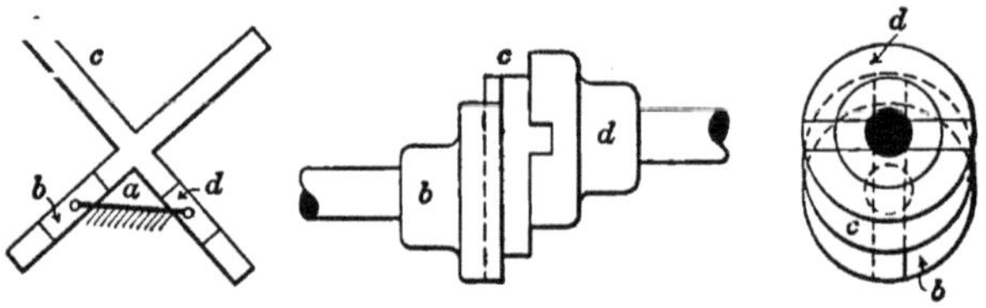

Fig. 15.—Oldham's coupling.

revolve, *c* works sideways and vertically, both shafts always turning at the same speed. Points on *c* describe ellipses and a modification of the device has been used on elliptical chucks and on instruments for drawing ellipses.

QUESTIONS ON CHAPTER I

1. Define the term machine and show that a gas engine, a stone crusher and a planer are machines. Is a plough or a hay rake or hay fork a machine? Why?

2. What are the methods of constrainment employed in the following:

Line shafting, loose pulley, sprocket chain, engine crankshaft, lathe spindle, eccentric sheave, automobile clutch, change gear, belt. Which are by force and which by chain closure?

3. Make a classification of the following with regard to constrainment and the form of closure: Gas-engine piston, lathe carriage, milling-machine head, ordinary *D*-slide valve, locomotive crosshead, valve rod, locomotive link. Give a sketch to illustrate each. Why would force closure not do for a connecting rod?

4. What form of pairing is used in the cases given in the above two questions? Is lower or higher pairing used in the following, and what is the type of contact: Roller bearing, ball bearing, vertical-step bearing, cam and roller in sewing machine, gear teeth, piston?

5. Define plane, helical and spherical motion. What form is used in the parts above mentioned, and in a pair of bevel gears?

6. In helical motion if the pitch of the helix is zero, what form of motion results; also what form for infinite pitch?

7. What is the resulting form of motion if the radius for a spherical motion becomes infinitely great?

8. Show that all the motions in an ordinary engine but that of the governor balls are plane. What form of motion do the latter have?

9. Define and illustrate the following terms: Element, lower pair, higher pair, link, chain, mechanism and compound chain.

10. List the links and their elements and give the form of motion and method of constrainment in the parts of a locomotive side rod, beam engine, stone crusher (Fig. 95) and shear (Fig. 94).

11. Explain and illustrate the inversion of the chain. Show that the epicyclic gear train is an inversion of the ordinary train.

CHAPTER II

MOTION IN MACHINES

27. Plane Motion.—It is now desirable to study briefly certain of the characteristics of plane motion, a term which may be defined by stating that a body has plane motion when it moves in such a way that any given point in it always remains in one and the same plane, and further, that the planes of motion of all points in the body are parallel. Thus, if any body has plane motion relative to the paper, then any point in the body must remain in a plane parallel to the plane of the paper during the motion of the body.

A little consideration will show that in the case of plane motion the location of a body is known when the location of any line in the body is known, provided this line lies in a plane parallel to the plane of motion or else in the plane of motion itself. The explanation is, that since all points in the body have plane motion, then the projection of the body on the plane is always the same for all positions and hence the line in it simply locates the body. For example, if a chair were pushed about upon the floor and had points marked R and L upon the bottoms of two of the legs, then the location of the chair is always known if the positions of R and L, that is, of the (imaginary) line RL is known. If, however, the chair were free to go up and down from the floor it would be necessary to know the position of the projection of RL on the floor and also the height of the line above the floor at any instant. Further, if it were possible for the chair to be tilted backward about the (imaginary) line RL, the position of the latter would tell very little about the position of the chair, as the tips of its legs might be kept stationary while tilting the chair back and forth, the position of RL being the same for various angular positions of the chair.

If the case where a body has not plane motion is considered, then the line will tell very little about the position of the body. In the case of an airship, for example, the ship may stand at various angles about a given line, say the axis of a pair of the wheels, the ship dipping downward or rising at the will of the operator.

28. Motion Determined by that of a *Line*.—Since the location of a body having plane motion is known when the location of any line in the body is known, then the motion of the body will be completely known, if the motion of any line in the body is known. Thus let C, Fig. 16, represent the projection on the plane of the paper of any body having plane motion, AB being any line in this body, and let AB be assumed to be in the plane of the paper, which is used as the plane of reference. Suppose now it is known that while C moves to C', the points A and B move over the paths AA' and BB', then the motion of C during the change is completely known. Thus at some intermediate position the line is at A_1B_1 and the figure of C can at once be drawn about this line, and this locates the position of the body corresponding to the location A_1B_1 of the line AB. It will therefore follow that the motion of a body is completely known provided only that the motion of any line in the body is known. This proposition is of much importance and should be carefully studied and understood.

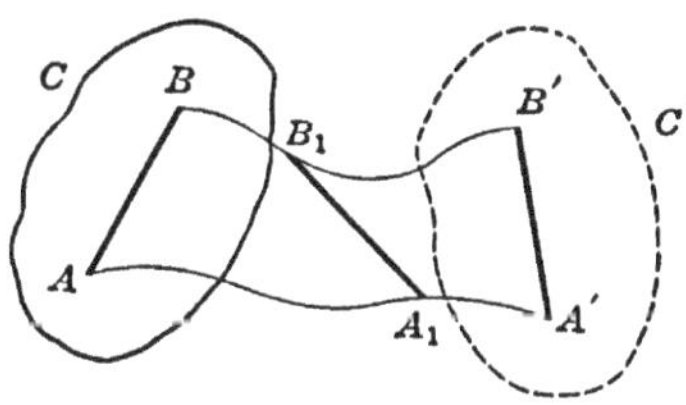

FIG. 16.

29. Relative Motion.—It will be necessary at this point to grasp some idea of the meaning of relative motion. We have practically no idea of any other kind of motion than that referred to some other body which moves in space, the moon is said to move simply because it changes its position as seen from the earth, or a train is said to move as it passes people standing on a railroad crossing. Again, one sees passengers in a railroad car as the train moves out and says they are moving, while each passenger in turn looks at other passengers sitting in the same car and says the latter are still. Again, a brakeman may walk backward on a flat car at exactly the same rate as the car goes forward, and a person on the ground who could just see his head would say he was stationary, while the engine driver would say he was moving at several miles per hour. If one stood on shore and saw a ship go out one would say that the funnel was moving, and yet a person on the ship would say that it was stationary.

These conflicting statements, which are, however, very common, would lead to endless confusion unless the essential differences in the various cases were grasped, and it will be seen that the real difference of view results from the fact that different persons have **entirely different standards** of comparison. Standing on the ground the standard of rest is the earth, and anything that moves relative to it is said to be moving. The man on the flat car would be described as stationary because he does not move with regard to the chosen standard—the earth, but the engine driver would be thinking of the train, and he would say the man moved because he moved relative to his standard—the train. It is easy to multiply these illustrations indefinitely, but they would always lead to the same result, that whether a body moves or remains at rest depends altogether upon the standard of comparison, and it is usual to say that a body is at rest when it has the same motion as the body on which the observer stands, and that it is in motion when its motion is different to that of the body on which the observer stands. On a railroad train one speaks of the poles flying past, whereas a man on the ground says they are fixed.

30. Absolute and Relative Motion.—When the standard which is used is the earth it is usual to speak of the motions of other bodies as **absolute** (although this is incorrect, for the earth itself moves) and when any standard which moves on the earth is used, the motions of the other bodies are said to be **relative. Thus the absolute motion of a body is its motion with regard to the earth, and the relative motion is the motion as compared with another body which is itself moving on the earth.** Unless these ideas are fully appreciated the reader will undoubtedly meet with much difficulty with what follows, for the notion of relative motion is troublesome.

In this connection it should be pointed out that a body secured to the earth may have motion relative to another body which is not so secured. Thus when a ship is coming into port the dock appears to move toward the passengers, but to the person on shore the ship appears to come toward the shore, thus the motion of the ship relative to the dock is equal and opposite to the motion of the dock relative to the ship.

31. Propositions Regarding Relative Motion.—Certain propositions will now be self-evident, the first being that if two bodies have no relative motion they have the same motion relative to every other body. Thus, two passengers sitting in a train have no relative motion, or do not change their positions relative to one another, and hence they have the same motion or cnange of position relative to the earth, or to another train or to any other body: the converse of this proposition is also true, or two bodies which have the same change of position relative to other bodies have no relative motion.

32. Another very important proposition may be stated as follows: The relative motions of two bodies are not affected by any motion which they have in common. Thus the motion of the connecting rod of an engine relative to the frame is the same whether the engine is a stationary one, or is on a steamboat or locomotive, simply because in the latter cases the motion of the locomotive or ship is common to the connecting rod and frame and does not affect their relative motions.

The latter proposition leads to the statement that if it be desired to study the relative motions in any machine it will not produce any change upon them to add the same motion to all parts. For example, if a bicycle were moving along a road it would be found almost impossible to study the relative motions of the various parts, but it is known that if to all parts a motion be added sufficient to bring the frame to rest it will not in any way affect the relative motions of the parts of the bicycle. Or if it be desired to study the motions in a locomotive engine, then to all parts a common motion is added which will bring one part, usually the frame, to rest relatively to the observer, or to the observer and to all parts of the machine such a motion is added as to bring him to rest relative to them, in fact, he stands upon the engine, having added to himself the motion which all parts of the engine have in common. So that, whenever it is found necessary to study the motions of machines all parts of which are moving, it will always be found convenient to add to the observer the common motion of all the links, which will bring onc of them to rest, relative to him.

To give a further illustration, let two gear wheels a and b run together and turn in opposite sense about fixed axes. Let a run at $+ 50$ revolutions per minute, and b at $- 80$ revolutions per minute; it is required to study the motion of b relative to a. To do this add to each such a motion as to bring a to rest, that is, $- 50$ revolutions per minute, the result being that a turns $+ 50 - 50 = 0$, while b turns $- 80 - 50 = - 130$ revolutions per minute or b turns relative to a at a speed of 130 revolutions per minute and in opposite sense to a. Here there has simply been added to each wheel the same motion, which does not affect their relative motions but has the effect of bringing one of the wheels to rest. To find the motion of a relative to b, bring b to rest by adding $+ 80$ revolutions per minute, so that a goes $+ 50 + 80 = 130$ revolutions per minute, or the motion of b relative to a is equal and opposite to that of a relative to b.

33. The Instantaneous or Virtual Center.—It has already been pointed out in Sec. 27 that when a body has plane motion, the motion of the body is completely known provided the motion of any line in the body in the plane of motion is known, that is, provided the motions or paths of any two points in the body are known. Now let c, Fig. 17, represent any body moving in the plane of the paper at any instant, the line AB being also in the plane of the paper, and let FA and BE represent short lengths of the paths of A and B respectively at this instant. The direction of motion of A is tangent to the path FA at A, and that of B is tangent to the path BE at B, the paths of A and B giving at once the direction of the motions of these points at the instant. Through A draw a normal AO to the direction of motion of A, then, if a pin is stuck through any point on the line AO into the plane of reference and c is turned very slightly about the pin it will give to A the direction of motion it actually has at the instant. The same argument applies to BO a normal to the path at B, and hence to the point O where AO and BO intersect, that is, if a pin is put through the point O in the body c and into the plane of reference, where the body is in the position shown, the actual motion of the body is the same as if it rotated for an instant about this pin. O is called the instantaneous or virtual center, because it is the point in the body c about which the latter is virtually turning, with regard to the paper, at the instant.

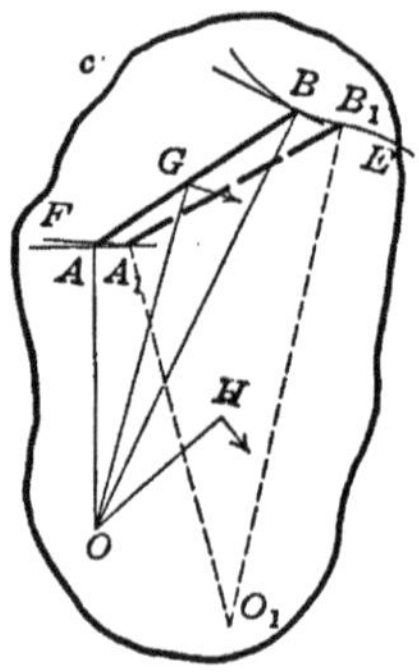

FIG. 17.

In going over this discussion it will appear that O may be found provided only the directions of motion of A and B are known at the instant. The only purpose for which the paths of these points have been used was to get the directions in which A and B were moving at the instant, and the actual path is unimportant in so far as the finding of O is concerned. It will further appear that the point O will, in general, change for each new position of the body, because the directions of motion of A and B will be such as to change the location of O. Should it happen, however, that A and B moved in parallel straight lines, O would be at infinity or the body c would have a motion of translation; on the other hand, if the points A and B moved around concentric circles, O would be fixed in position, being the common center of the two circles, and c would simply rotate about the fixed point O.

34. Directions of Motion of Various Points.—The virtual center so found gives much information about the motion of the body at the given instant. In the first place it shows that the direction of motion of G, with respect to the paper, which has been selected as the reference plane, is perpendicular at OG and that of H is perpendicular to OH, since the direction of motion of any point in a rotating body is perpendicular to the radius to the point; thus, when the virtual center is known, the direction of motion of every point in the body is known. It is not possible to put down at random the direction of motion of G as well as those of A and B because that of G is fixed when those of A and B are given; the virtual center does not, however, give the path of G but only its direction.

35. Linear Velocities.—In the next place the virtual center gives the relative linear velocities of all points in c at the instant. Let the body c be turning at the rate of n revolutions per minute, corresponding to ω radians per second, the relation being $\omega = \frac{2\pi n}{60}$. At the instant the velocity v of a point situated r ft. from O will evidently be $v = \frac{2\pi nr}{60} = r\omega$ ft. per second, and, since ω is the same for the whole body, the linear velocity of any point is proportional to its distance from the center O.

Thus if v_A, v_B, v_G be used to denote the velocities of the points A, B and G respectively then it follows that $v_A = OA \cdot \omega$, $v_B = OB \cdot \omega$ and $v_G = OG \cdot \omega$, and it will also be clear that even though ω is unknown the relations between the three velocities are known and also the sense of the motion.

36. Information Given by Virtual Centers.—The virtual center for a body may, therefore, be found, provided only that the directions (not necessarily the paths) of motion of two points in it are known, and having found this center the directions of motion of all points in the body are known, and their relative velocities; and also the actual velocities in magnitude, sense and direction will be known if the angular velocity is known. (This should be compared with the phorograph discussed in a later chapter.) It is to be further noted that the virtual center O is a double point; it is a point in the paper and also in c, and the motion of any point in c with regard to the paper being perpendicular to the radius from O to that point so also the motion of any point in the paper with regard to c is perpendicular to the line joining this point to O.

Another point is to be noticed, that if the various virtual centers O are known, then at once the relative motion of c to the paper is known. Thus the virtual center of one body with regard to another gives always the motion of the one body with regard to the other.

37. The Permanent Center.—It has already been pointed out that the instantaneous or virtual center is the center for rotation of any one body with regard to another at a given instant, and that the location of this center is changing from one instant to the next. There are, however, very many cases where one body is joined to another by means of a regular bearing, as in the case of the crankshaft of an engine and the frame, or a wagon wheel and the body of the axle, or the connecting rod and crankpin of an engine. A little reflection will show that in each of these cases the one body is always turning with regard to the other, and that the center or axis of revolution has a fixed position with regard to each of the bodies concerned, thus in these cases the virtual center remains relatively fixed and may be termed the **permanent center.**

38. The Fixed Center.—The permanent center must not be confused with the **fixed center,** which term would be applied to a center fixed in place on the earth, but is intended to include only the case where the virtual center for the rotation of one body with regard to another is a point which remains at the same place in each body and does not change from one instant to another. The center between the connecting rod and crank and between the crankshaft and frame are both permanent, the latter being also fixed usually.

39. Theorem of the Three Centers.—Before applying the virtual centers in the solution of problems of various kinds, a very important property connected with them will be proved. Let a, b and c, Fig. 18, represent three bodies all of which have plane motion of any nature whatever, and which motion is for the time being unknown. Now, generally a has motion relative to b, and b has motion relative to c, and similarly c with regard to a, in brief all three bodies move in different ways, hence from what has been said in Sec. 33, there is a virtual center of $a \sim b$[1] which may be called ab, and this is of course also the center of $b \sim a$. Further, there is a virtual center of $b \sim c$, that is bc, and also a center of $c \sim a$, which is ca, thus for the three bodies there are three virtual centers. Now it will be assumed that enough information has been given about the motions of a, b and c to determine ab and ac only, and it is required to find bc.

Since bc is a point common to both bodies b and c, let it be supposed to lie at P', then P' is a point in both the bodies b and c. As a point in b its motion with regard to a will be normal to $P' - ab$, that is, in the direction $P'A$, because the motion of a point in one body with regard to another body is normal to the line joining this point to the virtual center for the two bodies (Sec. 34). As a point in c, the motion of $P' \sim a$ is normal to $P' - ac$ or in the direction $P'B$, so that P' has two different motions with regard to a at the same time, which is impossible or P' cannot be the virtual center of $b \sim c$. Since, however, this is not the point, it shows at once that the point bc is located somewhere along the line $ab - ca$, or say at P, because it is only such a point as P which has the same motion with regard to a whether considered as a point in b or in c; thus the center bc must lie on the same straight line as the centers ab and ac. It is not possible to find the exact position of bc, however, without further information, all that is known is the line on which it lies.

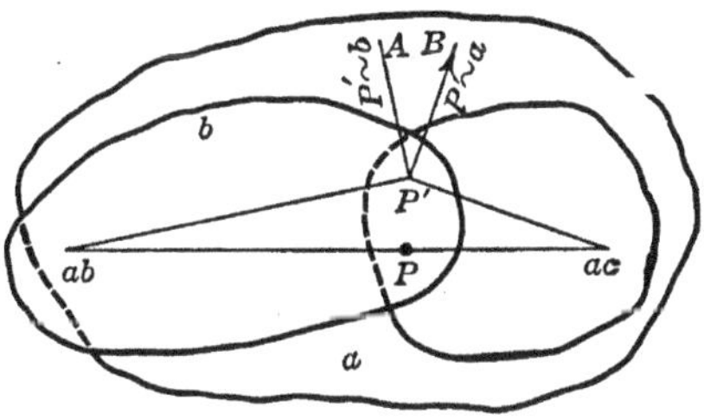

FIG. 18.

This proposition may be thus stated: **If in any mechanism there are any three links a, f, g, all having plane motion, then for the three links there are three virtual centers af, fg and ag, and these three centers must all lie on one straight line.**

Two of the centers may be permanent but not the third; in

[1] The sign $\sim$ means "with regard to."

the steam engine taking the crank *a*, the connecting rod *b* and the frame *d*, the centers *ab* and *ad* are permanent, but *bd* is not.

40. The Locating of the Virtual Centers.—The chapter will be concluded by finding the virtual centers in a few mechanisms simply to illustrate the method, the application being given in the next chapter. As an example, consider the chain with four turning pairs, which is first taken on account of its simplicity. It is shown in Fig. 19, and consists of four links, *a*, *b*, *c* and *d*, of different lengths, *d* being fixed, and by inspection the four permanent centers *ab*, *bc*, *cd* and *ad*, at the four corners of the chain, are at once located. It is also seen that there are six possible centers in the mechanism, viz., *ab*, *bc*, *cd*, *da*, *bd* and *ac*,

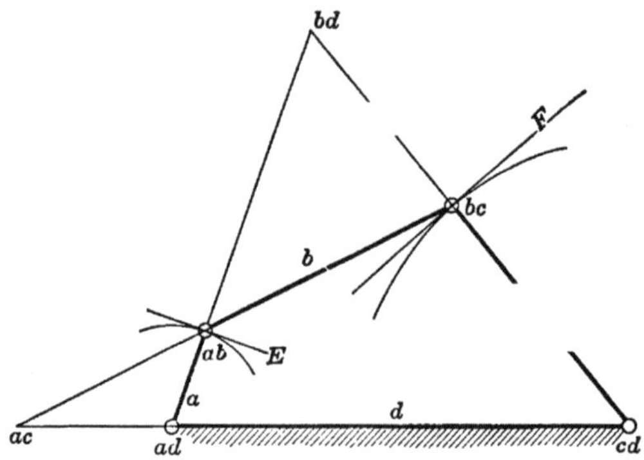

FIG. 19.

these being all the possible combinations of the links in the chain when taken in pairs, and of these six, the four permanent ones are found already, and only two others, *ac* and *bd*, remain. There are two methods of finding them, the first of which is the most instructive, and will be given first for that reason.

By the principle of the virtual center *bd* may be found if the **directions of motion** of any two points in $b \sim d$ are known. On examining *ab* remember that it is a point in *a* and also in *b*; as a point in *a* it moves with regard to *d* about the center *ad* and thus in a direction normal to $ad - ab$ or to *a* itself. And as a point in *b* it must have the same motion with regard to *d* as it has when considered as a point in *a*; that is, the motion of *ab* in *b* with regard to *d* is in the direction perpendicular to *a*. Hence, from Sec. 33, the virtual center will lie on the line through *ab* in the direction of *a*, that is, in *a* produced. Again *bc* is a point in *b* and *c*, and as a point in *c* it moves with regard to *d* in a direction perpendicular to *cd* − *bc*, or in the direction *bc* − *F*, and this must also be the direction of *bc* as a point in *b* ∼ *d*, so that the virtual center of *b* ∼ *d* must also lie in the line through *bc* normal to *bc* − *F*, or in *c* produced. Hence, *bd* is at the intersection of *a* and *c* produced.

This could have been solved by the theorem of the three centers, for there will be three centers, *ad*, *ab* and *bd*, for the three bodies *a*, *b* and *d*, and these must lie in one straight line, and as both *ad* and *ab* are known, this gives the line on which *bd* lies. Similarly, by considering the three bodies, *b*, *c* and *d*, and knowing the centers *bc* and *cd*, another line on which *bd* lies is isolated, and hence *bd* is readily found. To find the center *ac* it is possible to proceed in either of the ways already explained, and thus find *ac* at the intersection of the lines *b* and *d* produced.

41. Sliding Pairs.—One other example may be solved, and in order to include a sliding pair consider the case shown in Fig. 20, in which *a* is the crank, *b* the connecting rod, *c* the crosshead, piston, etc., and *d* the fixed frame. As before there are six centers *ad*, *ab*, *bc*, *cd*, *ac*, *bd*, of which *ad*, *ab*, and *bc* are permanent and found by inspection.

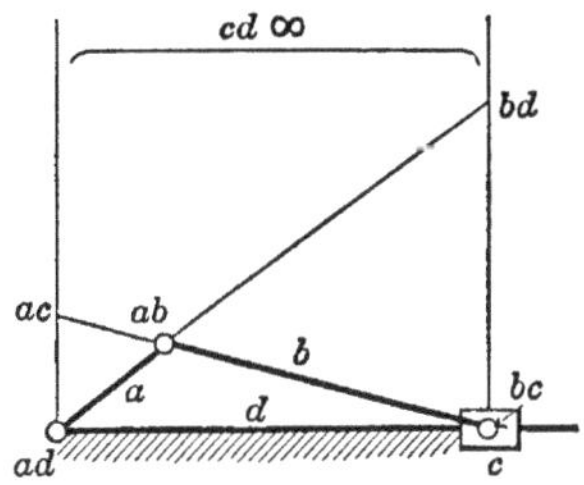

FIG. 20.

To find the center *cd* it is noticed that *c* slides in a horizontal direction with regard to *d*, that is, *c* has a motion of translation in a horizontal straight line, or, what is the same thing, it moves in a circle of infinite radius, and the center of this circle must, as before, lie in a line normal to the direction of motion of *c* ∼ *d*. Hence *cd* lies in a vertical line through *bc* or through any other point in the mechanism such as *ad*, and at an infinite distance away. The figure shows *cd* above the mechanism, but it might be below just as well.

Having found *cd*, the other centers *ac* and *bd* may be found by the theorem of the three centers. Thus *bd* lies on *bc* − *cd* and on *ad* − *ab* and is therefore at their intersection, and similarly *ac* lies at the intersection of *bc* − *ab* and *ad* − *cd*.

QUESTIONS ON CHAPTER II

1. Define motion. What data define the position and motion of a point, line, plane figure and body, the latter three having plane motion?

2. Two observers are looking at the same object; one sees it move while to the other it may appear stationary; explain how this is possible.

3. When a person in an automobile, which is gaining on a street car, looks at the latter without looking at the ground, the car appears to be coming toward him. Why?

4. Explain the difference between relative and absolute motion and state the propositions referring to these motions.

5. The speeds of two pulleys are 100 revolutions per minute in the clockwise sense and 125 revolutions per minute in the opposite sense, respectively; what is the relative speed of the former to the latter?

6. In a geared pump the pinion makes 90 revolutions per minute and the pump crankshaft 30 revolutions in opposite sense; what is the motion of the pinion relative to the shaft?

7. Distinguish between the instantaneous and complete motion of a body. What information gives the former completely? What is the virtual center?

8. What is the virtual center of a wagon wheel (*a*) with regard to the earth, (*b*) with regard to the wagon?

9. A vehicle with 36-in. wheels is moving at 10 miles an hour; what are the velocities in space and the directions of motion of a point at the top of the tire and also of points at the ends of a horizontal diameter? Is the motion the same for the latter two points? If not, find their relative motion.

10. A wheel turns at 150 revolutions per minute; what is its angular velocity in radians per second? Also, if it is 20 in. diameter, what is the linear velocity of the rim?

11. Give the information necessary to locate the virtual center between two bodies.

12. What is the difference between the virtual, permanent and fixed center? State and prove the theorem of the three centers.

13. Find the virtual centers for the stone crusher or any other somewhat complicated machine.

CHAPTER III

VELOCITY DIAGRAMS

42. Applications of Virtual Center.—Some of the main applications of the virtual center discussed in the last chapter are to the determination of the velocities of the various points and links in mechanisms, and also of the forces acting throughout the mechanism due to external forces. The latter question will be discussed in a subsequent chapter and the present chapter will be confined to the determination of velocities and to the representation of these velocities.

43. Linear and Angular Velocities.—There are two kinds of velocities which are required in machines, the linear velocities of the various points and the angular velocities of the various links, and it will be best to begin with the determination of linear velocities.

44. Linear Velocities of Points in Mechanisms.—The linear velocities of the various points may be required in one of two forms, either the absolute velocities may be required or else it may be only desired to compare the velocities of two points, that is, to determine their relative velocities. The latter problem may be always solved without knowing the velocity of any point in the machine, the only thing necessary being the shape of the mechanism and which link is fixed, while for the determination of the absolute velocity of a point in a mechanism that of some point or link must be known.

Again, the two points to be compared may be in one link, or in different links, and the solution will be made for each case and an effort will be made to obtain solutions which are quite general.

45. Points in the Same Link.—The first case will be that of the four-link mechanism, frequently referred to, containing four turning pairs and shown in Fig. 21, and the letter *d* will be used to indicate the fixed link. As a first problem, let the velocity of any point A_1 in *a* be given and that of another point A_2 in the same link be required. The six virtual centers have been found and marked on the drawing, and the link *a* has been selected for the first example because it has a permanent center which is *ad*.

Now the velocity of A_1 which is assumed given, is the absolute velocity, that is the velocity with regard to the earth. From A_1 lay off A_1E in any direction to represent the known velocity of A_1 and join $ad-E$ and produce this line outward to meet the line A_2F, parallel to A_1E, in the point F. Then A_2F will represent the linear velocity of A_2 on the same scale that A_1E represents the velocity of A_1. (It is assumed in this construction that ad, A_1 and A_2 are in the same straight line.) The reasoning is simple, for a turns with regard to the earth about the center ad and hence, since A_1 and A_2 are on the same link, their linear velocities are directly proportional to their distances from ad (Sec. 35) so that,

$$\frac{\text{Linear velocity of } A_1}{\text{Linear velocity of } A_2} = \frac{ad - A_1}{ad - A_2} = \frac{A_1E}{A_2F}.$$

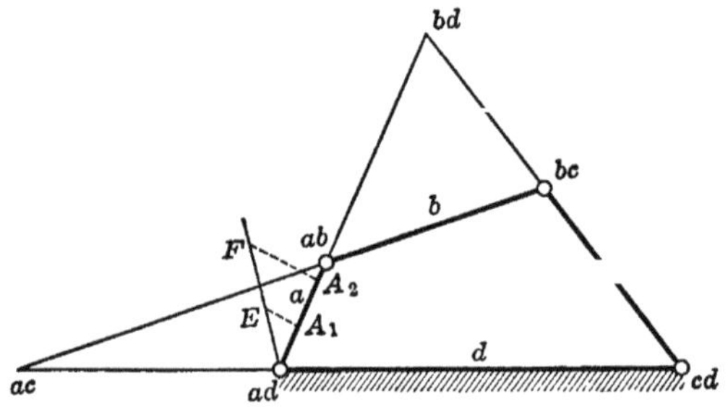

FIG. 21.

If the linear velocity of A_1 is given, so that A_1E can be drawn to scale, then the construction gives the numerical value of the velocity of A_2, but if the velocity of A_1 is not given then the above method simply gives the relative velocities of A_1 and A_2.

Next, let it be required to find the velocity of a point B_2 in b, Fig. 22, the velocity of B_1 in the same link being given and let d be the fixed link as before. Now since it is the absolute velocity of B_1 that is given, the first point is to find the center of bd about which b is turning with regard to the earth. The velocity of B_1 then bears the same relation to that of B_2 as the respective distances of these points from bd, or

$$\frac{\text{Linear velocity of } B_1}{\text{Linear velocity of } B_2} = \frac{bd - B_1}{bd - B_2}.$$

It is then only necessary to get a simple graphical method of obtaining this ratio and the figure indicates one way. First, with center bd draw arcs of circles through B_1 and B_2 cutting $bd-ab$ in B'_1 and B'_2. Then if B'_1G be drawn in any direction to represent the given velocity of B_1, it may be readily shown that B'_2H parallel to B'_1G, will represent the linear velocity of B_2, or the ratio of B'_1G to B'_2H is the ratio of the velocities of B_1 and B_2.

The only difference between this and the last case is that in the former case the center ad used was permanent, whereas in this case the center bd used is a virtual center.

46. Points in Different Links.—If it were required to compare the linear velocity of the point A_1 in a with that of B_1 in b the method would be as indicated in Fig. 22. The two links concerned are a and b and d is the fixed link and these links have the three centers ad, ab, bd, all on one line, also ab is a point common to a and b, being a point on each link. Treating it as a point in a, proceed as in the first example to find its velocity. Thus set off A_1E in any direction to represent the linear velocity of A_1, then $ab-F$ parallel to A_1E will represent the velocity of ab to the same scale. Now treat ab as a point in b and its velocity is given as $ab-F$, so that the matter now resolves itself into finding the velocity of a point B_1 in b, the velocity of the point ab in the same link being given.

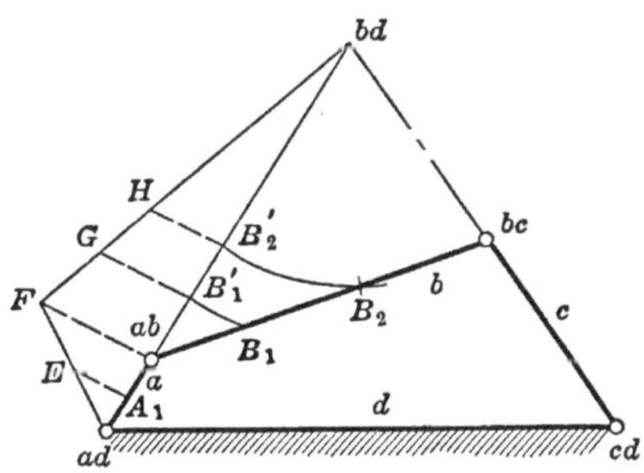

Fig. 22.

It must again be remembered that $ab-F$ represents the absolute velocity of the point ab, that is, the velocity of this point, using the fixed frame of the machine as the standard. With regard to the frame the link b is turning about the center bd, thus for the instant b turns relative to d about bd, and the velocities of all points in it at this instant are simply proportional to their distances from bd. The velocity of B_1 is to the velocity of ab in the ratio $bd-B_1$ to $bd-ab$, and in order to get this ratio conveniently, draw the arc $B_1B'_1$ with center bd, then join $bd-F$ and draw B'_1G parallel to $ab-F$ to meet $bd-F$ in G, then B'_1G represents the velocity of B_1 in the link b on the same scale that A_1E represents the velocity of A_1.

Notice that in dealing with the various links in finding relative velocities it is necessary to use the centers of the links under consideration with regard to the fixed link; thus the centers *ad* and *bd* and the common center *ab* are used. The reason *ad* and *bd* are employed, is because the velocities under consideration are all absolute.

To compare the velocity of any point A_1 in *a* with that of C_1 in *c*, Fig. 23, it would be necessary to use the centers *ad*, *ac* and *cd*. Proceeding as in the former case the velocity of *ac* is found by drawing the arc A_1L with center *ad* and drawing LN in any direction to represent the velocity of A_1 on a chosen scale, than the line $ac-M$ parallel to LN meeting $ad-N$ produced in M will represent the velocity of *ac*. Join $cd-M$, and draw the arc $C_1C'_1$ with center *cd*, then C'_1K parallel to $ac-M$, will represent the linear velocity of C_1.

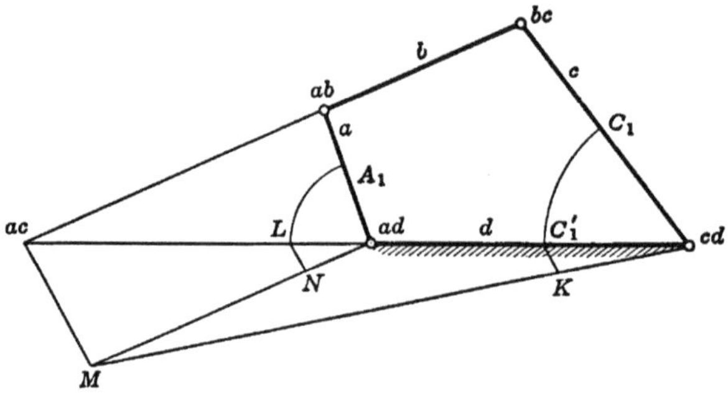

FIG. 23.

A general proposition may be stated as follows: **The velocity of any point** *A* **in link** *a* **being given to find the velocity of F in** *f*, **the fixed link being** *d*. **Find the centers** *ad*, *fd* **and** *af*, **and using** *ad* **and the velocity of** *A*, **find the velocity of** *af*, **and then treating** *af* **as a point in** *f* **and using the center** *fd*, **find the velocity of F.**

47. Relative Angular Velocities.—Similar methods to the preceding may be employed for finding angular velocities in mechanisms.

Let any body having plane motion turn through an angle θ about any axis, either on or off the body, in time t, then the angular velocity of the body is defined by the relation $\omega = \frac{\theta}{t}$. As all links in a mechanism move except the fixed link, there are in general as many different angular velocities as there are moving links. The angular velocities of the various links *a*, *b*, *c*, etc., will be designated by ω_a, ω_b, ω_c, etc., respectively, the unit being the radian per second.

As in the case of linear velocities, angular velocities may be expressed either as a ratio, in which case the result is a pure number, or as a number of radians per second, the method depending on the kind of information sought and also upon the data given. Unless the data includes the absolute angular velocity of one link it is quite impossible to obtain the absolute velocity of any other link and it is only the ratio between these velocities which may be found.

48. Methods of Expressing Velocities.—In finding the relative angular velocities between two bodies it is most usual to express the result as a ratio, thus $\frac{\omega_a}{\omega_b}$, which result is, of course, a pure number, such a method is very commonly employed in connection with gears, pulleys and other devices. If a belt connects two pulleys of 30 in. and 20 in. diameter their velocity ratio will be $^{20}/_{30} = ^{2}/_{3}$, that is, when standing on the ground and counting the revolutions with a speed counter, one of the wheels will be found to make two-thirds the number of revolutions the other one does, and this ratio is always the same irrespective of the absolute speed of either pulley.

It happens, however, that it may be necessary to know the relative angular velocities in a different form, that is, it may be desired to know how fast one of the wheels goes considering the other as a standard; the result would then be expressed in radians per second. Suppose a gear *a* turns at 20 revolutions per minute, $\omega_a = 2.09$ radians per second, and meshes with a gear *b* running at 30 revolutions per minute, $\omega_b = 3.14$ radians per second, the two wheels turning in opposite sense, then the velocity of *a* with regard to *b* is $\omega_a - \omega_b = 2.09 - (-3.14) = +5.23$ radians per second, that is, if one stood on gear *b* and looked at gear *a*, the latter would appear to turn in the opposite sense to *b* and at a rate of 5.23 radians per second. If, on the other hand, one stood on *a*, then since $\omega_b - \omega_a = -3.14 - 2.09 = -5.23$, *b* would appear to turn **backward** at a rate of 5.23 radians per second, the relative motion of *a* to *b* being equal and opposite to that of *b* with regard to *a*.

The first method of reckoning these velocities will alone be employed in this discussion and the construction will now be explained.

49. Relative Velocities of Links.—Given the angular velocity of a link a to find that of any other link b. Find the three centers ad, bd and ab, then as a point in a, ab has the linear velocity $(ad - ab)\ \omega_a$ and as a point in b, ab has the velocity $(bd - ab)\ \omega_b$. But as ab must have the same velocity whether considered as a point in a or in b, then $(ad - ab)\ \omega_a = (bd - ab)\ \omega_b$, or $\frac{\omega_b}{\omega_a} = \frac{ad - ab}{bd - ab}$. The construction is shown in Fig. 24 and will require very little explanation. Draw a circle with center ab and radius $ab - ad$, which cuts $ab - bd$ in G, lay off bd — F in any direc-

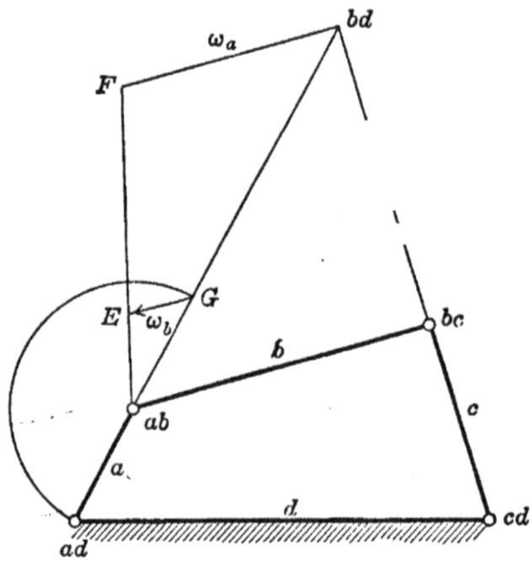

FIG. 24.

tion to represent ω_a on chosen scale, then draw GE parallel to bd — F to meet ab — F in E, and GE will represent the angular velocity of b or ω_b on the same scale.

Similar processes may be employed for the other links b and c, and no further discussion of the point will be given here. The general constructions are very similar to those for finding linear velocities.

As in the case of the linear velocities the following general method may be conveniently stated: **The angular velocity of any link a being given to find the angular velocity of a link f, d being the fixed link. Find the centers ad, fd and af, then the angular velocity ω_a is to ω_f in the same ratio that $fd - af$ is to $ad - af$.**

50. Discussion on the Method.—Although the determination of the linear and angular velocities by means of the virtual center is simple enough in the cases just considered, yet when it is employed in practice there is frequently much difficulty in getting convenient constructions. Many of the lines locating virtual centers are nearly parallel and do not intersect within the limits of the drafting board, and hence special and often troublesome methods must be employed to bring the constructions within ordinary bounds. Further, although it is common to have given the motion of one link such as *a*, and often only the motion of one other point or link say *f*, elsewhere in the mechanism is desired, requiring the finding of only three virtual centers, *ad*, *af* and *df*, yet frequently in practice these cannot be obtained without locating almost all the other virtual centers in the mechanism first. This may involve an immense amount of labor and patience, and in some cases makes the method unworkable.

51. Application to a Mechanism.—A practical example of a more complicated mechanism in common use will be worked out here to illustrate the method, only two more centers *af* and *bf* being found than those necessary for the solution of the problem. Fig. 25 shows the Joy valve gear as frequently used on locomotives and other reversing engines, more especially in England: *a* represents the engine crank, *b* the connecting rod, and *c* the piston, etc., as in the ordinary engine, the frame being *d*. One end of a link *e* is connected to the rod *b* and the other end to a link *f*, the latter link being also connected to the engine frame, while to the link *e* a rod *g* is jointed, which rod is also jointed to a sliding block *h*, and at its extreme upper end to the slide valve stem *V*. The part *m* on which *h* slides is controlled in direction by the engineer who moves it into the position shown or else into the the dotted position, according to the sense of rotation desired in the crankshaft, but once this piece *m* is set, it is left stationary and virtually becomes fixed for the time.

A very useful problem in such a case is to find the velocity of the valve and stem *V* for a given position and speed of the crankshaft. The problem concerns three links, *a*, *d* and *g*, the upper end of the latter link giving the valve stem its motion, so that the three centers *ad*, *ag*, and *dg* are required. First write on all the centers which it is possible to find by inspection, such as *ad*, *ab*, *be*, *bc*, *cd*, *ac*, *ef*, etc., and then proceed to find the required centers by the theorem of three centers given in Sec. 39. The centers *ag* and *dg* cannot be found at once and it will simplify the work to set down roughly in a circle (not necessarily accurately) anywhere on the sheet points which are approximately equidistant, there being one point for each link, in this case eight. Now letter these points *a, b, c, d, e, f, g, h,* to correspond with the links. As a center such as *ab* is found join the points *a* and *b* in the lower diagram and it is possible to join a fairly large number

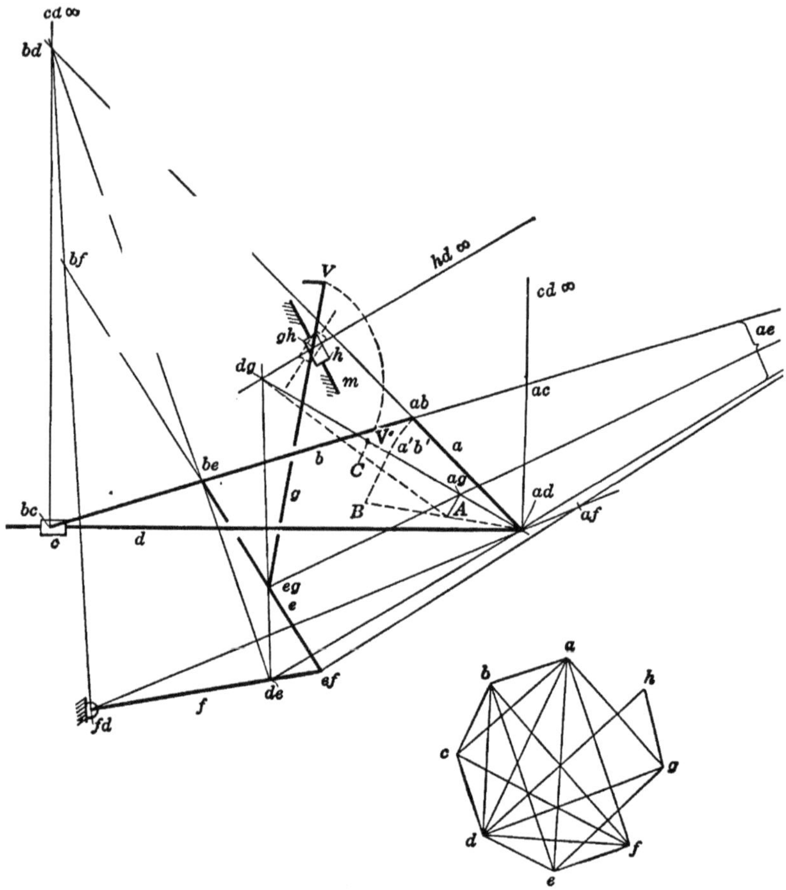

FIG. 25.—Joy valve gear.

of the points at once, then any two points not joined will represent a center still to be found. The figure shows by the plain lines the stage of the problem after the centers *ab, ac, ad, ae, af, ag, bc, bd, df, be, cd, de, df, dg, dh, ef, eg, gh,* have been found, which represents the work necessary to find the above three centers *ag, ad* and *dg.*

When all points on the lower figure are joined, all the centers have been found, and the figure shows by inspection what centers can be found at any time, for it is possible to find any center provided there are two paths between the two points corresponding to the center. It may happen that there will appear to be two paths between a given pair of points, but on examination it may be found that the paths are really coincident lines, in which case they will not fix the center and another path is necessary. The lower diagram in Fig. 25 shows that the centers *ab*, *ac*, *ad*, *de*, *df*, are known, while the centers *ah*, *bg*, *ch*, are not known, and that the center *fg* can probably be found as there are the two paths *fa* – *ag* and *fd* – *dg* between them as well as the path *fe* – *eg*. The center *gc* could not, however, be found before *gd*, as there would then be only one path *ga* – *ac* between the points.

Having now found the centers *ad*, *dg*, and *ag*, proceed as in the previous cases to find the velocity of *V* or the valve from the known velocity of *a*. If the velocity of the crankpin *ab* is given, revolve *ad* – *ab* into the line *ad* – *dg* and lay off *a′b′* – *B* to represent the velocity of *ab* on any scale. Join *ad* – *B*, then *ag* – *A* parallel to *a′b′* – *B* gives the velocity of *ag*. Next join *dg* – *A* and with center *dg* draw the arc *VV′*, then *V′C* parallel to *ag* – *A* will represent the velocity of the valve *V*. The whole process is evidently very cumbersome and laborious and is frequently too lengthy to be adopted. The reader's attention is called to the solution of the same problem by a different method in Chapter IV.

52. It must not be assumed that the methods here described are not used, because, in spite of the labor involved they are frequently more simple than any other method and a number of applications of the virtual center are given farther on in the present treatise. It is always necessary to do whatever work is required in solving problems, the importance of which frequently justifies large expenditure of time. In many cases the method described in the next chapter simplifies the work, and the reader's judgment will tell him in each case which method is likely to suit best.

53. Graphical Representation of Velocities.—It is frequently desirable to have a diagram to represent the velocities of the various points in a machine for one of its complete cycles, as the study of such diagrams gives very much information about the nature of the machine and of the forces acting on it. Two methods are in fairly common use (1) by means of a polar diagram, (2) by means of a diagram on a straight base.

To illustrate these a very simple case, the slider-crank mechanism, Fig. 26, will be selected, and the linear velocities of the piston will be determined, a problem which may be very conveniently solved by the method of virtual centers. Let the speed of the engine be known, and calculate the linear velocity of the crankpin *ab;* for example, let *a* be 5 in. long, and let the

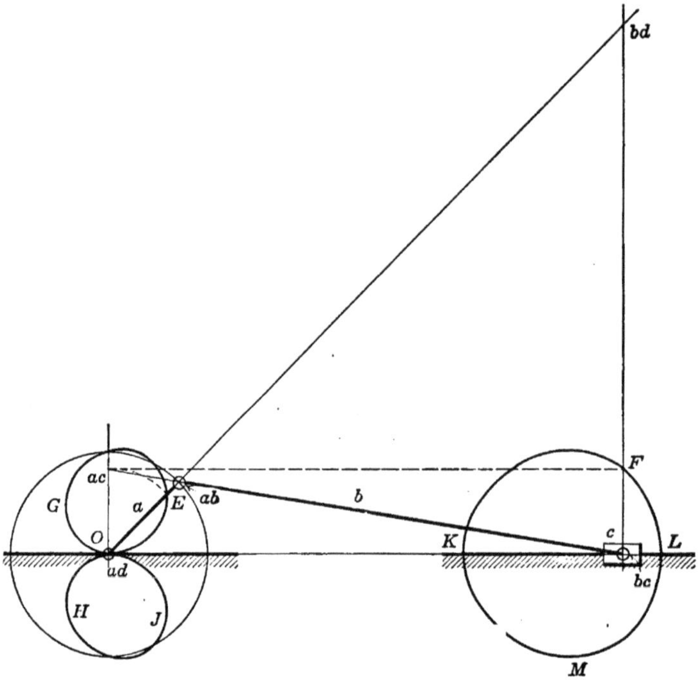

FIG. 26.—Piston velocities.

speed be 300 revolutions per minute, then the velocity of the crankpin $= 2\pi \times \frac{300}{60} \times \frac{5}{12} = 13.1$ ft. per second. Now *bc* is a point on both the piston *c* and on the rod *b* and clearly the velocity of *bc* is the same as that of *c*, the latter link having only a motion of translation, and further, the velocity of the crankpin *ab* is known, which is also the same as that of the forward end of the connecting rod. The problem then is: given the velocity of a point *ab* in *b* to find the velocity of *bc* in the same link, and from what has already been said (Sec. 35), the relation may be written:

$$\frac{\text{velocity of piston}}{\text{velocity of crankpin}} = \frac{\text{velocity of } bc}{\text{velocity of } ab} = \frac{bd - bc}{bd - ab}.$$

But by similar triangles

$$\frac{bd - bc}{bd - ab} = \frac{ac - ad}{ad - ab},$$

so that

$$\frac{\text{velocity of piston}}{\text{velocity of crankpin}} = \frac{ad - ac}{ad - ab},$$

and as $ad - ab$ is constant for all positions of the machine, it is evident that $ad - ac$ represents the velocity of the piston on the same scale as the length of a represents the linear velocity of the crankpin. Or, in the case chosen, if the mechanism is drawn full size then $ad - ab = 5$ in., and the scale will be 5 in. = 13.1 ft. per second or 1 in. = 2.62 ft. per second.

54. Polar Diagram.—Now it is convenient to plot this velocity of the piston either along a as $ad - E$ if the diagram is to show the result for the different crank positions, or vertically above the piston as bc — F, if it is desired to represent the velocity for different positions of the piston. If this determination for the complete revolution is made, there are obtained the two diagrams shown, the one *OEGOHJO* is called a **polar diagram,** *O* being the pole. The diagram consists of two closed curves passing through *O* and both curves are similar; in fact the lower one can be obtained from the upper by making a tracing of the latter and turning it over the horizontal line $ad - bc$. The longer the connecting rod the more nearly are the curves symmetrical about the vertical through *O*, and for an infinitely long rod the curves are circles, tangent to the horizontal line at *O*.

55. Diagram on a Straight Base.—The diagram found by laying off the velocities above and below the piston positions is *KFLMK*, and, as the figure shows, is egg-shaped with the small end of the egg toward *O*, and the whole curve symmetrical about the line of travel of the piston. Increasing the length of the rod makes the curve more nearly elliptical, and with the infinitely long rod it is a true ellipse.

If the direction of motion of the piston does not pass through *ad*, then the curve *FKML* is not symmetrical about the line of motion of the piston, but takes the form shown at Fig. 27, where the piston's direction passes above *ad*, a device in which it is clear from the velocity diagram that the mean velocity of the piston on its return stroke is greater than on the out stroke, and which may, therefore, be used as a quick-return motion in a shaper or other similar machine. Engines are sometimes made in this way, but with the cylinders only slightly **offset,** and not as much as shown in the figure.

56. Pump Discharge.—One very useful application of such diagrams as those just described may be found in the case of pumping engines. Let A be the area of the pump cylinder in square feet, and let the velocity of the plunger or piston in a given position be v ft. per second, as found by the preceding method, let Q cu. ft. per second be the rate at which the pump is discharging water at any instant, then evidently $Q = Av$ and as A is the same for all piston positions, Q is simply proportional

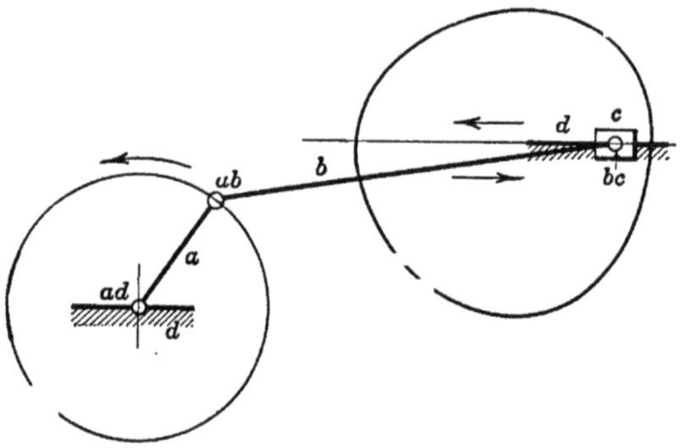

Fig. 27.—Off-set cylinder.

to v, or the height of the piston velocity diagram represents the rate of delivery of the pump for the corresponding piston position.

If a pipe were connected directly to the cylinder, the water in it would vary in velocity in the way shown in the velocity diagram (a), Fig. 28, the heights on this diagram representing piston velocities and hence velocities in the pipe, while horizontal distances show the distances traversed by the piston. The effect of both ends of a double-acting pump is shown; this variation in velocity would produce so much shock on the pipe as to injure it and hence a large **air chamber** would be put on to equalize the velocity.

Curve (b) shows two pumps delivering into the same pipe, their cranks being 90° apart, the heavy line showing that the variation of velocity in the pipe line is less than before and requires a much smaller air chamber. At (c) is shown a diagram corresponding to three cranks at 120° or a three-throw pump, in which case the variation in velocity in the pipe line would be much smaller still, this velocity being represented by the height up to the heavy line. All the curves are drawn for the case of a very long connecting rod, or of a pump like Fig. 6.

Thus the velocity diagram enables the study of such a problem to be made very accurately, and there are many other useful purposes to which it may be put, and which will appear in the course of the engineer's experience. Angular velocities may, of course, be plotted the same way as linear velocities.

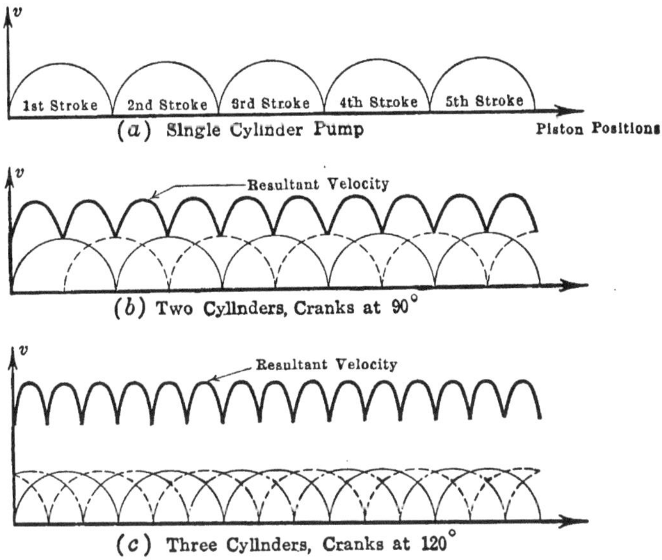

FIG. 28.—Rate of discharge from pumps.

Another method of finding both linear and angular velocities is described in the next chapter, and a few suggestions are made as to further uses of these velocities in practice.

QUESTIONS ON CHAPTER III

1. In the mechanism of Fig. 21, *a*, *b*, *c* and *d* are respectively 3, 15, 10 and 18 in. long, *d* is fixed and *a* turns at 160 revolutions per minute. Find the velocity of the center of each link with *a* at 45° to *d*.

2. Find also the angular velocities of the links in the same case as above.

3. An 8-in by 10-in. engine has a connecting rod 20 in. long and a speed of 250 revolutions per minute. Find the velocity of the center of the rod and the angular velocity of the latter in radians per second (*a*) for the dead points, (*b*) when the crank has moved 45° from the inner dead point, (*c*) for 90° and 135° crank angles.

4. Find the linear velocity of the left end of the jaw in Fig. 84, knowing the angular velocity of the camshaft.

5. Plot a diagram on a straight base for every 15° of crank angle for the piston and center of the rod in question 3; also plot a polar diagram of the angular velocity of the rod.

6. In the mechanism of Fig. 27 with $a = 6$, $b = 18$ in., and the line of travel of c, 9 in. above ad, plot the velocity diagram for c.

7. What maximum speed will be obtained with a Whitworth quick-return motion, Fig. 37, with $d = \frac{1}{2}$ in., $a = 2$ in., $b = 2\frac{3}{4}$ in. and $e = 13$ in., the line of motion of the table passing through ad?

8. Find the velocity of the tool in one of the riveters given in Chapter IX, assuming the velocity of the piston at the instant to be known.

9. Draw the polar diagram for the angular velocity of the valve stem in the mechanism of Fig. 40.

CHAPTER IV

THE MOTION DIAGRAM

57. Uses of Velocity Diagrams.—In the previous chapter something has been said about the methods of finding the velocities of various points and links in mechanisms, and a few applications of the methods were given. As a matter of fact there are a very great number of cases in which such velocity diagrams are of great value in studying the conditions existing in machines. Such problems, for example, as the value of a quick-return motion, or of a given type of valve gear or link motion; or again, problems involving the action of forces in machines, such as the turning moment produced on the crankshaft by a given piston pressure, or the belt pull necessary to crush a certain kind of stone in a stone crusher, and many other similar matters.

All of the above problems may be solved by the determination of the velocities of various parts and hence the matter of finding these velocities deserves some further consideration, more especially in view of the fact that a somewhat simpler method than that described in Chapter III, and which enables the rapid solution of all such problems, is available and may now be discussed. The graphical method of solution is usually the best, because it is simple and because the designing engineer always has drafting instruments available for such a method, and further because motions in machines are frequently so complex as to render mathematical solutions altogether too cumbersome.

58. Method to be Used.—In all machines there is one part which has a definitely known motion, and frequently this motion is one of rotation about a fixed center at uniform speed, as in the case of the flywheel of an engine, or the belt wheel of a stone crusher or punch or machine tool, this part is called the **link of reference.** Provided the motion of this link is known, it is possible to definitely determine the motions of all other parts, but if its motion is not known, then all that is possible is the determination of the relative motions of the various parts; the method described here may be used in either case.

The construction about to be explained has been called by its discoverer the phorograph method, and, as the name suggests, is a method for graphically representing the motions. It is a vector method of a kind similar to that used in determining the stresses in bridges and roofs with the important differences that the vector used in representing stresses are always parallel to the member affected, while the vector representing velocity is in many cases normal to the direction of the link concerned and further that the diagram is drawn on an arbitrarily selected **link of reference** which is itself moving.

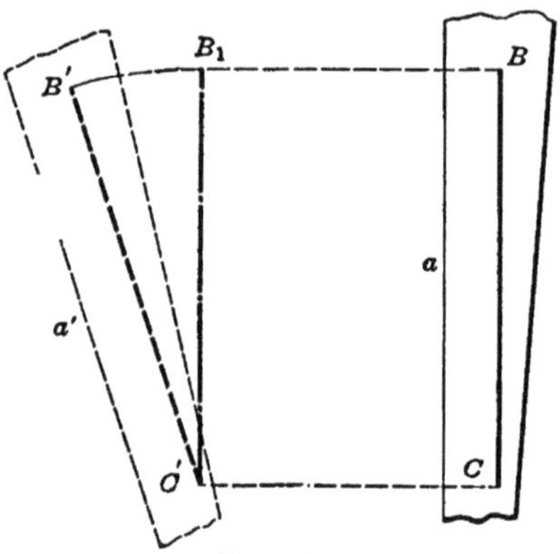

FIG. 29.

59. The Phorograph.—The phorograph is a construction by which the motions of all points on a machine may be represented in a convenient graphical manner. As discussed here the only application made is to plane motion although the construction may readily be modified so as to make it apply to non-plane motion, but in most cases of the latter kind any graphical construction becomes complicated. The method is based on very few important principles and these will first be explained.

60. Relative Motion of Points in Bodies. First Principle.—The first principle is that any one point in a rigid body can move relatively to any other point in the same body only in a direction at right angles to the straight line joining them; that is to say, if the whole body moves from one position to another, then the only motion which the one point has that the other has not is in a direction normal to the line joining them. To illustrate this take the connecting rod of an engine, a part of which is shown at a in Fig. 29 and let the two points B and C, and hence the line BC, lie in the plane of the paper. Let the rod move from a to a', the line taking up the corresponding new position $B'C'$.

During the motion above described C has moved to C' and B

to B'. Now draw BB_1, parallel to CC' and $C'B_1$, parallel to BC, then an inspection of the figure shows at once that if the rod had only moved to B_1C' the points B and C would have had exactly the same net motion, that is, one of translation through $CC' = BB_1$ in the same direction and sense, and hence B and C would have had no relative motion. But when the rod has moved to a', B has had a further motion which C has not had, namely B_1B'.

Thus during the motion of a, B has had only one motion not shared by C, or B has moved relatively to C through the arc B_1B', and at each stage of the motion the direction of this arc was evidently at right angles to the radius from C', or at right angles to the line joining B and C.

Thus when a body has plane motion any point in the body can move relatively to any other point in the body only at right angles to the line joining the two points. It follows from this that if the line joining the two points should be normal to the plane of motion, then the two points could have no relative motion.

61. Second Principle.—Let Fig. 30 represent a mechanism having four links, a, b, c and d, joined together by four turning

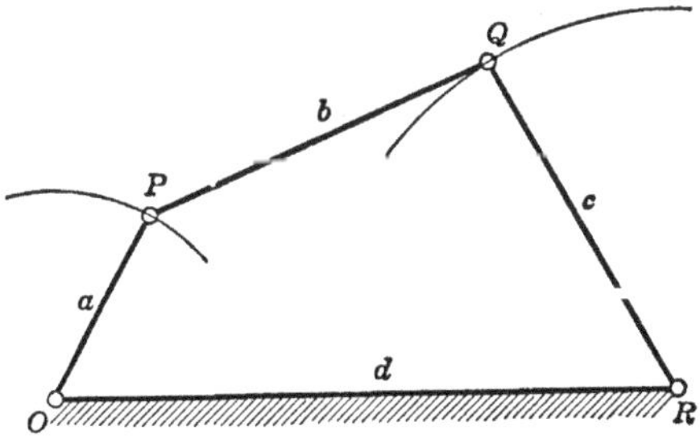

FIG. 30.

pairs O, P, Q and R as indicated. This mechanism is selected because of its common application and the reader will find it used in many complicated mechanisms. For example, it forms half the mechanism used in the beam engine, when the links are somewhat differently proportioned, a being the crank, b the connecting rod and c one-half of the walking beam. The same chain is also used in the stone crusher shown at Fig. 95 and in many other places.

The second principle upon which the phorograph depends may now be explained by illustrating with the above mechanism.

In this mechanism the fixed link is d which will be briefly referred to as the frame. Thus O and R are fixed bearings or permanent centers, while P and Q move in arcs of circles about O and R respectively. Choose one of the links as the link of reference, usually a or c will be most suitable as they both have a permanent center while b has not; the one actually selected is a. Imagine that to a there is attached an immense sheet of cardboard extending indefinitely in all directions from O, and for brevity the whole sheet will be referred to as a.

A consideration of the matter will show that on the cardboard on the link a there are points having all conceivable motions and velocities in magnitude, direction and sense. Thus, if a circle be drawn on a with center at O, all points on the circle will have velocities of the same magnitude, but the direction and sense will be different; or if a vertical line be drawn through O, all points on this line will move in the same direction, that is, horizontal, those above O moving in opposite sense to those below and all points having different velocities. If any point on a be selected, its velocity will depend on its distance from O, the direction of its motion will be normal to the radial line joining it to O, and its sense will depend upon the relative positions of the point and O on the radial line. The above statements are true whether a has constant angular velocity or not, and are also true even though O moves.

From the foregoing it follows that it will be possible to find a point on a having the same motion as that of any point Q in any link or part of the machine, which motion it is desired to study; and thus to collect on a a set of points each representing the motion of a given point on the machine at the given instant. Since the points above described are all on the link a, their relative motions will be easily determined, and this therefore affords a very direct method of comparing the velocities of the various points and links at a given instant. If the motion of a is known, as is frequently the case, then the motions of all points on a are known, and hence the motion of any point in the mechanism to which the determined point on a corresponds; whereas, if the motion of a is unknown, only the relative motions of the different points at the instant are known.

A collection of points on a certain link, arbitrarily chosen as the link of reference, which points have the same motions as points on the mechanism to which they correspond, and about

which information is desired, is called the phorograph of the mechanism, because it represents graphically (vectorially) the relative motions of the different points in the mechanism.

62. Third Principle.—The third point upon which this graphical method depends is that the very construction of the mechanism supplies the information necessary for finding in a simple way the representative point on the reference link corresponding to a given point on the mechanism; this representative point may be conveniently called the image of the actual point. Looking at the mechanism of Fig. 30, and remembering the first principle as enunciated in Sec. 60, it is clear that if it is desired to study the motion and velocity of such a point as Q, then the mechanism gives the following information at once:

1. The motion of Q relative to P is normal to QP since P and Q are both in the link b, and as P is also a point in link a,

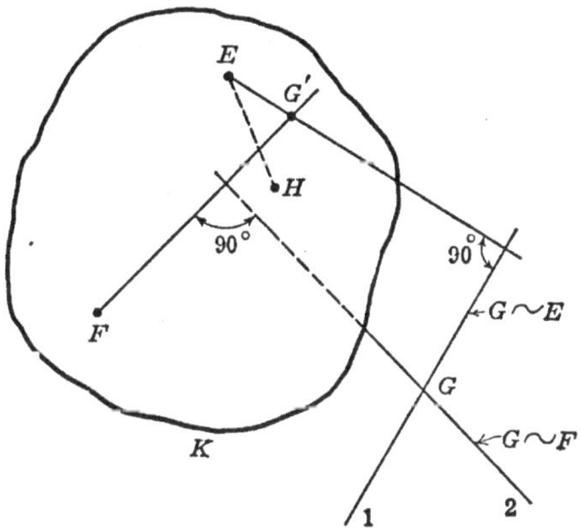

FIG. 31.

the motion of $Q \sim P$ in a is normal to QP, or the motion of Q in b with regard to a point P in a is known.

2. Since Q is also a point in c the motion of $Q \sim R$ is normal to QR. But R is a point in the fixed link d and hence R is stationary as O is, so that the motion of $Q \sim R$ is the same as the motion of $Q \sim O$. Hence the motion of Q with regard to a second point in a is known.

As will now be shown these facts are sufficient to determine on a point Q' in a having the same motion as Q and the method of doing this will now be demonstrated in a general way.

63. Images of Points.—Let there be a body K, Fig. 31, containing two points E and F, and let K have plane motion of any nature whatsoever, the exact nature of its motion being at present unknown. On some other body there is a point G also moving in the same plane as K; the location of G is unknown and the only information given about it is that its instantaneous motion relative to E is in the direction $G - 1$ and its motion relative to F is in the direction $G - 2$. It is required to find a point G' on K which has the same motion as G; the point G' is called the image of G.

Referring to the first principle it is seen that the motion of any point in $K \sim E$ is perpendicular to the line joining this point to E, for example the motion of F $\sim E$ is perpendicular to FE. Now a point G' is to be found in K having the same motion as G, and as the direction of motion of $G \sim E$ is given, this gives at once the position of the line joining E to the required point; it must be perpendicular to $G - 1$ and pass through E. The point could not lie at H for instance, because then the direction of motion would be perpendicular to HE, which is different to the specified direction $G - 1$. Thus G' lies on a line EG' perpendicular to $G - 1$ through E.

Similarly it may be shown that G' must lie on a line through F perpendicular to $G - 2$, and hence it must lie at the intersection of the lines through E and F or at G' as shown in Fig. 31. Then G' is a point on K having the same motion as G in some external body.

64. Possible Data.—A little consideration will show that it is not possible to assume at random the sense or magnitude of the motions, but only the two directions. The point G' could, however, be found by assuming the data in another form; for example, if the angular velocity of K were known and also the magnitude direction and sense of motion of $G \sim E$, G' could be located, and then the motion of $G \sim F$ could be determined, the reader will readily see how this is done. In general the data is given in the form stated first, as will appear later.

Now as to the information given, the discussion farther on will show this more clearly but to introduce the subject in a simple way let the virtual center of the body at the given instant, with regard to the earth, be H and let the body be rotating at this instant in a clockwise sense at the rate of n revolutions per minute or $\omega = \frac{2\pi n}{60}$ radians per second. Then the motion of G in space is in the direction normal to $G'H$, and it is moving to the right with a velocity $G'H.\omega$ ft. per second, where $G'H$ is in feet. Further, the motion of $G \sim E$ is in the sense $G - 1$ and the velocity of $G \sim E$ is $EG'.\omega$ ft. per second.

65. Application to Mechanisms.—The application of the above principles to the solution of problems in machinery will illustrate the method very well, and in doing this the principles upon which the construction depends should be carefully studied, and attention paid to the fact that if too much is assumed the different items may not be consistent.

The simple mechanism with four links and four turning pairs will be again selected as the first example, and is shown in Fig. 32, the letters a, b, c, d, O, P, Q and R having the same significance as in former figures and a is chosen as the link of reference, or more conveniently, the primary link, a rough outline being shown to indicate its wide extent. In future this outline will be omitted. It is required to find the linear velocities of the point S the center of b, of T in c and of Q, also the angular velocities of b and c compared to a while the mechanism is passing through the position shown in the figure.

Points will first be found on a having the same motions as Q and R, these points being the **images** of Q and R, and are indicated by accents; thus Q' is a point on a having the same motion in every respect as Q actually has.

Inspection at once shows that since P is a point in a therefore P^1 the image of P will coincide with the latter, and if ω be the angular velocity of a (where ω may be constant or variable), then the linear velocity of P at the instant is $OP.\omega = a\omega$ ft. per second, where a is the length in feet of the link a. The direction of motion of P is perpendicular to OP and its sense must correspond with ω. Such being the case, the length OP or a represents $a\omega$ ft. per second, so that the velocity scale is $\omega : 1$. Again since R is stationary it is essential that R' be located at O the only stationary point in the link a.

The point Q' may be found thus: The direction of motion of $Q \sim P$ is perpendicular to QP or b, and hence, from the proposition given in Sec. 60 to 63, Q' must lie in a line through P' (which coincides with P) perpendicular to the motion of $Q \sim P$, that is in a line through P' in the direction of b, or on b produced. Again, the direction of motion of $Q \sim R$ is perpendicular to QR or c, and since R' at O has the same motion as R, both being fixed. this is also the direction of motion of $Q \sim R'$, so that Q' lies on a line through R' perpendicular to the motion of $Q \sim R$, that is, on the line through R' in the direction of c. Now as Q' has been shown to lie on b or on b produced, and also on the line through O parallel to c, therefore it lies at the point shown on the diagram at the intersection of these two lines.

66. Images are Points on the Primary *Link*.—It may be well again to remind the reader that the point Q' is a point on a but that its motion is identical with that of Q at the junction of the links b and c. If the angular velocity of a is ω radians per second, then the linear velocity of Q' on a is $Q'O.\omega$ ft. per second and its direction in space is perpendicular to $Q'O$, and from the sense of rotation shown on Fig. 32 it moves to the left. Since the motion of Q' is the same as that of Q then Q also moves to the left in the direction normal to $Q'O$ and with the velocity $Q'O.\omega$ ft. per second.

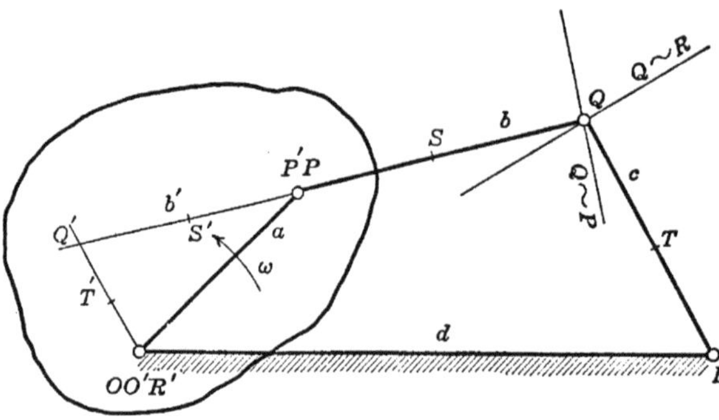

FIG. 32.

67. *Images of Links*.—Since P' and Q' are the images of P and Q on b, $P'Q'$ may be regarded as the image of b, and will in future be denoted by b'; similarly $R'Q'(OQ')$ will be denoted by c'. By a similar process of reasoning it may be shown that since S bisects PQ, so will S' bisect $P'Q'$ and also T' may be found from the relation $R'T' : T'Q' = RT : TQ$.

Since the latter point is of importance and of frequent occurrence, it may be well to prove the method of locating S'. The direction of motion of $S \sim P$ is clearly the same as that of $Q \sim P$, that is, perpendicular to PQ or b, but the linear velocity of $Q \sim P$ is twice that of $S \sim P$, both being on the same link and S bisecting PQ. But the motion of P' is the same as P and of Q' is the same as Q; hence the motion of $Q' \sim P'$ is exactly the same as that of $Q \sim P$, so that the velocity of $S' \sim P'$ is one-half that of $Q' \sim P'$, or S' will lie on $P'Q'$ and in the center of the latter line.

68. Angular Velocities.—The diagram may be put to further use in determining the angular velocities of b and c when that of a is known or the relation between them when that of a is not known. Let ω_b and ω_c, respectively, denote the angular velocities of b and c in space, the angular velocity of the primary link a being ω radians per second. Now Q and P are on one link b and the motion of $Q \sim P$ is perpendicular to QP, and hence the velocity of $Q \sim P$ is $QP.\omega_b = b.\omega_b$ ft. per second where ω_b is the angular velocity of b, and ω_b is as yet unknown. Again Q' and P' are points on the same link a, which turns with the known angular velocity ω, and hence the velocity of $Q' \sim P'$ is $Q'P'.\omega = b'\omega$ ft. per second. But from the nature of the case, since Q' has the same motion as Q, and P' the same motion as P, the velocity of $Q \sim P$ is equal to that of $Q' \sim P'$, that is, $b\omega_b = b'\omega$ or

$$\omega_b = \frac{b'}{b} \cdot \omega = b' \times \frac{\omega}{b}.$$

Similarly

$$\omega_c = \frac{c'}{c}\,\omega = c' \times \frac{\omega}{c}.$$

69. *Image of Link Represents Its Angular Velocity.*—The above discussion shows that if the angular velocity of a is constant then the lengths of the images b' and c' represent the angular velocities of the links b and c to the scale $\frac{\omega}{b}$ and $\frac{\omega}{c}$ respectively, since b and c are the same for all positions of the mechanism. On the other hand, even though ω is variable, at any instant $\frac{\omega_b}{\omega} = \frac{b'}{b}$, etc., so that there is a direct method of getting the relation between the angular velocities in such cases.

70. Sense of Rotation of *Links.*—The diagram further shows the sense in which the various links are turning, and by the formulas for the angular velocities these are readily inferred. Thus $\omega_b = \frac{b'}{b}\omega$, and starting at the point P, $P'Q' = b'$ is drawn to the left and $PQ = b$ to the right, hence the ratio $\frac{b'}{b}$ is negative, or the link b is at the instant turning in opposite sense to a or in a clockwise sense. In the case of the link c the lines $R'Q'$ and

RQ are drawn upward from R and R', that is $\frac{c'}{c}$ is positive and hence a and c are turning in the same sense.

71. Phorograph a Vector Diagram.—The figure $OP'Q'R'$ is evidently a vector diagram for the mechanism, the distance of any point on this diagram from the pole O being a measure of the velocity of the corresponding point in the mechanism. The direction of the motion in space is normal to the line joining the image of the point to O, and the sense of the motion is known from the sense of rotation of the primary link. Further, the lengths of the sides of this vector diagram, $b'(P'Q')$, $c'(R'Q')$ and $d'(OR')$ are measures of the angular velocities of these links the sense of motion being determined as explained. As d is at rest, OR' has no length.

In Fig. 33 other positions and proportions of a similar mechanism are shown, in which the solution is given and the results will be as follows:

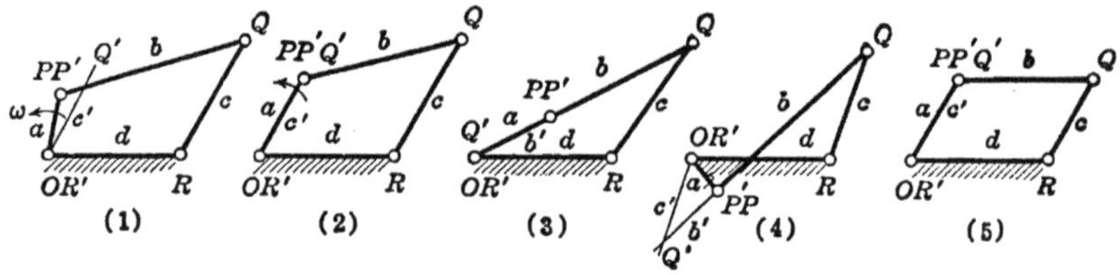

FIG. 33.

At (1) the ratio $\frac{b'}{b}$ is positive as is also $\frac{c'}{c}$ or all links are turning in the same sense; at (2) the link a is parallel to c and hence Q^1 and P^1 lie at P, so that $b' = 0$ or $\omega_b = \frac{b'}{b}\omega = 0$, that is, at this instant the link b has no angular velocity and is either at rest or has a motion of translation. Evidently it is not at rest since the velocity of all points on it are not zero but are $OP.\omega = a\omega$ ft. per second. As shown at (3) the links a and b are in one straight line and in the phorograph Q' and R' both lie at O, so that $Q'R' = c' = 0$, and hence $\omega_c = 0$, in which case the link c is for the instant at rest, since both Q' and R' are at O, the only point at rest in the figure. At (4) both the ratios $\frac{b'}{b}$ and $\frac{c'}{c}$ are negative so that b and c both turn in opposite sense to a and therefore in the same sense as one another. At (5) the parallelogram used commonly on the side rods of locomotives, is shown and the phorograph shows that Q' and P' coincide with P so that $b' = 0$ or the side rod b has a motion of translation, as is well known. There is the further well-known conclusion that since $c^1 = c$ the links a and c turn with the same angular velocity.

It is to be noted that if the image of any link reduces to a single point two explanations are possible: (a) if this point falls at O the link is stationary for the instant as for d in the former figure, and also as for c in (3) of Fig. 33; but (b) if the point is not at O, the inference is that all points in the link move in the same way or the link has a motion of translation at the instant, as for the link b in (2).

The method will now be employed in a few typical cases.

72. Further Example.—The mechanism shown in Fig. 34 is a little more complicated than the previous ones. Here P', Q'

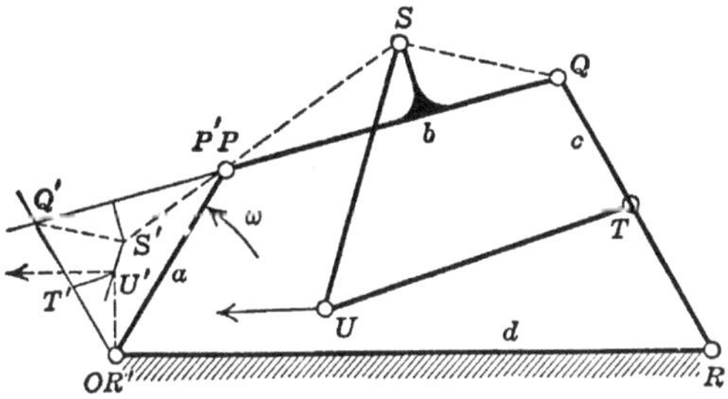

FIG. 34.

and R' are found as before, and since the motions of $S \sim Q$ and $S \sim P$ are perpendicular respectively to SQ and SP, therefore $S'P'$ and $S'Q'$ are drawn parallel respectively to SP and SQ, thus locating S'. Also T' is located from the relation $R'T' : T'Q' = RT : TQ$ (Sec. 67). Next since the motions of $U \sim S$ and $U \sim T$ are given, draw $S'U'$ parallel to SU and $U'T'$ parallel to UT, their intersection locating U'. Assuming a to turn at angular velocity ω in the sense shown, then the angular velocity of SU is $\frac{S'U'}{SU} \cdot \omega$ in the same sense as a and that of UT is $\frac{U'T'}{UT}\,\omega$ in opposite sense to a (Sec. 70). The linear velocity of U is $OU'.\omega$ ft. per second.

73. Image is Exact Copy of Link.—There is an important point which should be emphasized here and it is illustrated in finding the image of the link b. The method shows that the relation of S' to $P'Q'$ is the same as that of S to PQ, or the image of the link is an exact copy of the link itself, and although it may be inverted and is usually of different size, all lines on the image are parallel with the corresponding lines on the mechanism. Whenever the image of the link is inverted it simply means that the link, of which this is the image, is turning at the given instant in the opposite sense to the link of reference; if the image is the same size as the original link, then the link has the same angular velocity as the link of reference.

74. Valve Gear.—The mechanism shown in Fig. 35 is very commonly used by some engine builders for operating the slide valve, *OP* being the eccentric, *RS* the rocker arm pivoted to the frame at *R*, *ST* the valve rod and *T* the end of the valve stem which has a motion of sliding. *OP* has been selected as the link of reference and *P′*, *Q′*, *R′* and *S′* are found as before.

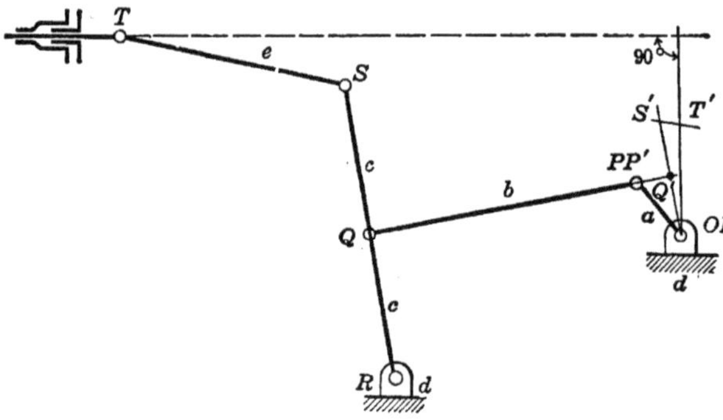

FIG. 35.—Valve gear.

The construction forces *T* to move horizontally in space, and therefore *T′* must lie on a line through *O* perpendicular to the motion of *T*, that is *T′* is on a vertical line through *O*, and further *T′* lies on a line through *S′* parallel to *ST*, which fixes *T′*.

Following the instructions given regarding former cases, it is evident that the velocity of the valve is $OT'.\omega$ ft. per second, ω being the angular velocity of *OP*. While the other velocities are not of much importance yet the figure gives the angular velocity of *ST* as $\frac{S'T'}{ST}\cdot\omega$ in the opposite sense to *a* and the linear velocity of *S* is greater than that of *T* in the ratio $OS':OT'$.

75. Steam Engine.—The steam engine mechanism is shown in Fig. 36, (*a*) where the piston direction passes through *O* and (*b*) where it passes above *O* with the cylinder offset, but the same letters and description will apply to both. Evidently Q' lies on $P'Q'$ through P', parallel to PQ, that is, on QP produced, and also since the motion of Q in space is horizontal, Q' will lie on the vertical through O.

The velocity of the piston is $OQ'.\omega$ in both cases and the angular velocity of the connecting rod is $\frac{b'}{b}\cdot\omega$ in the opposite sense to that of the crank, since $P'Q'$ lies to the left of P' while PQ lies to the right of P, and it is interesting to note that in both cases when the crank is to the left of the vertical line through O, the crank and rod turn in the same sense; further that the rod is not turning when the crank is vertical because Q' and P' coincide and hence $b' = 0$. Again, the piston velocity will be zero when Q' lies at O, which will occur when a and b are in a straight line; the maximum piston velocity will be when OQ' is greatest and this will not occur for the same crank angle in the two cases (a) and (b).

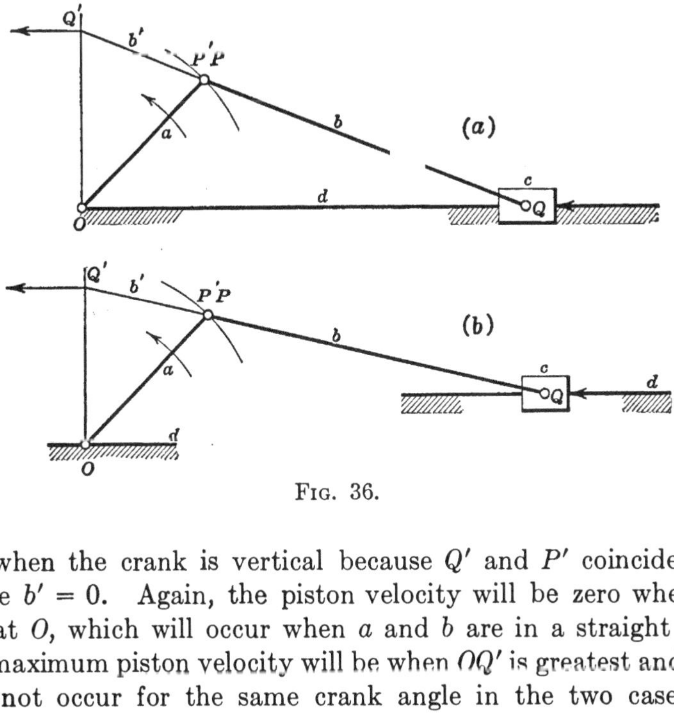

FIG. 36.

If a comparison is made between the two figures of Fig. 36 it will be clear that the length OQ' in the upper figure is greater than the corresponding length in the lower one, or the upper piston is moving at this instant at a higher rate than the lower one, but if the whole revolution be examined the reverse will be true of other crank positions; in fact, the lower construction is frequently used as a quick-return motion (see Fig. 27).

76. Whitworth Quick-return Motion.—The Whitworth quick-return motion, already described and shown at Fig. 11, is illustrated at Fig. 37. There are four links a, b, d and e and two sliding blocks, c and f, d being the fixed and a the driving link which rotates at speed ω, and is selected as the primary link. P' and Q' are found by inspection. Further, S' lies on a vertical line through O, and R' on a line through Q' parallel to QR, but the exact positions of S' and R' are unknown.

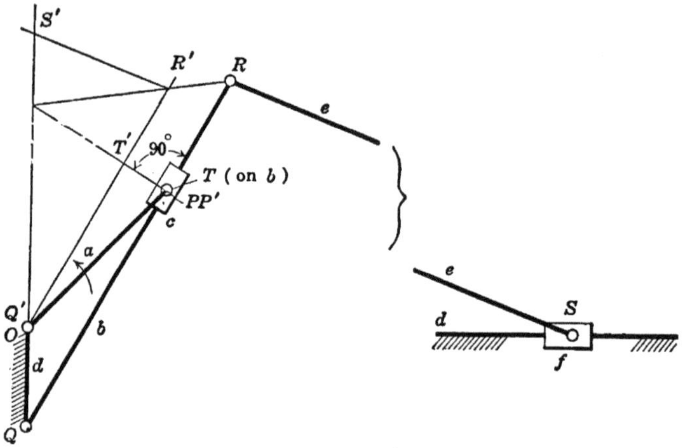

FIG. 37.—Whitworth quick-return motion.

Now P is a point on both a and c. Choose T on b exactly below P on a, T thus having a different position on b for each position of a. As all links have plane motion, the only motion which T can have relative to P is one of sliding in the direction of b, or the motion of $T \sim P$ is in the direction of b; hence T' lies on a line through P' perpendicular to b. But T is a point on the link b and hence T' must be on a line through Q' parallel with b; thus T' is determined, and having found T' it is very easy to find R^1 by dividing $Q'T'$ externally at R' so that

$$\frac{Q'R'}{Q'T'} = \frac{QR.}{QT}$$

The dotted lines show a simple geometrical method for finding this ratio, and it is always well to look for some such method as it enormously reduces the time involved in the problem.

Now since S moves horizontally, S' will be located on the vertical line through O and also on the line $R'S'$ through R' parallel to RS. This return motion is frequently used on shapers and other machines, the tool holder being attached to the block f, so that the tool holder is moving with linear velocity $OS'.\omega$ and the angular velocity of the link b is $\frac{Q'R'}{QR}\cdot\omega$ in the same sense as a.

It must be noticed that although P and T coincide, their images do not, for T has a sliding motion with regard to P at the rate $P'T'.\omega$ ft. per second, and hence both cannot have the same velocity. If P' and T' coincided then P and T would have the same velocity. The velocity diagram for S for the complete revolution of a is shown in Fig. 38.

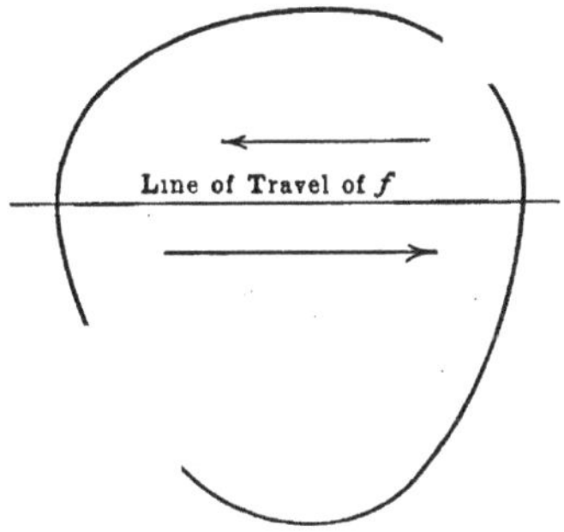

FIG. 38.—Velocity of tool in Whitworth motion.

77. Stephenson Link Motion.—The Stephenson link motion shown in Fig. 39 involves a slightly different method of attack. The proportions have been considerably distorted to avoid confusion of lines. The primary link is the crankshaft containing the crank C and the eccentrics E and F, and the scale here will be altered so that $OC' = 2 \times OC$, OE' and OF' being similarly treated. The scale will then be $OC' = OC.\omega$ ft. per second or $\frac{1}{2}\omega : 1$.

The points C', E', F', H', D' and J' are readily located. Further choose M on the curved link AGB directly below K on the rocker arm LDK and draw lines $E'A', F'B', H'G'$ and $D'K'$, of unknown lengths but parallel respectively to EA, FB, HG and DK. It has already been seen (Sec. 73) that the image of each link is similar to and similarly divided to the link itself (it is, in fact, a photographic image of it); hence the link AGB must have an image similar to it, that is, the (imaginary) straight lines $A'G'$ and $G'B'$ must be parallel to AG and GB and the triangle $A'G'B'$ must be similar to AGB. But the lines on which A', G' and B' lie are known; hence the problem is simply the geometrical one of drawing a triangle $A'G'B'$ similar and parallel to AGB with its vertices on three known lines. The reader may easily invent a geometrical method of doing this with very little effort, the process being as simple as the one shown in Fig. 37, but the construction is not shown because the figure is already complicated.

Having now located the points A', G' and B' the curved link $A'G'B'$ may be made by copying AGB on an enlarged scale and on it the point M' may be located similarly to M in the actual link. For the purpose of illustrating the problem the image of the curved link has been drawn in on the figure although this is not at all necessary in locating M'. Since K slides with regard to M, then $K'M'$ is drawn normal to the curved link at M' which locates K' and then L' is found from the relation $LD : DK = L'D' : D'K'$.

The construction shows that the curved link is turning in the same sense as the crank with angular velocity $\frac{1}{2} \cdot \frac{A'B'}{AB} \cdot \omega$ since the scale is such that $OE' = 2OE$. Again, the valve is moving to

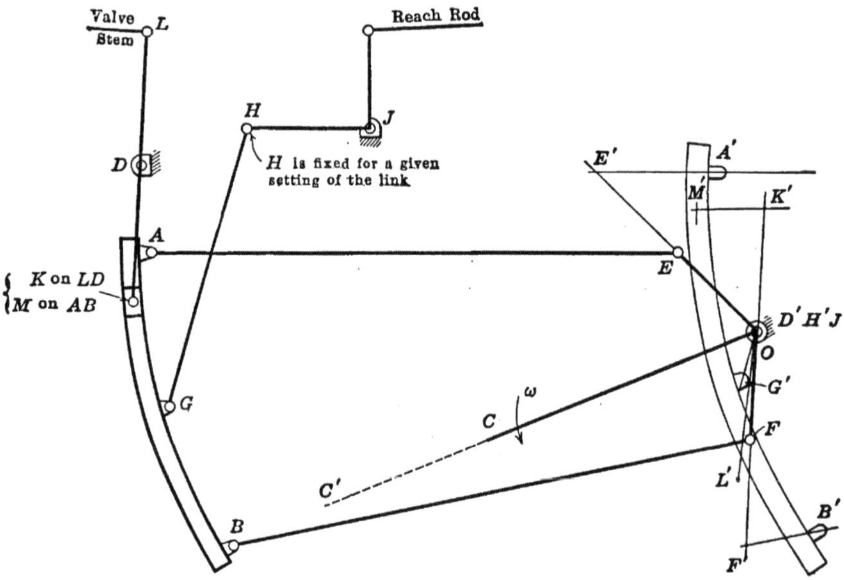

FIG. 39.—Stephenson link motion.

the right at the velocity represented by OL', and further the velocity of sliding of the block in the curved link is represented by $K'M'$, both of these on the same scale that $OC' = 2.OC$ represents the linear velocity of the crankpin. The method gives a very direct means of studying the whole link motion.

78. Reeves Valve Gear.—The Corliss valve gear used on the Reeves engine is shown diagrammatically in Fig 40, there being no great attempt at proportions. Very little explanation is necessary; O, R and S are fixed in space, S being the end of the rocking valve stem; hence R' and S' are at O, and the link OP is driven direct from the crankshaft through the eccentric connec-

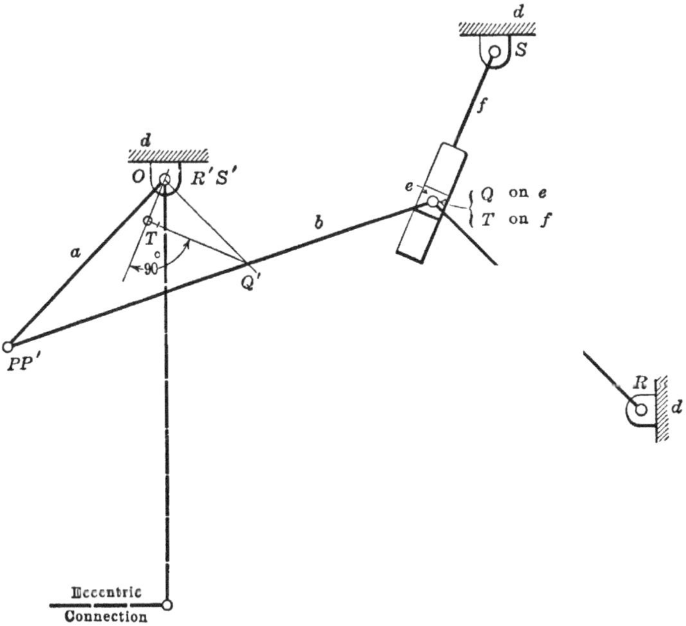

FIG. 40.—Reeves valve gear.

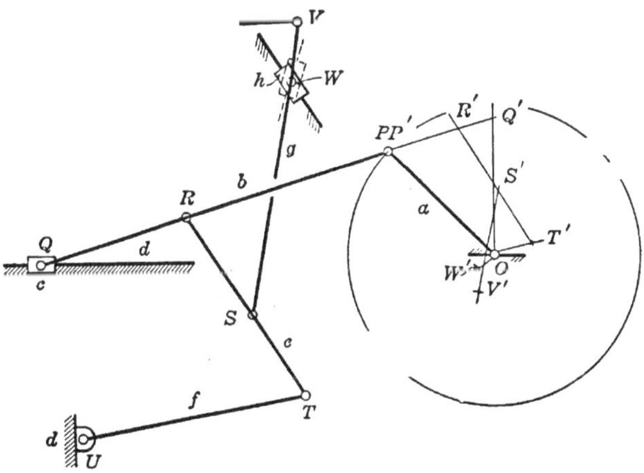

FIG. 41.

tion. The point Q on the sliding block e is directly over a movable point T on the lever f which lever is keyed to the valve stem. The image of the link f is found by drawing $S'T'$ parallel to ST and T' is located by drawing $Q'T'$ perpendicular to $S'T'$ or to ST. In this position the angular velocity of the valve is $\frac{S'T'}{ST}$ times that of the link OP, and from this data the linear velocity of the valve face is readily found

79. Joy Valve Gear.—This chapter will be concluded by one other example here, although the method has been used a good deal throughout the book and other examples appear later on. The example chosen is the Joy valve gear, shown in Fig. 41, this gear having been largely used for locomotives and reversing engines. Referring to the figure, a is the crank, b the connecting rod, c the crosshead, e or RST, f and g or SWV are links connected as shown. The link g, to which the valve stem is connected at V, is pivoted to a block h, which ordinarily slides in a slotted link fixed in position, but in order to reverse the engine the slotted block is thrown over to the dotted position.

There should be no difficulty in solving this problem, the only point causing any hesitation being in drawing OW', which should be normal to the direction of motion of the block h. The velocity of the valve is $OV^1 \cdot \omega$ in opposite sense to P and of such a point as S is $OS^1 \cdot \omega$ in a direction at right angles to OS^1.

80. Important Principles.—It may be well to call attention to certain fundamental points connected with the construction discussed. In the first place, it will be seen that the method is a purely vector one for representing velocities, and is thus quite analogous to the methods of graphic statics. As in the latter case the idea seems rather hard to grasp but the application will be found quite simple, and even in complicated mechanisms there is little difficulty. Graphical methods for dividing up lines and determining given ratios are worth the time spent in devising them.

It should be remembered that in the phorograph **the image of a link is a true image of the actual link, that is, it is exactly similar to it, and similarly divided, and is always parallel to the link but may be inverted relative to it. The image may be the same size or larger or smaller than the link depending on how fast it is revolving; it is in fact exactly what might be seen by looking through a lens at the link.** If this statement is kept in mind, it will greatly aid in the solution of problems and the understanding of the method.

QUESTIONS ON CHAPTER IV

1. Prove that any point in a body can only move relatively to any other point, in a direction perpendicular to the line joining them.

2. Define the phorograph and state the principles involved.

3. A body *a* rotates about a fixed center *O;* show that all points in it have different velocities, either in magnitude, sense or direction.

4. Show that the phorograph is a vector diagram. What quantities may be determined directly by vectors from it?

5. If the image of a link is equal in length to the link and in the same sense, what is the conclusion? What would it be if the image was a point?

6. In the mechanism (3) of Fig. 9, let *d* turn at constant speed, as in the Gnome motor; find the phorograph and the angular velocity of the rod *b*.

7. In an engine of 30 in. stroke the connecting rod is 90 in. long, and the engine runs at 90 revolutions per minute. Find the magnitude and sense of the angular velocity of the rod for crank angles 45°, 135°, 225° and 315°.

8. Make a diagram of a Walschaert valve gear and find the velocity and direction of motion of the valve for a given crank position.

9. Plot the angular velocity of the jaw and the linear velocity of the center *G* of the crusher given in Chapter XV, Fig. 168. See also Fig. 95.

CHAPTER V

TOOTHED GEARING

81. Forms of Drives.—In machinery it is frequently necessary to transmit power from one shaft to another, the ratio of the angular velocities of the shafts being known, and in very many cases this ratio is constant; thus it may be desired to transmit power from a shaft running at 120 revolutions per minute to another running at 200 revolutions per minute. Various methods are possible, for example, pulleys of proper size may be attached to the shafts and connected by a belt, or sprocket wheels may be used and connected by a chain, as in a bicycle, or pulleys may be placed on the shafts and the faces of the pulleys pressed together, so that the friction between them may be sufficient to transmit the power, a drive used sometimes in trucks, or, again, toothed wheels called gear wheels may be used on the two shafts, as in street cars and in most automobiles.

Any of these methods is possible in a few cases, but usually the location of the shafts, their speeds, etc., make some one of the methods the more preferable. If the shafts are far apart, a belt and pulleys may be used, but as the drive is not positive the belt may slip, and thus the relative speeds may change, the speed of the driven wheel often being 5 per cent. lower than the diameters of the pulleys would indicate. Where the shafts are fairly close together a belt does not work with satisfaction, and then a chain and sprockets are sometimes used which cannot slip, and hence the speed ratio required may be maintained. For shafts which are still closer together either friction gears or toothed gears are generally used. Thus the nature of the drive will depend upon various circumstances, one of the most important being the distance apart of the shafts concerned in it, another being the question as to whether the velocity is to be accurately or only approximately maintained, and another being the power to be transmitted.

82. Spur Gearing.—The discussion here deals only with drives of the class which use toothed gears, these being generally used between shafts which must turn with an exact velocity ratio which must be known at any instant, and they are generally used when the shafts are fairly close together. It will be convenient to deal first with parallel shafts, which turn in opposite senses, the gear wheels connecting which are called **spur wheels,** the larger one being commonly known as the **gear,** and the smaller one as the **pinion.** Kinematically, spur gears are the exact equivalent of a pair of smooth round wheels of the same mean diameter, and which are pressed together so as to drive one another by friction. Thus if two shafts 15 in. apart are to rotate at 100 revolutions per minute and 200 revolutions per minute, respectively, they may be connected by two smooth wheels 10 in and 20 in. in diameter, one on each shaft, which are pressed together so that they will not slip, or by a pair of spur wheels of the same mean diameter, both methods producing the desired results. But if the power to be transmitted is great the friction wheels are inadmissible on account of the great pressure between them necessary to prevent slipping. If slipping occurs the velocity ratio is variable, and such an arrangement would be of no value in such a drive as is used on a street car, for instance, on account of the jerky motion it would produce in the latter.

83. Sizes of Gears.—In order to begin the problem in the simplest possible way consider first the very common case of a pair of spur gears connecting two shafts which are to have a constant velocity ratio. This is, the ratio between the speeds n_1 and n_2 is to be constant at every instant that the shafts are revolving. Let l be the distance from center to center of the shafts. Then, if friction wheels were used, the velocities at their rims will be $\pi d_1 n_1$ and $\pi d_2 n_2$ in. per minute, where d_1 and d_2 are the diameters of the wheels in inches, and it will be clear that the velocity of the rim of each will be the same since there is to be no slipping.

Therefore $\quad d_1 n_1 = d_2 n_2$

since $\quad \pi d_1 n_1 = \pi d_2 n_2$

and $\quad r_1 n_1 = r_2 n_2$ where r_1 and r_2 are the radii.

But $\quad r_1 + r_2 = l$

Hence $\quad \dfrac{n_2}{n_1} r_2 + r_2 = l$

or
$$r_2 = \frac{n_1}{n_1 + n_2} \cdot l \text{ in.}$$

and
$$r_1 = \frac{n_2}{n_1 + n_2} \cdot l \text{ in.}$$

Now, whatever actual shape is given of these wheels, the motion of the shafts must be the same as if two smooth wheels, of sizes as determined above, rolled together without slipping. In other words, whatever shape the wheels actually have, the resulting motion must be equivalent to that obtained by the rolling together of two cylinders centered on the shafts. In gear wheels these cylinders are called **pitch cylinders,** and their projections on a plane normal to their axes, **pitch circles,** and the circles evidently, touch on a line joining their centers, which point is called the **pitch point.**

84. Proper Outlines of Bodies in Contact.—Let a small part of the actual outline of each wheel be as shown in the hatched lines of Fig. 42, the projections on the wheels being required to prevent slipping of the pitch lines. It is required to find the necessary shape which these projections must have.

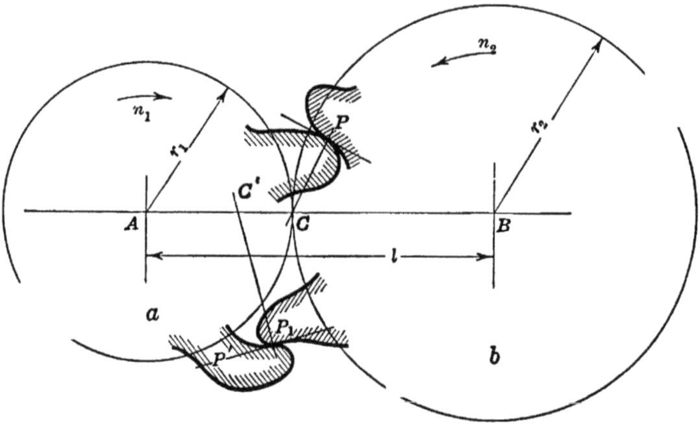

FIG. 42.

Let the actual outlines of the two wheels touch at P and let P be joined to the pitch point C; it has been already explained that there must be no slipping of the circles at C. Now P is a point in both wheels, and as a point in the gear b it moves with regard to C on the pinion a at right angles to PC, while as a point on the pinion a it moves with regard to the gear b also at right angles to PC. Whether, therefore, P is considered as a point on a or b its motion must be normal to the line joining it to C. A little consideration will show that, in order that this condition may be fulfilled, the shape of both wheels at P must be normal to PC.

In order to see this more clearly, examine the case shown in the lower part of the figure, where the projections are not normal to P_1C at the point P_1 where they touch. From the very nature of the case sliding must occur at P_1, and where two bodies slide on one another the direction of sliding must always be along the common tangent to their surfaces at the point of contact, that is, the direction of sliding must in this case be P_1P'. But P_1 is the point of contact and is therefore a point in each wheel, and the motion of the two wheels must be the same as if the two pitch circles rolled together, having contact at C. Such being the case, if the two projections shown are placed on the wheels, the direction of motion at their point of contact should be perpendicular to P_1C, whereas here it is perpendicular to P_1C'. This would cause slipping at C, and would give the proper shape for pitch circles of radii AC' and BC', which would correspond to a different velocity ratio. Thus C' should lie at C and P_1P' should be normal to P_1C.

Another method of dealing with this matter is by means of the virtual center. Calling the frame which supports the bearings of a and b, the link d, then A is the center ad and B is bd while the pitch point C is ab. It is shown at Sec. 33 that the motion of b with regard to a at the given instant is one of rotation about the center ab and hence the motion of P in b is normal to PC. Where the two wheels are in contact at P there is relative sliding perpendicular to PC, that is, at P the surfaces must have a common tangent perpendicular to PC. The shape shown at P_1 is incorrect because from Sec. 33 the center ab must lie in a normal through P_1 and also on $ad - bd$ from the theorem of the three centers, Sec. 39; so that it would lie at C'. But if ab were at C_1' then $\frac{n_1}{n_2} = \frac{BC'}{AC'}$ which does not give the ratio required.

85. Conditions to be Fulfilled.—From the foregoing the following important statements follow: **The shapes of the projections or teeth on the wheels must be such that at any point of contact they will have a common normal passing through the fixed pitch point, and while the pitch circles roll on one another the projections or teeth will have a sliding motion.** These projections on gear wheels are called **teeth**, and for convenience in manufacturing, all the teeth on each gear have the same shape, although this is not at all necessary to the motion. The teeth on the pinion are not the same shape as those on the gear with which it **meshes.**

There are a great many shapes of teeth, which will satisfy the necessary condition set forth in the previous paragraph, but by far the most common of these are the cycloidal and the involute teeth, so called because the curves forming them are cycloids and involutes respectively.

86. Cycloidal Teeth.—Select two circles *PC* and *P'C*, Fig. 43, and suppose these to be mounted on fixed shafts, so that the

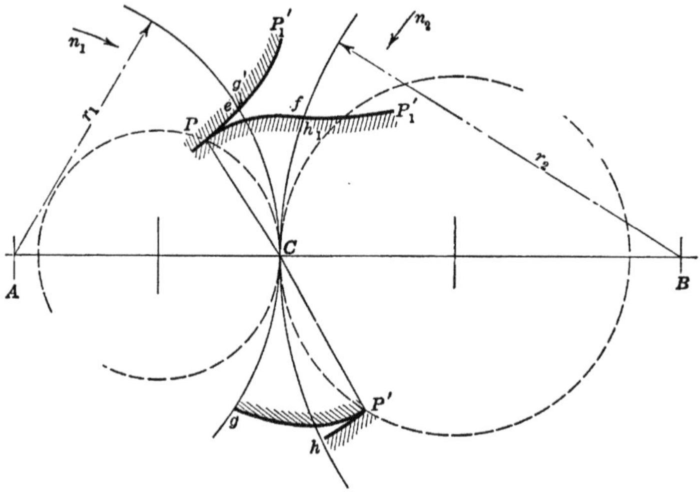

FIG. 43.—Cycloidal teeth.

centers *A* and *B* of the pitch circles, and the centers of the **describing circles** *PC* and *P'C*, as well as the pitch point *C*, all lie in the same straight line, which means that the four circles are tangent at *C*. Now place a pencil at *P* on the circle *PC* and let all four circles run in contact without slipping, that is, the circumferential velocity of all circles at any instant is the same. As the motion continues *P* approaches the pitch circles *eC* and *fC*, and if the right-hand wheel is extended beyond the circle *fCh*, the pencil at *P* will describe two curves, a shorter one *Pe* on the wheel *eCg* and a longer one *Pf* on the wheel *fCh*, the points *e* and *f* being reached when *P* reaches the point *C*, and from the conditions of motion arc PC = arc $eC = fC$.

Now *P* is a common point on the curves *Pe* and *Pf* and also a point on the circle *PC*, which has the common point *C* with the remaining three circles. Hence the motion of *P* with regard to eCg is perpendicular to PC, and of P with regard to fCh is perpendicular to PC, that is, the tangents to Pe and Pf at P are normal to PC, or the two curves have a common tangent, and hence a common normal PC at their point of contact, and this normal will pass through the pitch point C. Thus Pe and Pf fulfil the necessary conditions for the shapes of gear teeth. Evidently the points of contact along these two curves lie along PC, since both curves are described simultaneously by a point which always remains on the circle PC. Now these curves are first in contact at P and the point of contact travels down the arc PC relative to the paper till it finally reaches C where the points, P, C, e and f coincide, so that since Pe is shorter than Pf, the curve Pe slips on the curve Pf through the distance $Pf - Pe$ during the motion from P to C.

Below C the pencil at P would simply describe the same curves over again only reversed, and to further extend these curves, a second pencil must be placed at P' on the right-hand circle $P'C$, which pencil will, in moving downward from C, draw the curves $P'g$ and $P'h$, also fulfilling the necessary conditions, the points of contact lying along the arc CP' and the amount of slipping being $P'g - P'h$.

Having thus described the four curves join the two formed on wheel eCg, that is gP' and Pe, forming the curve $Peg'P'_1$ and the two curves on fCh as shown at $Pfh_1P'_1$, and in this way long curves are obtained which will remain in contact from P to P', the point of contact moving relative to the paper, down the arcs PCP', and the common normal at the point of contact always passing through C. The total relative amount of slipping is $Pfh_1P'_1 - Peg'P'_1$. If now two pieces of wood are cut out, one having its side shaped like the curve PeP'_1 and pivoted at A, while the other is shaped like PfP'_1 and pivoted at B; then from what has been said, the former may be used to drive the latter, and the motion will be the same as that produced by the rolling of the two pitch circles together; hence these shapes will be the proper ones for the profiles of gear teeth.

87. Cycloidal Curves.—The curves Pe, Pf, $P'g$ and $P'h$, which are produced by the rolling of one circle inside or outside of another, are called **cycloidal** curves, the two Pe and $P'h$ being known as hypocycloids, since they are formed by the describing circle rolling inside the pitch circle, while the two curves Pf and $P'g$ are known as epicycloidal curves, as they lie outside the pitch circles. Gears having these curves as the profiles of the teeth are said to have **cycloidal teeth** (sometimes erroneously called epicycloidal teeth), a form which is in very common use. So far only one side of the tooth has been drawn, but it will be evident that the other side is simply obtained by making a tracing of the curve PeP'_1 on a piece of tracing cloth, with center A also marked, then by turning the tracing over and bringing the point A on the tracing to the original center A on the draw-

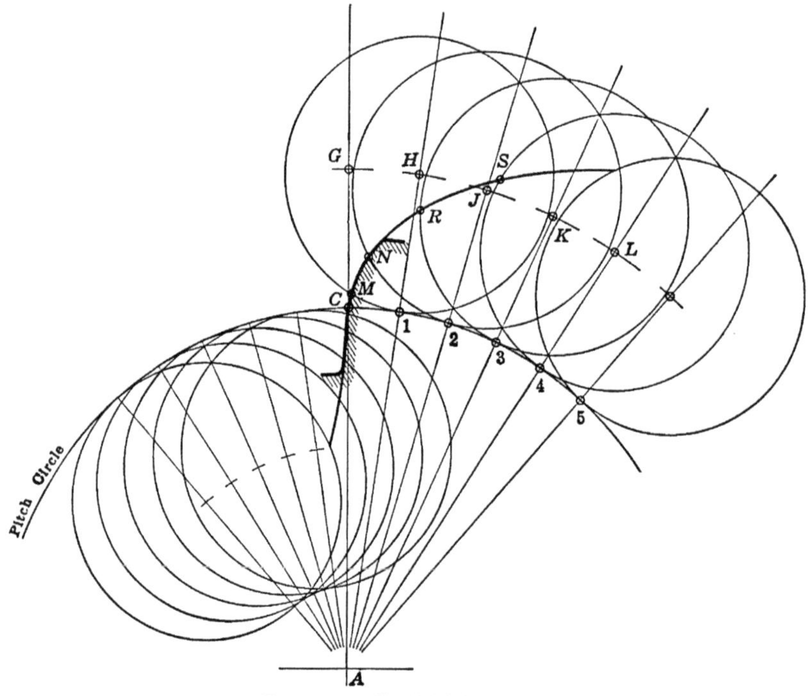

FIG. 44.—Cycloidal teeth.

ing, the other side of the tooth on the wheel eCg may be pricked through with a needle. The same method may be employed for the teeth on wheel fCh.

The method of drawing these curves on the drafting board is not difficult, and may be described. Let C . . . 5 in Fig. 44 represent one of the pitch circles and the smaller circles the describing circle. Choose the arc $C5$ of any convenient length and divide it into an equal number of parts the arcs $C-1$, $1-2$, etc., each being so short as to equal in length the corresponding chords. Draw radial lines from A as shown, and locate points G,H,J,K,L at distance from the pitch circle equal to the radius of the describing circle, and from these points draw in a number of circles equal in size to the describing circle. Now lay off the arc $1M$ = arc $1C$, and arc $2N$ equal the arc $C2$ or twice the arc $C1$, and the arc $3R$ equal three times arc $C1$, etc., in this way finding the points C, M, N, R, S, which are points on the desired epicycloidal curve. Similarly the hypocycloidal curve below the pitch circle may be drawn.

88. Size of Describing Circle.—Nothing has so far been said of the sizes of the describing circles, and, indeed, it is evident that any size of describing circle, so long as it is somewhat smaller than the pitch circle, may be used, and will produce a curve ful-

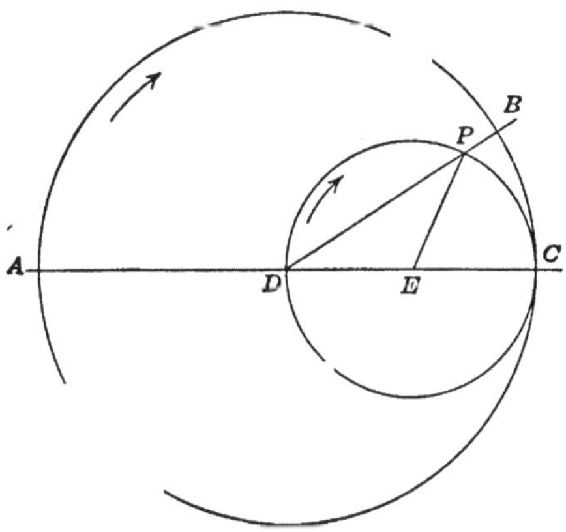

Fig. 45.

filling the desired conditions, but it may be shown that when the describing circle is one-half the diameter of the corresponding pitch circle the hypocycloid becomes a radial line in the pitch circle, and for reasons to be explained later this is undesirable. The maximum size of the describing circle is for this reason limited to one-half that of the corresponding pitch circle and when a set of gears are to run together, the describing circle for the set is usually half the size of a gear having from 12 to 15 teeth. This will enable any two wheels of the set to work properly together.

The proof that the hypocycloid is a radial line if the describing circle is half the pitch circle, may be given as follows: Let ABC, Fig. 45, be the pitch circle and DPC the describing circle, P being the pencil, and BP the line described by P as P and B approach C. The arc BC is equal to the arc PC by construction, and hence the angle PEC at the center E of DPC is twice the angle BDC, because the radius in the latter case is twice that in the former. But the angles BDC and PEC are both in the smaller circle, the one at the circumference and the other at the center, and since the latter is double the former they must stand on the same arc PC. In other words BP is a radial line in the larger circle since DP and DB must coincide.

89. Teeth of Wheels.—In the actual gear the tooth profiles are not very long, but are limited between two circles concentric

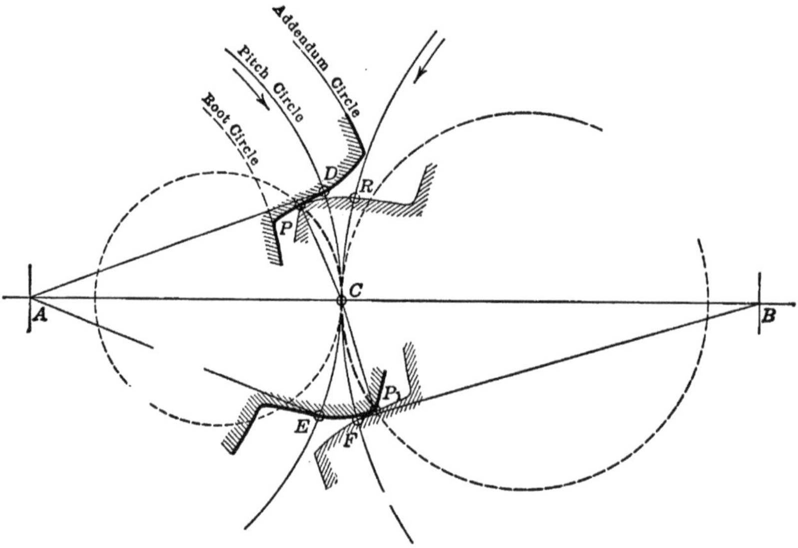

FIG. 46.

with the pitch circle in each gear, and called the **addendum** and **root circles** respectively, for the tops and bottoms of the teeth, the distances between these circles and the pitch circle being quite arbitrarily chosen by the manufacturers, although certain proportions, as given later, have been generally adopted. These circles are shown on Fig. 46 and they limit the path of contact to the reversed curve PCP_1 and the amount of slipping of each pair of teeth to $PR - PD + P_1E - P_1F = PR + P_1E - (PD + P_1F)$, the distances being measured along the profiles of the teeth in all cases. Further, since the common normal to the teeth always passes through C, then the direction of pressure between a given pair of teeth is always along the line joining their point of contact to C, friction being neglected, the limiting directions of this line of pressure thus being PC and P_1C.

The arc PC is called the **arc of approach,** being the locus of the points of contact down to the pitch point C, while the arc CP_1 is the **arc of recess,** P_1 being the last point of contact. Similarly, the angles DAC and CAE are called the **angle of approach** and **angle of recess,** respectively, for the left-hand gear. The reversed curve PCP_1 is the **arc of contact** and its length depends to some extent on the size of the describing circles among other things, being longer as the relative size of the describing circle increases. If this arc of contact is shorter than the distance between the centers of two adjacent teeth on the one gear, then only one pair of teeth can be in contact at once and the running is uneven and unsatisfactory, while if this arc is just equal to the distance between the centers of a given pair of teeth on one gear, or the **circular pitch,** as it is called (see Fig. 52), one pair of teeth will just be going out of contact as the second pair is coming in, which will also cause jarring. It is usual to make PCP_1 at least 1.5 times the pitch of the teeth. This will, of course, increase the amount of slipping of the teeth.

With the usual proportions it is found that when the number of teeth in a wheel is less than 12 the teeth are not well shaped for strength or wear, and hence, although they will fulfil the kinematic conditions, they are not to be recommended in practice.

90. Involute Teeth.—The second and perhaps the most common method of forming the curves for gear teeth is by means of involute curves. Let A and B, Fig. 47, represent the axes of the gears, the pitch circles of which touch at C, and through C draw a secant DCE at any angle θ to the normal to AB, and with centers A and B draw circles to touch the secant in D and E. Now (Sec. 83) $\frac{n_1}{n_2} = \frac{BC}{AC} = \frac{BD}{AE}$, so that the new circles have the same speed ratios as the original pitch circles. If then a string is run from one dotted circle to the other and used as a belt between these dotted or **base circles** as they are called, the proper speed ratio will be maintained and the two pitch circles will still roll upon one another without slipping, having contact at C.

Now, choose any point P on the belt DE and attach at this point a pencil, and as the wheels revolve it will evidently mark on the original wheels having centers at A and B, two curves Pa and Pb respectively, a being reached when the pencil gets down to E, and b being the starting point just as the pencil leaves

D, and since the point P traces the curves simultaneously they will always be in contact relation to the paper at some point along DE, the point of contact traveling downward with the pencil at P. Since P can only have a motion with regard to the

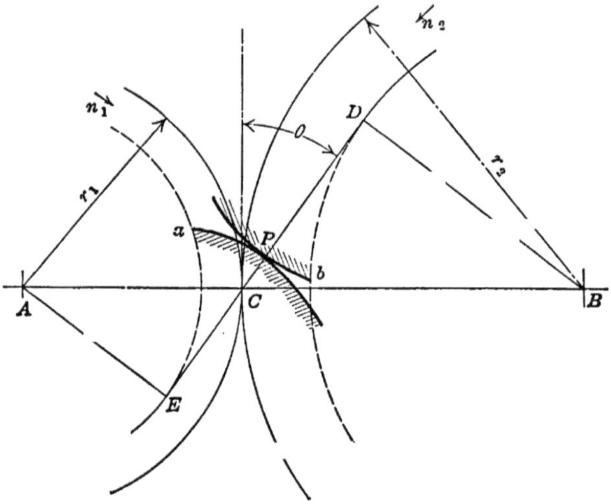

FIG. 47.—Involute teeth.

wheel aE normal to the string PE, and its motion with regard to the wheel Db is at right angles to PD, it will be at once evident

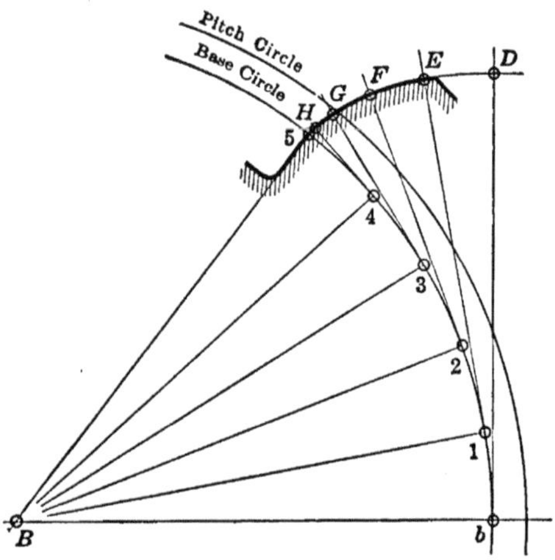

FIG. 48.—Involute teeth.

that these two curves have a common normal at the point where they are in contact, and this normal evidently passes through C. Hence the curves may be used as the profiles of gear teeth (Sec. 85).

The method of describing these curves on the drafting board is as follows: Draw the base circle $b-5$ with center B, Fig. 48, and lay off the short arcs $b-1$, $1-2$, $2-3$, $3-4$, etc., all of equal length and so short that the arc may be regarded as equal in length to the chord. Draw the radial lines Bb, $B-1$, $B-2$, etc., and the tangents bD, $1-E$, $2-F$, $3-G$, $4-H$ any length, and lay off $4-H$ = arc $5-4$, $3-G$ = arc $3-5$ which equals twice arc $5-4$, $2-F$ equal three times arc $5-4$, $1-E$ equal four times arc $5-4$, etc.; then D, E, F, G, H and 5 are all points on the desired curve and the latter may now be drawn in and extended, if desired, by choosing more points below b.

91. Involute **Curves.**—The curves Pa and Pb, Fig. 47, are called involute **curves,** and when they are used as the profiles of gear teeth the latter are **involute teeth.** The angle θ is the **angle of obliquity,** and evidently gives the direction of pressure between the teeth, so that the smaller this angle becomes the less will be the pressure between the teeth for a given amount of power transmitted. If, on the other hand, this angle is unduly small, the base circles approach so nearly to the pitch circles in size that the curves Pa and Pb have very short lengths below the pitch circles. Many firms adopt for θ the angle $14\frac{1}{2}°$, in which case the diameter of the base circle is 0.968 (about $\frac{31}{32}$) that of the pitch circle. If the teeth are to be extended inside the base circles, as is usual, the inner part is made radial. With teeth of this form the distance between the centers A and B may be somewhat increased without affecting in any way the regularity of the motion. Recently some makers of gears for automobile work have increased the angle of obliquity to 20°, in this way making the teeth much broader and stronger. **Stub teeth** to be discussed later, are frequently made in this way, largely for use on automobiles. A discussion of the forms of teeth appears in a later section.

92. Sets of Wheels with Involute Teeth.—Gears with involute teeth are now in very common use, and if a set of these is to be made, any two of which are capable of working together, then all must have the same angle of obliquity. The arc of contact is usually about twice the circular pitch and the number of teeth in a pinion should not be less than 12 as the teeth are liable to be weak at the root unless the angle of obliquity is increased.

A more complete drawing of a pair of gears having involute teeth is shown in Fig. 49. Taking the upper gear as the driver, the line of contact will be along $DPCE$, but the addendum circles usually limit the length of this contact to some extent, contact taking place only on the part of the obliquity line DE inside the addendum line. The larger the addendum circles the longer the lines of contact will be and the proportions are such in Fig 49 that contact occurs along the entire line DE. No contacts can possibly occur inside the base circles.

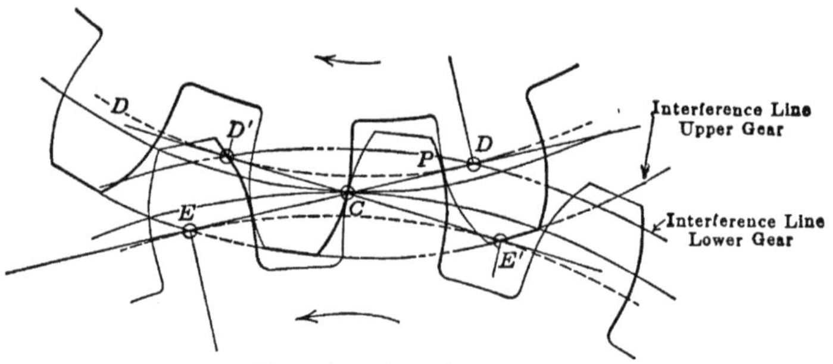

FIG. 49.—Involute teeth.

93. Racks.—When the radius of one of the gears becomes infinitely large the pitch line of it becomes a straight line tangent to the pitch line of the other gear and it is then called a **rack.** The teeth of the rack in the cycloidal system are made in exactly the same manner as those of an ordinary gear, but both the describing circles roll along a straight pitch line, generating **cycloidal curves,** having the same properties as those on the ordinary gear.

For the involute system the teeth on the rack simply have straight sides normal to the angle of obliquity, each side of such teeth forming the angle θ with the radius line AC drawn from the center of the pinion to the pitch point.

94. Annular or Internal Gears.—In all cases already discusesd the pair of gears working together have been assumed to turn in opposite sense, resulting in the use of **spur gears,** but it not infrequently happens that it is desired to have the two turn in the same sense, in which case the larger one of the gears must have teeth inside the rim and is called an **annular or internal gear.** An annular gear meshes with a spur pinion, and it will be self-evident that the annular gear must always be somewhat larger than the pinion.

A small part of annular gears both on the cycloidal and involute systems is shown at Fig. 50 and the odd appearance of the involute internal gear teeth is evident; such gears are frequently avoided by the use of an extra spur gear.

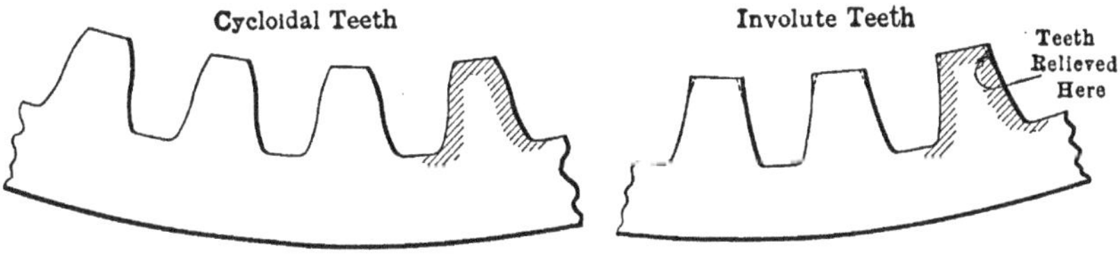

FIG. 50.—Internal gears.

95. Interference.—The previous discussion deals with the correct theoretical form of teeth required to give a uniform velocity

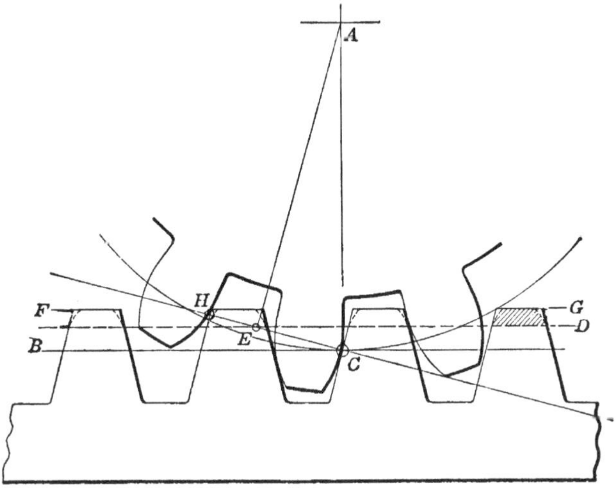

(a)

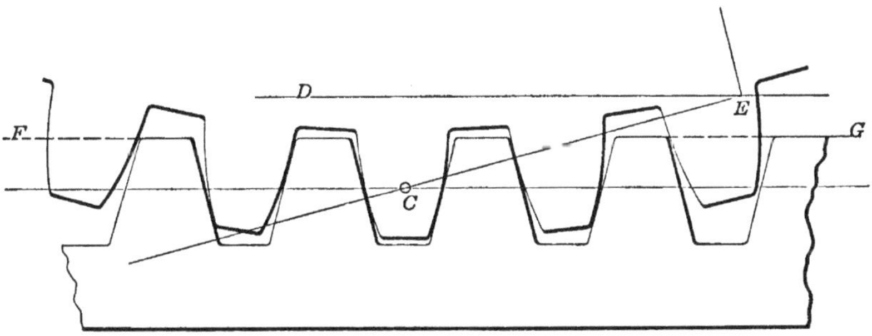

(b)

FIG. 51.—Interference.

ratio, but with the usual proportions adopted in practice for the addendum, pitch and root circles, it is found that in certain cases parts of teeth on one of the gears would cut into the teeth on the other gear, causing **interference.** This is most common with the involute system and occurs most where the difference in size of the gears in contact is greatest; thus interference is worst where a small pinion and a rack work together, but it may occur, to some extent with all sizes of gears.

An example will make this more clear. The drawing in Fig. 51 represents one of the smaller pinions geared with a rack in the involute system and it is readily seen that the point of the rack tooth cuts into the root of the tooth on the gear at H and that in order that the pair may work together it will be necessary either to cut away the bottom of the pinion tooth or the top of the rack tooth. This conflict between the two sets of teeth is called **interference.**

Looking at the figure, and remembering the former discussion (Sec. 90) on involute teeth, it is seen that contact will be along the line of obliquity from C to E and that points on this line CE produced have no meaning in this regard, so that if BC denote the pitch line of the rack, the teeth of the rack can only be usefully extended up to the line ED, whereas the actual addendum line is FG. Thus, the part of the rack teeth between ED and FG, as shown hatched on one tooth on the right, cannot be made the same shape as the involute would require but must be modified in order to clear the teeth of the pinion. The usual practice is to modify the teeth on the rack, leaving the lower parts of the teeth on the pinion unchanged, and the figure shows dotted how the teeth of the rack are trimmed off at the top to make proper allowance for this.

Interference will occur where the point E, Fig. 51 (*a*), falls below the addendum line FG, the one tooth cutting into the other at H on the line of obliquity. Where a pinion meshes with a gear which is not too large, then the curvature of the addendum line of the gear may be sufficient to prevent contact at the point H, in which case interference will not occur. As has already been explained, interference occurs most when a pinion, meshes with a gear which is very much larger, or with a rack. Where a large gear meshes with a rack as in the diagram at Fig. 51 (*b*), the interference line DE is above the addendum line FG and hence no modification is necessary.

In Fig. 49 the interference line for the lower gear is inside the addendum line and hence the points of these teeth must be cut away, but the points of the teeth on the upper gear would be correct as the interference line for it coincides with its addendum line.

96. Methods of Making Gears.—Gear wheels are made in various ways, such as casting from a solid pattern, or from a pattern on a moulding machine containing only a few teeth, neither of which processes give the most accurate form of tooth. The only method which has been devised of making the teeth with great accuracy is by cutting them from the solid casting, and the present discussion deals only with cut teeth. In order to produce these, a casting or forging is first accurately turned to the outside diameter of the teeth, that is to the diameter of the addendum line, and the metal forming the spaces between the teeth is then carefully cut out by machine, leaving accurately formed teeth if the work is well done. Space does not permit the discussion of the machinery for doing this class of work, for various principles are used in them and a number of makes of the machines will produce theoretically correct tooth outlines. The reader will be able to secure information from the builders of these machines himself.

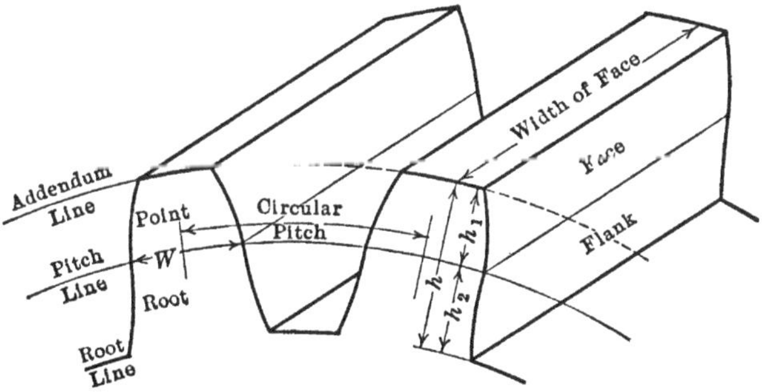

FIG. 52.

97. Parts of Teeth.—The various terms applied to gear teeth, either of the involute or cycloidal form, will appear from Fig. 52. The **addendum** line is the circle whose diameter is that of the outside of the gear, the **dedendum** line is a circle indicating the depth to which the tooth on the other gear extends; usually the addendum and dedendum lines are equidistant from the pitch line. The teeth usually are cut away to the **root circle** which is slighly inside the dedendum circle to allow for some **clearance,** so that the total depth of the teeth somewhat exceeds the working depth or distance between the addendum and dedendum circles. The dimension or length of the tooth parallel to the shaft is the **width of face of the gear,** or often only the **face of the gear,** while the face of the tooth is the surface of the latter above the pitch line and the **flank of the tooth** is the surface of the tooth below the pitch line. The solid part of the tooth outside the pitch line is the **point** and the solid part below the pitch line is the **root.**

Two systems of designating cut teeth are now in use, the one most commonly used being by Brown and Sharpe and it will first be described.

Let d be the pitch diameter of a gear having t teeth, h_1 the depth of the tooth between pitch and addendum circles, and h_2 the depth below the pitch circle, so that the whole depth of the tooth is $h = h_1 + h_2$, while the working depth is $2h_1$. The distance measured along the circumference of the pitch circle from center to center of teeth is called the **circumferential** or **circular pitch** which is denoted by p and it is evident that $pt = \pi d$. In the case of cut teeth the width W of the tooth and also of the space along the pitch circle are equal, that is, the width of the tooth measured around the circumference of the pitch circle is equal to one-half the circular pitch. The statements in the present paragraph are true for all systems.

98. System of Teeth Used by Brown and Sharpe.—Brown and Sharpe have used very largely the term **diametral pitch** which is defined as the number of teeth divided by the diameter in inches of the pitch circle, and the diametral pitches have been largely confined to whole numbers though some fractional numbers have been introduced. Thus a gear of 5 diametral pitch means one in which the number of teeth is five times the pitch diameter in inches, that is such a gear having a pitch diameter of 4 in. would have 20 teeth. Denoting the diametral pitch by q then $q = \frac{t}{d}$ and from this it follows that $pq = \pi$ or the product of the diametral and circular pitches is 3.1416 always. The circular pitch is a number of inches, the diametral pitch is not.

The standard angle of obliquity used by Brown and Sharpe is 14½° and further $h_1 = \frac{1}{q}$ in., $h_2 = \frac{1}{q} + \frac{p}{20}$ in., clearance $= \frac{p}{20}$ in., and the width W of the tooth is $\frac{p}{2}$, so that there is no side clearance or **back lash** between the sides of the teeth.

99. Stub Tooth System.—Recently the very great use of gears for automobiles and the severe service to which these gears have been put has caused manufacturers to introduce what is often called the "**Stub Tooth**" system in which the teeth are not proportioned as adopted by Brown and Sharpe. Stub teeth are made on the involute system with an obliquity of 20° usually, and are not cut as deep as the teeth already described. The dimensions of the teeth are designated by a fraction, the numerator of which indicates the diametral pitch used, while the denominator shows the depth of tooth above the pitch line. A $5/7$ gear is one of 5 diametral pitch and having a tooth of depth $h_1 = 1/7$ in. above the pitch line (in the Brown and Sharpe system h_1 would be $1/5$ in. for the same gear).

The usual pitches with stub tooth gears are $4/5$, $5/7$, $6/8$, $7/9$, $8/10$, $9/11$, $10/12$ and $12/14$. Some little difference of opinion appears to exist with regard to the clearance between the tops and roots of the teeth, the Fellows Gear Shaper Co. making the clearance equal to one-quarter of the depth h_1. Thus, a $5/7$ gear would have the same h_1 as is used in the Brown and Sharpe system for a 7 diametral pitch gear, that is $h_1 = 0.1429$ in., and a clearance equal to $0.25 \times 1/7 = 0.0357$ in., which is much greater than the 0.0224 in. which would be used in the Brown and Sharpe system on a 7 pitch gear.

100. The Module.—In addition to the methods already explained of indicating the size of gear teeth, by means of the circular and diametral pitch, the **module** has also to some extent been adopted, more especially where the metric system of measurement is in use. The module is the number of inches of diameter per tooth, and thus corresponds with the circular pitch, or number of inches of circumferences per tooth, and is clearly the reciprocal of the diametral pitch. Using the symbol m for the module the three numbers indicating the pitch are related as follows:

$$m = \frac{1}{q}; \text{ also } q = \frac{1}{m} = \frac{\pi}{p}.$$

The module is rarely expressed in other units than millimeters.

101. Examples.—A few illustrations will make the use of the formulas clear, and before working these it is necessary to remember that any pair of gears working together must have the same pitch and a set of wheels constructed so that any two may work together must have the same pitch and be designed on the same system.

Let d_1 and d_2 be the pitch diameters of two gears of radii r_1 and r_2 respectively and let these be placed on shafts l in. apart and turning at n_1 and n_2 revolutions per minute. Then from Sec. 83, where spur gears only are used,

$$r_1 = \frac{n_2}{n_1 + n_2} \cdot l$$

and

$$_2 = \frac{n_1}{n_1 + n_2} \cdot l$$

also

$$\frac{r_1}{r_2} = \frac{t_1}{t_2} = \frac{n_2}{n_1}$$

Suppose $l = 9$ in. between centers of shafts which turn at 100 revolutions and 200 revolutions per minute; then, substituting in the above formula $r_1 = \frac{200}{100 + 200} \times 9 = 6$ in. and $r_2 = \frac{100}{100 + 200} \times 9 = 3$ in., or the gears will be 12 in. and 6 in. diameter respectively. If cut to 4 diametral pitch the numbers of teeth will be $t_1 = 4 \times 12 = 48$ and $t_2 = 4 \times 6 = 24$. The circular pitch is $\frac{\pi}{4} = 0.7854$ in. Further, $h_1 = \frac{1}{4}$ in. and the outside diameters of the gears are $12\frac{1}{2}$ in. and $6\frac{1}{2}$ in., the tooth clearances $= \frac{0.7854}{20} = 0.0393$ in. The module would be $\frac{1}{4}$ in. $= \frac{6 \text{ in.}}{24 \text{ teeth}}$.

If the gears have stub teeth of four-fifths size, then the numbers of teeth will be 48 and 24 as before, but h_1 will be $\frac{1}{5} = 0.2$ in., so that the outside diameters will be 12.4 in. and 6.4 in. respectively, the clearance will be $\frac{1}{4} \times \frac{1}{4} = 0.0625$ in. and the total depth of the teeth 0.4625 in. as compared with 0.5393 in. for the teeth on the Brown and Sharpe system.

Inasmuch as it is rather more usual to use the outside diameter of a gear than the pitch diameter in shops where they are made, it is very desirable that the reader become so familiar with the proportions as to be able to know instantly the relations between the different dimensions of the gears in terms of the outside diameter and the pitch.

102. Discussion of the Gear Systems.—The involute form of tooth is now more generally used than the cycloidal form. In the first place the profile is a single curve instead of the double one required with the cycloidal shape. Again, because of its construction, it is possible to separate the centers of involute gears without causing any unevenness of running, that is, if the gears are designed for shafts at certain distance apart this distance may be slightly increased without in the least altering the velocity ratio or disturbing the evenness of the running, this is an advantage not possessed by cycloidal teeth.

In cycloidal teeth the direction of pressure between a given pair of teeth is variable, being always along the line joining the pitch point to the point of contact, and when the point of contact is the pitch point, the direction of pressure is normal to the line joining the shaft centers. In the involute teeth the pressure is always in the same direction being along the line of obliquity, and thus the pressure between the teeth and the force tending to separate the shafts is somewhat greater in the involute form, although there is no very great advantage in cycloidal teeth from this point of view. The statements in this paragraph assume that there is no friction between the teeth.

Interference is somewhat greater in involute teeth.

As regards the Brown and Sharpe proportions and the stub teeth, of course the large angle of obliquity of the latter teeth increases the pressure for the power transmitted. The stub tooth gears are, however, stronger and there is very little interference owing to the shorter tooth. The teeth would possibly be a little cheaper to cut, and this as well as their greater strength would give them a considerable advantage in such machines as automobiles.

103. Helical Teeth.—A study of such drawings as are shown at Fig. 40, etc., shows that the smaller the depth of the teeth the less will be the amount of slipping and therefore the less the frictional loss. But this is also accompanied by a decrease in the arc of contact and hence the number of teeth in contact at any one time will, for a given pitch, be decreased, which may cause unevenness in the motion. If, however, the whole width of the gear be assumed made up of a lot of thin discs, and if, after the teeth had been cut across all the discs at once, they were then slightly twisted relatively to one another, then the whole width of the gear would be made up of a series of steps and if these steps were made small enough the teeth would run across the face of the gear as helices, and the gear so made would be called **a helical gear.** The advantage of such gears will appear very easily, for instead of contact taking place across the entire width of the face of a tooth at one instant, the tooth will only gradually come into contact, the action beginning at one end and working gradually over to the other, and in this way very great evenness of motion results, even with short teeth of considerable pitch.

The profile of the teeth of such gears is made the same, on a plane normal to the shaft, as it would be if they were ordinary spur gears.

Fig. 53.—Double helical gears.

Helical gears are a necessity in any case where high speed and velocity ratio are desired, and the modern reduction gear now being much used between steam turbines and turbine pumps and dynamos would be a failure, on account of the noise and vibration, if ordinary spur gears were used. Such reduction gears are always helical and frequently two are used with the teeth running across in opposite directions so as to avoid end thrust. Some turbines have been made in which such gears, running at speeds of over 400 revolutions per second, have worked without great noise.

A photograph of a De Laval double helical gear is shown in Fig. 53, this gear being used to transmit over 1,000 hp. without serious noise. In the figure the teeth run across the face of the gear at about 45°, and are arranged to run in oil so as to prevent undue friction.

QUESTIONS ON CHAPTER V

1. A shaft running at 320 revolutions per minute is to drive a second one 20 in. away at 80 revolutions per minute, by means of spur gears; find their pitch diameters.

2. If both shafts were to turn in the same sense at the speeds given and were 4 in. apart, what would be the sizes of the gears?

3. What is the purpose of gear teeth and what properties must they possess?

4. Define cycloid, hypocycloid and epicycloid.

5. Draw the gear teeth, cycloidal system, for a gear of 25 teeth, 2½ pitch with 3-in. describing circle, (*a*) for a spur gear, (*b*) for an annular gear.

6. A pair of gears are to connect two shafts 9 in. apart, ratio 4 to 5; the diametral pitch is to be 2 and the describing circles 1½ in. diameter. Draw the teeth.

7. Define the various terms used in connection with gears and gear teeth

8. Define involute curve, angle of obliquity, base circle. Show that all involute curves from the same base circle are identical.

9. Lay out the gears in problem 6, for involute teeth, obliquity 14½°.

10. What is meant by interference in gear teeth and what is the cause of it? Why does it occur only under some circumstances and not always?

11. How may interference be prevented? What modification is usually made in rack teeth?

12. What effect has variation of the angle of obliquity on involute teeth?

13. What are the relative merits of cycloidal and involute teeth?

14. Obtain all the dimensions of the following gears: (*a*) Two spur wheels, velocity ratio 2, pitch 2, shafts 9 in. apart. (*b*) Outside diameter of gear 4 in., diametral pitch 8. (*c*) Gear of 50 teeth, 4 pitch. (*d*) Wheel of 10½ in. outside diameter, 40 teeth. (*e*) Pair of gears, ratio 3 to 4, smaller 6 in. pitch diameter with 30 teeth.

15. What is a stub tooth, and what are the usual proportions? What advantages has it?

16. Describe the various methods of giving the sizes of gear teeth and find the relation between them.

17. Two gears for an automobile are to have a velocity ratio 4 to 5, shaft centers 4½ in., 6 pitch; draw the correct teeth on the 20° stub system.

18. Explain the construction of the helical gear and state its advantages

CHAPTER VI

BEVEL AND SPIRAL GEARING

104. Gears for Shafts not Parallel.—Frequently in practice the shafts on which gears are placed are not parallel, in which case the spur gears already described in the former chapter cannot be used and some other form is required. The type of gearing used depends, in the first place, on whether the axes of shafts intersect or not, the most common case being that of intersecting axes, such as occurs in the transmission of automobiles, the connection between the shaft of a vertical water turbine and the main horizontal shaft, in governors, and in very many other well-known cases.

On the other hand it not infrequently happens that the shafts do not intersect, as is true of the crank- and camshafts of many gas engines, and of the worm-gear transmissions in some motor trucks. In many of these cases the shafts are at right angles, as in the examples quoted, but the cases where they are not are by no means infrequent and the treatment of the present chapter has been made general.

105. Types of Gearing.—Where the axes of the shafts intersect the gears connecting them are called **bevel gears.** Where the axes of the shafts do not intersect the class of gearing depends upon the conditions to be fulfilled by it. If the work to be done by the gearing is of such a nature that point contact between the teeth is sufficient, then **screw or spiral gearing** is used, a form of transmission very largely used where the shafts are at rightangles, although it may also be used for shafts at other angles. One peculiarity of this class is that the diameters of the gears are not determined by the velocity ratio required, and in fact it would be quite possible to keep the velocity ratio between a given pair of shafts constant and yet to vary within wide limits the relative diameters of the two gears used.

Where it is desired to maintain **line contact** between the teeth of the gears on the two shafts, then the sizes of the gears are exactly determined, as for spur gears, by the velocity ratio and also the angle and distance between the shafts. Such gears are called **hyperboloidal** or **skew bevel gears** and are not nearly so common as the spiral gears, but are quite often used.

The different forms of this gearing will now be discussed, and although a general method of dealing with the question might be given at once, it would seem better for various reasons to defer the general case for a while and to deal in a special way with the simpler and more common case afterward giving the general treatment.

BEVEL GEARING

106. Bevel Gearing.—The first case is where the axes of the shafts intersect, involving the use of **bevel gearing.** The intersecting angle may have any value from nearly zero to nearly 180°, and it is usual to measure this angle on the side of the point of intersection on which the bevel gears are placed. A very common angle of intersection is 90° and if in such a case both shafts turn at the same speed the two wheels would be identical and are then called **mitre gears.** The type of bevel gearing corresponding to annular spur gearing is very unusual on account of the difficulty of construction, and because such gears are usually easily avoided, however they are occasionally used.

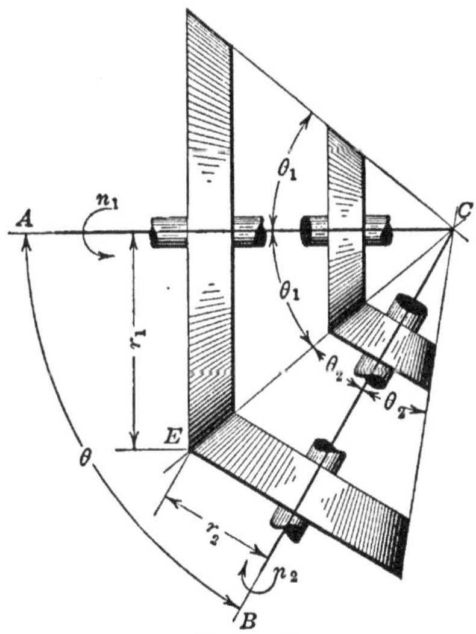

Fig. 54.

Let A and B, Fig. 54, represent the axes of two shafts intersecting at the point C at angle θ, the speeds of the shafts being, respectively, n_1 and n_2 revolutions per minute; it is required to find the sizes of the gears necessary to drive between them. Let E be a point cf contact of the pitch lines of the desired gears and let its distances from A and B be r_1 and r_2, these being the respective radii. Join EC.

Now from Sec. 83 it will be seen that $r_1n_1 = r_2n_2$ since the pitch circles must have the same velocity, there being no slipping between them, and hence $\frac{r_1}{r_2} = \frac{n_2}{n_1}$ is constant, that is at any point where the pitch surfaces of the gears touch, $\frac{r_1}{r_2}$ must be constant, a condition which will be fulfilled by any point on the line EC. In the case of bevel gearing, therefore, the pitch cylinders referred to in Sec. 83 will be replaced by **pitch cones** with apex at C, which cones are generated by the revolution of the line CE about the axes A and B. Two pairs of frustra are shaded in on Fig. 54 and both of these would fulfil the desired conditions for the pitch surfaces of the gear wheels, so that in the case of bevel gearing the diameters of the gears are not fixed as with spur wheels, but the ratio between these diameters alone is fixed by the velocity ratio desired. The angles at the apexes of the two pitch cones are $2\theta_1$ and $2\theta_2$ as indicated. If $\theta = 90°$ and $n_1 = n_2$, then $\theta_1 = \theta_2 = 45°$ and this gives the case of the mitre gears.

In going over the discussion it will be observed that when θ and n_1/n_2 are known, the angles θ_1 and θ_2 and hence the pitch cones are fully determined but the designer has still the option of selecting one of the radii, for example r_1, at his pleasure after which the other one, r_2, is determined.

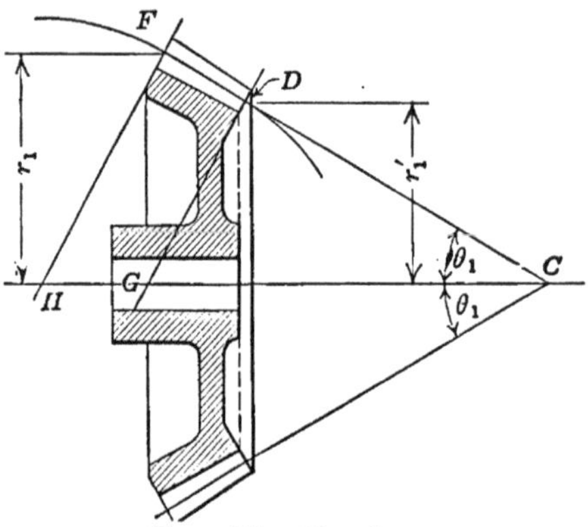

FIG. 55.—Bevel gears.

107. Proper Shape of Teeth on Bevel Gears.—It is much beyond the purpose of the present treatise to go into a discussion of the exact method of obtaining the form of teeth for bevel wheels, because such a method is indeed complicated, and the practical approximation produces very accurate teeth and will be described. Let Fig. 55 represent one of the wheels with angle 2θ, at the vertex of the pitch cone and radii r_1 and r_1' selected in accordance with the previous discussion, the power to be transmitted, and other details fixed by the place in which the gear is to be used. The diameter $2r_1$ is the pitch diameter of the gear, while the distance CF is called the **cone distance** and FH the **back cone distance,** θ_1 is the pitch angle and other terms are the same as are used for spur gears and already explained. The lines DG and FH are normal to CF and intersect the shaft at G and H respectively.

The practical method is to make the teeth at F the same shape and proportions as they would be on a spur gear of radius FH, while at D they correspond to the teeth on a spur gear of radius DG, and a similar method is used for any intermediate point. The teeth should taper from F to D and a straight edge passing through C would touch the tooth at any point for its entire length. Either the involute or cycloidal system may be used.

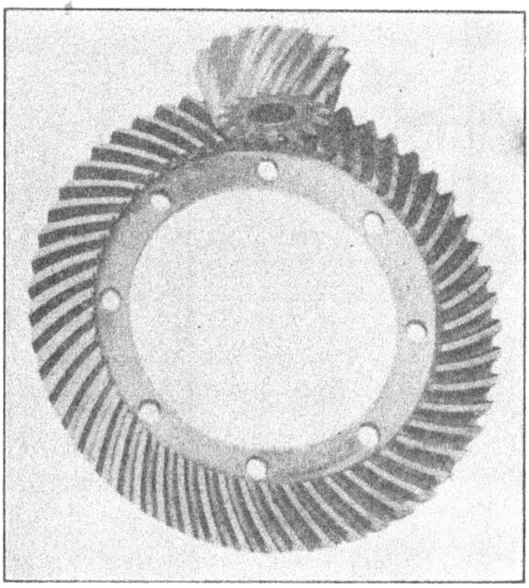

FIG. 56.—Spiral tooth bevel gears.

108. Spiral Tooth Bevel Gears.—Within the past few years the Gleason Works, and possibly others, have devised a method for cutting bevel gears with a form of "spiral" tooth of the same general nature as the helical teeth used with spur gears and described at Sec. 103. A cut of a pair of these from a photograph kindly supplied by the Gleason Works is given at Fig. 56, and shows the general appearance of the gears. Nothing appears to be gained in the way of reducing friction, but they run very smoothly and noiselessly and the greater steadiness of motion makes them of value in automobiles and other similar machines, in which they are mainly used at present.

HYPERBOLOIDAL OR SKEW BEVEL GEARING

The Teeth of which have Line Contact

109. Following the bevel gearing the next class logically is the hyperboloidal gearing and the treatment of this includes the general case of all gearing having line contact between the teeth. Let AO and BP, Fig. 57, represent the axes of two shafts which are to be geared together, the line OP being the shortest line between the axes and is therefore their common perpendicular. Let the axes of the shafts be projected in the ordinary way on two planes, one normal to OP and the other passing through OP and one of the axes AO, the projections on the former plane being AO, OP and PB while those on the second plane are $A'O'$, $O'P'$ and $P'B'$.

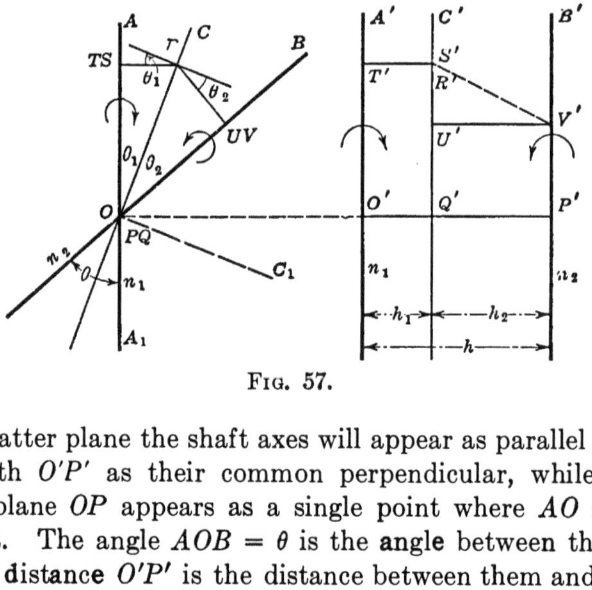

Fig. 57.

On the latter plane the shaft axes will appear as parallel straight lines with $O'P'$ as their common perpendicular, while on the former plane OP appears as a single point where AO and BP intersect. The angle $AOB = \theta$ is the angle between the shafts and the distance $O'P'$ is the distance between them and when θ and $O'P' = h$ are known the exact positions of the shafts are given. The speeds of the shafts n_1 and n_2 must also be known, as well as the sense in which they are to turn.

In stating the angle between the shafts it is always intended to mean the angle in which the line of contact must lie, thus in Fig. 57 the sense of rotation would indicate that the line of contact CQ must lie somewhere in the angle AOB and not in A_1OB so that the angle $\theta = AOB$ is used instead of A_1OB. Should the shaft AO turn in the opposite to that shown, then the line of contact would be in the angle, A_1OB, such as C_1Q (since annular gears are not used for this type) and then the angle A_1OB would be called θ.

110. Data Assumed.—It is assumed in the problem that the angle θ, the distance h, and the speeds n_1 and n_2 or the ratio n_1/n_2, are all given and it is required to design a pair of gears for the shafts, such that the contact between the teeth shall be along a straight line, the gears complying with the above data.

111. Determination of Pitch Surfaces.—Let the line of contact of the pitch surfaces be CQ and let it be assumed that this line passes through and is normal to OP, so that on the right-hand projections $A'O'$, $C'Q'$ and $B'P'$ are all parallel. The problem then is to locate CQ and the pitch surfaces to which it corresponds, the first part of the problem being therefore to determine h_1, h_2, θ_1 and θ_2, Fig. 57, and this will now be done.

Select any point R on CQ, Fig. 57, R being thus one point of contact between the required gears, and from R drop perpendiculars RT and RV on OA and BP respectively. These perpendiculars, which are radii of the desired gears, have the resolved parts ST and UV, respectively, parallel to OP, and the resolved parts RS and RU perpendicular both to OP and to the respective shafts. These resolved parts are clearly shown in the figure, and a most elementary knowledge of descriptive geometry will enable the reader to understand their locations. Further, it is clear the $RT^2 = RS^2 + S'T'^2$ and $RV^2 - RU^2 + U'V'^2$.

At the point of contact R, the correct velocity ratio must be maintained between the shafts, and as R is a point of contact it is a point common to both gears From the discussion in Sec. 84 it will be clear that, as a point on the gear located on OA the motion of R in a plane normal to the line of contact CQ must be identical with the motion of the same point R considered as a point on the wheel on BP, that is, in the plane normal to the line of contact CQ, the two wheels must have the same motion at the point of contact R. Sliding along CQ is not objectionable, however, except from the point of view of the wear on the teeth and causes no unevenness of motion any more than the axial motion of spur gears would do, it being evident that the endlong motion of spur gears will in no way affect the velocity ratio or the steadiness of the motion. In designing this class of gearing, therefore, no effort is made to prevent slipping along the line of contact CQ.

Imagine now that the motion of R in each wheel in the plane normal to CQ is divided into two parts, namely, those normal to each plane of reference in the drawing. Taking first the motion of R parallel to OP (that is, normal to the first plane) and in the plane normal to CQ, its motion as a point in the wheel on OA in the required direction is proportional to $RS \times n_1$ and as a point in the wheel on BP its motion in the same direction is proportional to $RU \times n_2$. So that the first condition to be fulfilled is that

$$RS \times n_1 = RU \times n_2$$

But $$RS = OR \sin \theta_1 \text{ and } RU = OR \sin \theta_2$$

Hence $$OR \sin \theta_1 \times n_1 = OR \sin \theta_2 \times n_2$$

or $$n_1 \sin \theta_1 = n_2 \sin \theta_2$$

In the second place, consider the motion of R in the plane normal to CQ but in the direction normal to OP. As a point in the wheel on OA its motion in the required direction is proportional to $S'T' \times n_1 \times \cos \theta_1$, while as a point in the wheel on PB its motion is proportional to $U'V' \times n_2 \times \cos \theta_2$.

The second condition therefore is

$$S'T' \times n_1 \times \cos \theta_1 = U'V' \times n_2 \times \cos \theta_2$$

$$h_1 n_1 \cos \theta_1 = h_2 n_2 \cos \theta_2.$$

112. Equations for Finding the Line of Contact.—Two other conditions may be written as self-evident, and assembling the four sets at one place, for convenience, gives

$$\theta_1 + \theta_2 = \theta \quad (1)$$

$$h_1 + h_2 = h \quad (2)$$

$$n_1 \sin \theta_1 = n_2 \sin \theta_2 \quad (3)$$

and $$h_1 n_1 \cos \theta_1 = h_2 n_2 \cos \theta_2 \quad (4)$$

These four equations are clearly independent, and since θ, h, n_1 and n_2 or n_1/n_2 are given, the values of θ_1, θ_2, h_1 and h_2 are known and hence the location of the line of contact CQ.

113. Graphical Solution for Line of Contact.—The most simple solution is graphical and the method is indicated in Fig. 58 where OA and OB represent the projections of the axes of the shafts on a plane normal to their common perpendicular. Lay off OM and ON along the directions of OB and OA respectively to represent to any scale the speeds n_2 and n_1, if the latter are given absolutely, but if not, make the ratio $OM/ON = n_2/n_1$ choosing one of the lines, say OM, of any convenient length. It is important to lay off OM and ON in the proper sense, and since the shafts turn in opposite sense, these are laid off in opposite sense from O. Join MN and draw OK perpendicular to MN. Then $NK/KM = h_2/h_1$, and to find their numerical values take any distance NL to represent h, join LM, and draw KJ parallel LM. Then $h_1 = JL$, $h_2 = JN$, $ONM = \theta_1$ and $OMN = \theta_2$.

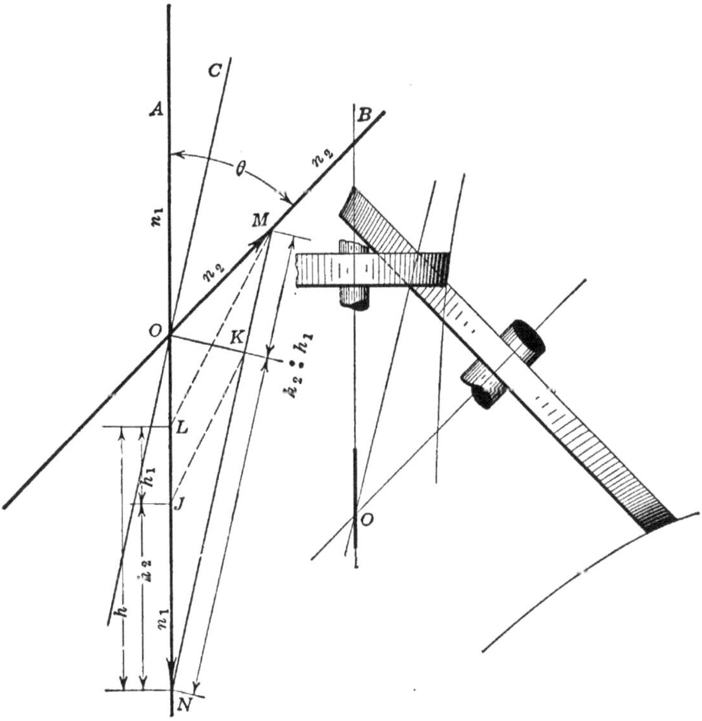

FIG. 58.

The proof of the construction is as follows: Since $\dfrac{OM}{ON} = \dfrac{n_2}{n_1}$ and since $OK = OM \sin OMK = ON \sin ONK$, it follows that

$$\frac{n_2}{n_1} = \frac{OM}{ON} = \frac{\sin ONK}{\sin OMK}.$$

Comparing this with equation (3), Sec. 112, it is clear that $ONK = \theta_1$ and $OMK = \theta_2$ so that OC is parallel to MN. Again, $NK = ON \cos \theta_1$ and $MK = OM \cos \theta_2$ from which

$$\frac{NK}{MK} = \frac{ON}{OM} \cdot \frac{\cos \theta_1}{\cos \theta_2} = \frac{n_1}{n_2} \cdot \frac{\cos \theta_1}{\cos \theta_2} = \frac{h_2}{h_1}$$

by comparing with equation (4). The construction for finding the numerical values of h_1 and h_2 requires no explanation.

114. General and Special Cases.—A few applications will show the general nature of the solution found.

Case 1.—**Shafts inclined at any angle θ and at distance h apart.** This is the general case already solved and θ_1, θ_2, h_1 and h_2 are found as indicated.

Case 2.—**Shafts inclined at angle $\theta = 90°$ and at distance h apart.** Care must be taken not to confuse the method and type of gear here described with the spiral gear to be discussed later. Choose the axes as shown in Fig. 59, lay off to scale $ON = n_1$ and $OM = n_2$ and join MN; then draw OK perpendicular to MN from which (Sec. 113) $NK:KM = h_2:h_1$. In this case $h_1n_1 \cos \theta_1 = h_2n_2 \cos \theta_2$ gives

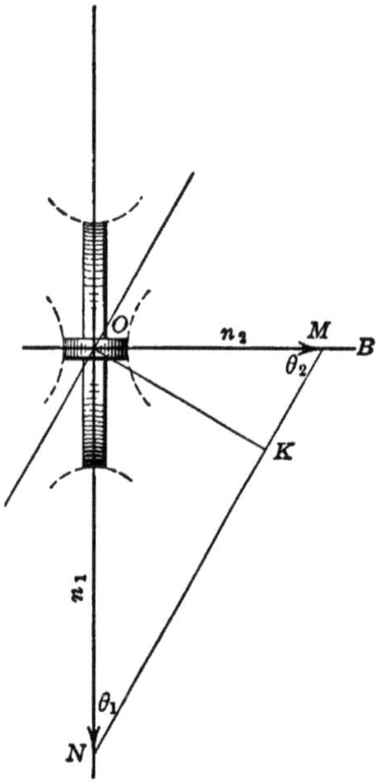

FIG. 59.

$$\frac{h_1n_1}{h_2n_2} = \frac{\cos \theta_2}{\cos \theta_1} = \frac{\sin \theta_1}{\cos \theta_1} = \tan \theta_1 = \frac{n_2}{n_1},$$

and hence

$$\frac{h_1}{h_2} = \left(\frac{n_2}{n_1}\right)^2.$$

To take a definite case, suppose $n_1 = 2n_2$ then

$$\frac{h_1}{h_2} = \left(\frac{n_2}{n_1}\right)^2 = \left(\frac{n_2}{2n_2}\right)^2 = \frac{1}{4}$$

and if the distance apart of the shafts, h, is 20 in. then $h_1 = 4$ in. and $h_2 = 16$ in., and the angle θ_1 is given by

$$\tan \theta_1 = \frac{n_2}{n_1} = \frac{1}{2} = 0.5,$$

or

$$\theta_1 = 26° \, 34' \text{ and } \theta_2 = 90 - \theta_1 = 63° \, 26',$$

so that the line of contact is located.

Case 3.—**Parallel shafts at distance h apart.** This gives the ordinary case of the spur gear. Here $\theta = 0$ and therefore $\theta_1 = 0 = \theta_2$, hence, $\sin \theta_1 = 0 = \sin \theta_2$ and $\cos \theta_1 = 1 = \cos \theta_2$, so that there are only two conditions to satisfy: $h_1 + h_2 = h$ and $h_1 n_1 = h_2 n_2$. Solving these gives

$$h_2 = \frac{n_1}{n_2} h_1$$

and substituting in $h_1 + h_2 = h$ gives

$$h_1 = \frac{n_2}{n_1 + n_2} \cdot h,$$

and

$$h_2 = \frac{n_1}{n_1 + n_2} h,$$

formulas which will be found to agree exactly with those of Sec. 83 for spur gears.

Case 4.—**Intersecting shafts.** Here $h = 0$, therefore $h_1 = 0$ and $h_2 = 0$. Referring to Fig. 60, draw $OM = n_2$ and $ON = n_1$

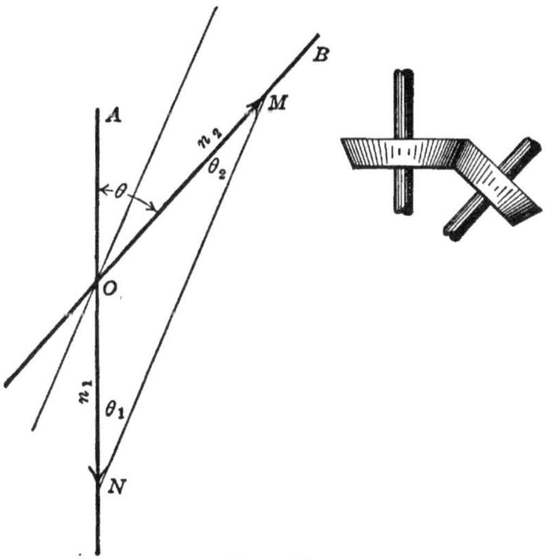

FIG. 60.

then MN is in the direction of the line of contact OC. There are only two equations here to satisfy: $\theta_1 + \theta_2 = \theta$ and $n_1 \sin \theta_1 = n_2 \sin \theta_2$ and these are satisfied by MN. Then draw OC parallel to MN (compare this with the case of the bevel gear taken up at the beginning of the chapter).

Case 5.—**Intersecting shafts at right angles.** Here $\theta = 90°$. Further let $n_2 = n_1$ then $\theta_2 = 45°$, thus the wheels would be equal and are **mitre wheels.**

115. Pitch Surfaces.—Returning to the general problem in which the location of the line of contact CQ is found by the method described for finding h_1, h_2, θ_1 and θ_2. Now, just as in the case of the spur and bevel gears, a short part of the line of contact is selected to use for the pitch surfaces of the gears, according to the width of face which is decided upon, the width of face largely depending upon the power to be transmitted, and therefore being beyond the scope of the present discussion.

It is known from geometry that if the line CQ were secured to AO, while the latter revolved the former line would describe a surface known as an hyperboloid of revolution and a second hyperboloid would be described by securing the line CQ to BP, the curved lines in the drawing, Fig. 61, showing sections of these hyperboloids by planes passed through the axes AO and BP. As the process of developing the hyperboloid is somewhat difficult and long, the reader is referred to books on descriptive geometry or to other works for the method. In the solution of such problems as the present one, however, it is quite unnecessary to draw the exact forms of the curves, and at any time the true radius to the curve may be computed as explained at Sec. 111. Or referring to Fig. 57 the radius of the wheel on OA at the point T on its axis is computed from the relation $RT^2 = RS^2 + ST^2$ where RT is the radius sought. In this way any number of radii may be computed and the true form of the wheels drawn in.

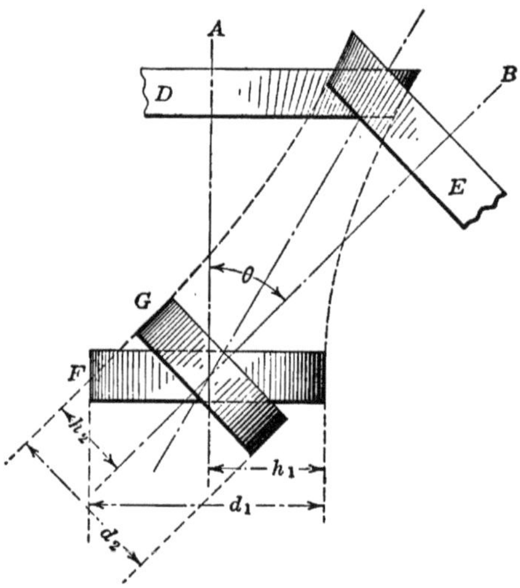

FIG. 61.

Should the distance h be small, then sections of the hyperboloids selected as shown at D and E must be used, the distances of these from the common normal depending upon the size of the teeth desired, the power to be transmitted, the velocity ratio, etc., in which case true curved surfaces will have to be used, more especially if the gears are to have a wide face. If the face is not wide, it may be possible to substitute frustra of approximately similar conical surfaces.

If the distance h is great enough, and other conditions permit of it, it is customary to use the gorges of the hyperboloids as shown at F and G and where the width of gear face is not great, cylindrical surfaces may often be substituted for the true curved surfaces. For the wheels F and G the angles θ_1 and θ_2 give the inclination of the teeth and the angles of the teeth for D and E may be computed from θ_1 and θ_2.

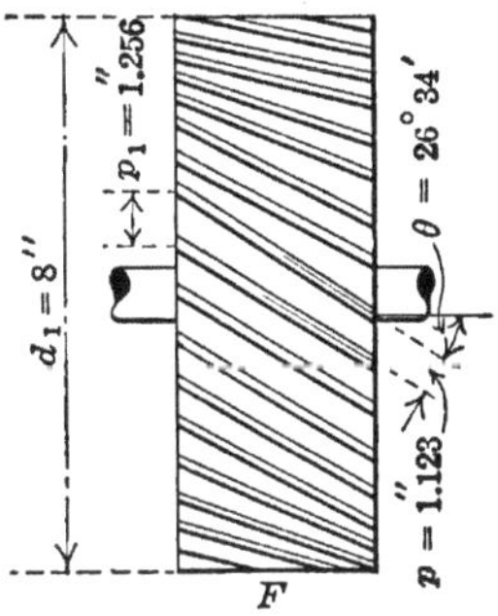

FIG. 62.

116. Example.—To explain more fully, take Case (2), Sec. 114, for which $\theta = 90°$, $n_1 = 2n_2$, and $h = 20$ in., then the formulas give $h_1 = 4$ in. and $h_2 = 16$ in. Let it be assumed that the drive is such as to allow the use of the gorge wheels corresponding to F and G, then the wheel on OA will have a diameter $d_1 = 2h_1 = 8$ in. and that on BP will be $d_2 = 32$ in. diameter. Further the angles have been determined to be $\theta_1 = 26° 34'$, $\theta_2 = 63° 26'$. As the numbers of teeth will depend on the power transmitted, etc., it will here be assumed that gear F has $t_1 = 20$ teeth. Then the **circular pitch,** measured on the end of the gear, from center to center of teeth along the pitch line is $p_1 = \frac{\pi \times 8}{20}$ $= 1.256$ in., which distance will be different from the corresponding pitch in the other gear G which has 40 teeth. As the gears are to work together the normal distance from center to center of teeth on the pitch surface must be the same in each gear. This distance is called the **normal pitch,** and is the shortest distance from center to center of teeth measured around the pitch surface; it is, in fact, the distance from center to center of the teeth around the pitch line measured on a plane normal to the line of contact (CQ in Fig. 57) and agrees with what has already been said, that the motion in this normal plane must be the same in each gear. Calling the normal pitch p, then for both gears $p = p_1 \cos \theta_1 = p_2 \cos \theta_2 = 1.256 \cos 26° 34' = 1.123$ in.

For the gear G the number of teeth $t_2 = 40$ since $n_1 = 2n_2$ and $p_1 = 2.513$ in. while $p = 1.123$ in. A sketch of the gear F is given at Fig. 62.

117. Form of Teeth.—Much discussion has arisen over the correct form of the teeth on such gears, and indeed it is almost impossible to make a tooth which will be theoretically correct, but here again one is to be guided by the fact the correct conditions must be fulfilled in the plane normal to the line of contact. Hence on this normal plane the teeth should have the correct involute or cycloidal profile.

In this type of gearing there is a good deal of slip along the line of contact (CQ) resulting in considerable frictional loss and wear, but such gearing, if well made will run very smoothly and quietly. Although it is difficult to construct there are cases where the positions of the shafts make its use imperative.

SPIRAL OR SCREW GEARING

The Teeth of which have Point Contact

118. Screw Gearing.—In speaking of gears for shafts which were not parallel and did not interest two classes were mentioned: (*a*) hyperboloidal gears, and (*b*) spiral or screw gears and this latter class will now be discussed, the former having just been dealt with. In screw gearing there is no necessary relation between the diameters of the wheels and the velocity ratio n_1/n_2 between the shafts; thus it is frequently found that while the camshaft of a gas engine runs at half the speed of the crankshaft, the two screw gears producing the drive are of the same diameters, while if skew bevel gears were used the ratio of diameters would be 1 to 4 (Sec. 114(*d*)) and bevel and spur gears for the same work would have a ratio 1 to 2.

119. Worm Gearing.—The most familiar form of this gearing is the well-known worm and worm wheel which is sketched in Fig. 63, and it is to be noticed that the one wheel takes the form of a screw, this wheel being distinguished by the name of the worm. The distance which any point on the pitch circle of the worm wheel is moved by one revolution of the worm is called the axial pitch of the worm, and if this pitch corresponds to the distance from thread to thread along the worm parallel to its axis, the thread is **single** pitch. If the distance from one thread to the next is one-half of the axial pitch the thread is **double** pitch, and if this ratio is one-third the pitch is triple, etc. The latter two cases are illustrated at (*a*) and (*b*), Fig. 64.

Fig. 63.—Worm gearing.

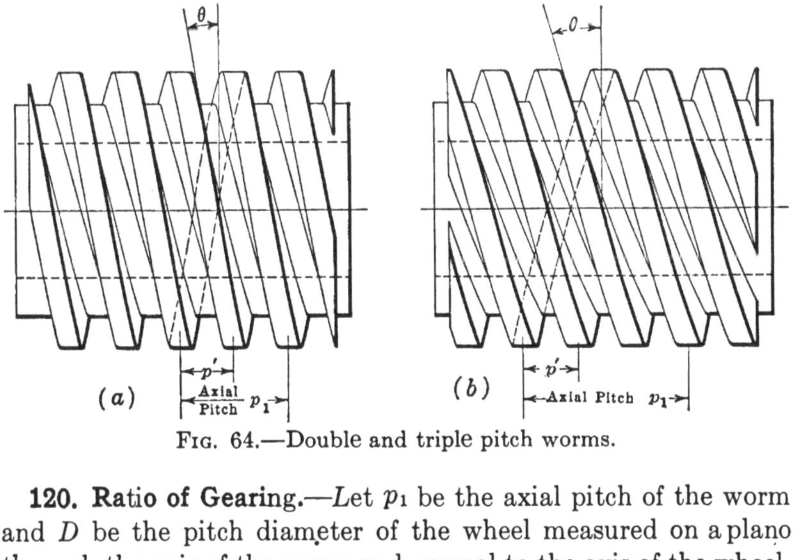

FIG. 64.—Double and triple pitch worms.

120. Ratio of Gearing.—*Let* p_1 be the axial pitch of the worm and D be the pitch diameter of the wheel measured on a plano through the axis of the worm and normal to the axis of the wheel.

Then the circumference of the wheel is πD, and since, by definition of the pitch, one revolution of the worm will move the gear forward p_1 in., hence there will be $\frac{\pi D}{p_1}$ revolutions of the worm for one revolution of the wheel, or this is the ratio of the gears. Let t be the number of teeth in the gear, then if the worm is single pitch $t = \frac{\pi D}{p_1}$ or the ratio of the gears is simply the number of teeth in the wheel. If the worm is double pitch, then p_1 the distance from center to center to teeth measured as before is given by $p_1 = 2p'$, where p' is the axial distance from the center of one thread to the center of the next one, and $t = \frac{\pi D}{p'}$ and as the ratio of the gears is $\frac{\pi D}{p_{-1}}$, in the double pitch worm this is equal to $\frac{t}{2}$, and for triple pitch it is $\frac{t}{3}$, etc.

121. Construction of the Worm.—A brief study of the matter will show that as the velocity ratio of the gearing is fixed by the pitch of the worm and the diameter of the wheel, hence no matter how large the worm may be made it is possible still to retain the same pitch, and hence the same velocity ratio, for the same worm wheel. The only change produced by changing the diameter of the worm is that the angle of inclination of the spiral thread is altered, being decreased as the diameter increases, and *vice versa*. The angle made by the teeth across the face of the wheel must be the same as that made by the spiral on the worm, and if the pitch of the worm be denoted by p_1 and the mean diameter of the thread on the latter by d, then the inclination of the thread is given by $\tan \theta = \frac{p_1}{\pi d}$, and this should also properly be the inclination of the wheel teeth. From the very nature of the case there will be a great deal of slipping between the two wheels, for while the wheel moves forward only a single tooth there will be slipping of amount πd, and hence considerable frictional loss, so that the diameter of the worm is usually made as small as possible consistent with reasonable values of θ. The worm is often immersed in oil and the friction loss is frequently below 5 per cent but in poorly made worms it may be much higher.

When both the worm and wheel are made parts of cylinders, Fig. 65, then there will only be a very small wearing surface on the wheel, but as this is unsatisfactory for power transmission, the worm and wheel are usually made as shown in section in the left-hand diagram in Fig. 65 where the construction increases the wearing surface. The usual method of construction is to turn the worm up in the lathe, cutting the threads as accurately as may be desired, then to turn the wheel to the proper outside finished dimensions. The cutting of the teeth in the wheel rim may then be done in various ways of which only one will be described, that by the use of a **hob.**

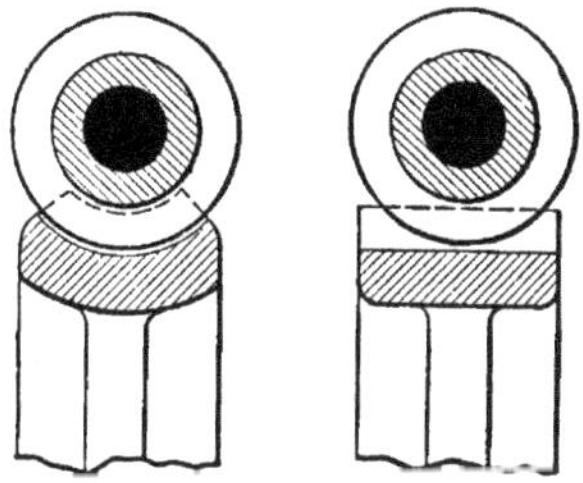

FIG. 65.

122. Worm and Worm-wheel Teeth.—A hob is constructed of steel and is an exact copy of the worm with which the wheel is to work, and grooves are cut longitudinally across the threads so as to make it after the fashion of a milling cutter; the hob is then hardened and ground and is ready for service. The teeth on the wheel may now be roughly milled out by a cutter, after which the hob and gear are brought into contact and run at proper relative speeds, the hob milling out the teeth and gradually being forced down on the wheel till it occupies the same relative position that the worm will even-

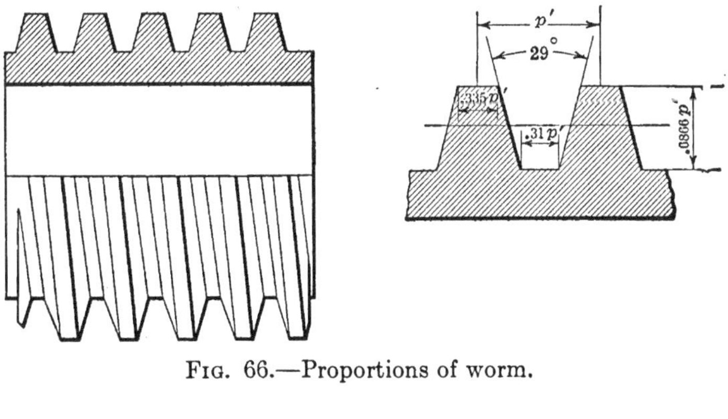

FIG. 66.—Proportions of worm.

tually take. In this way the best form of worm teeth are cut and the worm and wheel will work well together.

The shape of teeth on the worm wheel is determined by the worm, as explained above. The A.G.M.A. standard for single- and double-threaded worms is a $14\frac{1}{2}°$ angle, but angles of 20° or more are used on triple threads, etc., as this gives greater clearance for machining and grinding the worms. For automobile and high-speed work nearly 30° may be used. A section of the worm thread is shown at Fig. 66 in which the proportions used by Brown and Sharpe are the same as in a rack.

123. Large Ratio in this Gearing.—Although the frictional losses in screw gearing are large, even when the worm works immersed in oil, yet there are great advantages in being able to obtain high velocity ratios without excessively large wheels. Thus if a worm wheel has 40 teeth, and is geared with a single-threaded worm, the velocity ratio will be $\frac{1}{40}$, while with a double-threaded worm it will be $\frac{2}{40} = \frac{1}{20}$, so that it is very convenient for large ratios. It also finds favor because ordinarily it cannot be reversed, that is, the worm must always be used as the driver and cannot be driven by the wheel unless the angle θ is large. In cream separators, the wheel is made to drive the worm.

124. Screw Gearing.—Consider now the case of the worm and wheel shown in Fig. 65, in which both are cylinders, and suppose that with a worm of given size a change is made from a single to double thread, at the same time keeping the threads of the same size. The result will be that there will be an increase in the angle θ and hence the threads will run around the worm and the teeth will run across the wheel at greater angle than before. If the pitch be further increased there is a further increase in θ and this may be made as great as 45°, or even greater, and if at the same time the axial length of the worm be somewhat decreased, the threads will not run around the worm completely, but will run off the ends just in the same way as the teeth of wheels do.

By the method just described the diameter of the worm is unaltered, and yet the velocity ratio is gradually approaching unity, since the pitch is increasing, so that keeping to a given diameter of worm and wheel, the velocity ratio may be varied in any way whatever, or the velocity ratio is independent of the diameters of the worm and wheel. When the pitch of the worm is increased and its length made quite short it changes its appearance from what it originally had and takes the form of a gear wheel with teeth running in helices across the face. A photograph of a pair of these wheels used for driving the camshaft of a gas engine is shown in Fig. 67, and in this case the wheels give a velocity ratio of 2 to 1 between two shafts which do not intersect, but have an angle of 90° between planes passing through their axes. This form of gear is very extensively used for such purposes as aforesaid, giving quiet steady running, but, of course, the frictional loss is quite high.

Some of the points mentioned may be made clearer by an illustration. Let it be required to design a pair of gears of this type to drive the camshaft of a gas engine from the crankshaft, the velocity ratio in this case being 1 to 2, and let both gears be of the same diameter, the distance between centers being 12 in. From the data given the pitch diameter of each wheel will be

Fig. 67.—Screw gears.

12 in. and since for one revolution of the camshaft the crankshaft must turn twice, the pitch of the thread on the worm must be $\frac{1}{2} \times \pi \times 12 = 18.85$ in. For the gear on the crankshaft (corresponding to the worm) the "teeth" will run across its face at an angle given by $\tan \theta = \dfrac{18.85}{\pi \times 12} = 0.5$, or $\theta = 26° \ 34'$, and this angle is to be measured between the thread or tooth and the plane normal to the axis of rotation of the worm (see Fig. 64). The angle of the teeth of the gear on the camshaft (corresponding to

the worm wheel) will be $90 - 26° 34' = 63° 26'$ measured in the same way as before (compare this with the gear in Sec. 116).

It will be found that the number of teeth in one gear is double that in the other, also the normal pitch of both gears must be the same. The distance between adjacent teeth is made to suit the conditions of loading and will not be discussed.

Spiral gearing may be used for shafts at any angle to one another, although they are most common in practice where the angle is 90°. A more detailed discussion of the matter will not be attempted here and the reader is referred to other complete works on the subject.

125. General Remarks on Gearing.—In concluding this chapter it is well to point out the differences in the two types of gearing here discussed. In appearance in many cases it is rather difficult to tell the gears apart, but a close examination will show the decided difference that in **hyperboloidal gearing contact between the gears is along a straight line,** while in **spiral gearing contact is at a** point only. A study of gears which have been in operation shows this clearly, the ordinary spiral gear as used in a gas engine wearing only over a very small surface at the centers of the teeth. It is also to be noted that the teeth of hyperboloidal gears are straight and run perfectly straight across the face of the gear, while the teeth of spiral gears run across the face in helices.

Again in both classes of gears where the spiral gears have the form shown at Fig. 67, the ratio between the numbers of teeth on the gear and pinion is the velocity ratio transmitted, but in the case of the spiral gears the relative diameters may be selected as desired, while in the hyperboloidal gears the diameters are fixed when the angle between the shafts and the velocity ratio is given.

QUESTIONS ON CHAPTER VI

1. Two shafts intersect at 80°, one running at double the speed of the other. Design bevel gears for the purpose, the minimum number of teeth being 12 and the diametral pitch 3.

2. What would be the sizes of a pair of miter gears of 20 teeth and 1½ in. pitch? If the face is two and one-half times the pitch, find the radii of the spur gears at the two ends, from which the teeth are determined.

3. Two shafts cross at angle $\theta = 45°$ and are 10 in. apart, velocity ratio 2, locate the line of contact of the teeth.

4. Find the diameters of a pair of gorge wheels to suit question 3; also the sizes of the gears if the distance $OS = 12$ in.

5. If the angle between the shafts is 90° in the above case, find the sizes of the gorge wheels.

6. Explain fully the difference between spiral and skew bevel gears.

7. A worm gear is to be used for velocity ratio of 100, the worm to be 6 in. diameter, 1¼ in. pitch, and single-thread; find the size of the gear and the angle of the teeth.

8. What would be the dimensions above for a double-threaded worm?

CHAPTER VII

TRAINS OF GEARING

126. Trains of Gearing.—In ordinary practice gears are usually arranged in a series on several separate axles, such a series being called a **train** of gearing, so that a **train of gearing consists of two or more toothed wheels which all have relative motion at the same time, the relative angular velocities of all wheels being known when that of any one is given.** A train of gearing may always be replaced by a single pair of wheels of suitable diameters, but frequently the sizes of the two gears are such as to make the arrangement undesirable or impracticable.

When the train consists of four or more wheels, and when two of these of different sizes are keyed to the same intermediate shaft, the arrangement is a **compound train.** This agrees with the definition of a compound chain given in Chapter I, because one of the links contains over two elements, this being the pair of gears on the intermediate shaft. The compound trains are in very common use and are sometimes arranged so that the axes of the first and last gears coincide, in which case the train is said to be **reverted;** a very common illustration of this is the train of gears between the minute and hour hands of a clock, the axes of both hands coinciding.

127. Kinds of Gearing Trains.—If one of the gears in the train is prevented from turning, or is held stationary, and all of the other gears revolve relatively to it, usually by being carried bodily about the fixed gear as in the Weston triplex pulley block or the differential on an automobile when one wheel stops and the other spins in the mud, the arrangement is called an **epicyclic train.** Such a train may be used as a simple train of only two wheels, but is much more commonly compounded and reverted so that the axis of the last wheel coincides with that of the first.

For the ordinary train of gearing the **velocity ratio** is the number of turns of the last wheel divided by the number of turns of the first wheel in the same time, whereas in the epicyclic train the velocity ratio is the number of turns of the last wheel in the train divided by the number of turns in the same time of the frame carrying the moving wheels.

128. Ordinary Trains[1] of Gearing.—It will be well to begin this discussion with the most common class of gearing trains, that is those in which all the gears in the train revolve, and the frame carrying their axles remains fixed in space. The outline of such a train is shown in Fig. 68, where the frame carrying the axles is shown by a straight line while only the pitch circles of the gears are drawn in; there are no annular gears in the train shown, although these may be treated similarly to spur gears. Let n_2 be the number of revolutions per minute made by the

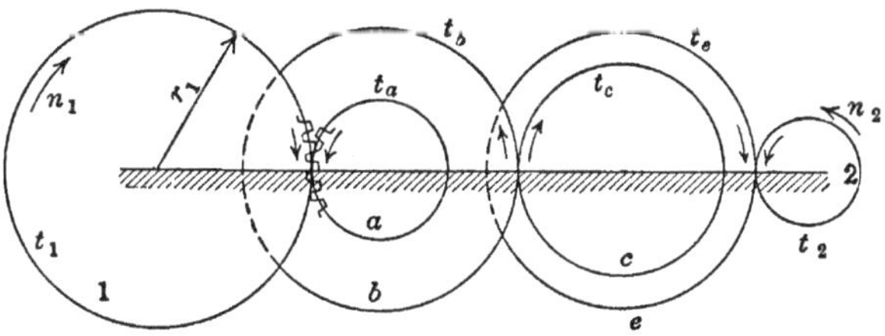

FIG. 68.

last gear and n_1 the corresponding number of revolutions per minute made by the first gear, then, from the definition already given, the ratio of the train is

$$R = \frac{n_2}{n_1} = \frac{\text{the number of revolutions per minute of the last gear}}{\text{the number of revolutions per minute of the first gear.}}$$

The figure shows a train consisting of six spur gears marked 1, a, b, c, e and 2, and let 1 be considered the first gear and 2 the last gear.

The following notation will be employed: n_1, $n_a = n_b$, $n_c = n_e$ and n_2 will represent the revolutions per minute, r_1, r_a, r_b, r_c, r_e and r_2 the radii in inches, and t_1, t_a, t_b, t_c, t_e and t_2 the numbers of teeth for the several gears used. The gears a and b and also the gears c and e are assumed to be fastened together, so that the train is compounded, a statement true of any train of over two gears. Any pair of gears such as b and c, which mesh with one another, must have the same type and pitch of teeth, but both the type and pitch may be different for any other pair which mesh together, such as e and 2; the only requirement is that each gear

[1]In what follows in this chapter reference is made to spur and bevel gears only.

must have teeth corresponding with those in the gear with which it meshes.

129. Ratio of the Train.—Now, from the results given in Sec. 101, evidently

$$\frac{n_a}{n_1} = \frac{r_1}{r_a} = \frac{t_1}{t_a}, \text{ and } \frac{n_c}{n_b} = \frac{r_b}{r_c} = \frac{t_b}{t_c}$$

also

$$\frac{n_2}{n_e} = \frac{r_e}{r_2} = \frac{t_e}{t_2}.$$

Therefore, the ratio of the train

$$R = \frac{n_2}{n_1} = \frac{n_a}{n_1} \cdot \frac{n_c}{n_a} \cdot \frac{n_2}{n_c}$$

$$= \frac{n_a}{n_1} \cdot \frac{n_c}{n_b} \cdot \frac{n_2}{n_e} \text{ since } n_a = n_b \text{ and } n_c = n.$$

and therefore

$$R = \frac{r_1}{r_a} \times \frac{r_b}{r_c} \times \frac{r_e}{r_2} = \frac{r_1 \times r_b \times r_e}{r_a \times r_c \times r_2}$$

$$= \frac{t_1}{t_a} \times \frac{t_b}{t_c} \times \frac{t_e}{t_2} = \frac{t_1 \times t_b \times t_e}{t_a \times t_c \times t_2}.$$

Calling the first wheel in each pair (*i.e.*, 1, *b* and *e*) the driver, then the formula for the ratio of the train may be written thus, **The ratio of any gear train is the product of the radii of the drivers divided by the product of the radii of the driven wheels, or the ratio of the train is the product of the numbers of teeth in the drivers divided by the product of the numbers of teeth in the driven wheels.**

The same law may be readily shown to apply although some of the gears are annular, and indeed is true when a pair of gears is replaced by an open or a crossed belt or a pair of sprockets and a chain.

To take an illustration, let the train shown in Fig. 68 have gears of the following sizes:
$r_1 = 6$ in., $r_a = 3\frac{3}{4}$ in., $r_b = 4$ in. $r_c = 2\frac{2}{3}$ in., $r_e = 5$ in. and $r_2 = 3$ in. and let the diametral pitches be 4, 6 and 8 for the pairs 1 and *a*, *b* and *c*, and *e* and 2 respectively. Then $t_1 = 48$, $t_a = 30$, $t_b = 48$, $t_c = 32$, $t_e = 80$ and $t_2 = 48$ teeth.

The ratio of the train is then

$$R = \frac{6 \times 4 \times 5}{3\frac{3}{4} \times 2\frac{2}{3} \times 3} = 4 \text{ from the radii}$$

or

$$R = \frac{48 \times 48 \times 80}{30 \times 32 \times 48} = 4 \text{ from the numbers of teeth.}$$

Further, if wheel 1 turn at a speed of $n_1 = 50$ revolutions per minute the speeds of the other gears will be $n_a = n_b = 80$ revolutions, $n_c = n_e = 120$ revolutions and $n_2 = 200$ revolutions per minute.

If the distance between the axes of gears 1 and 2 were fixed by some external conditions at the distance apart corresponding to the above train, then the whole train could be replaced by a pair of gears having radii of 19.53 in. and 4.88 in., and these would give the same velocity ratio as the train, but would often be objectionable on account of the large size of the larger gear.

The sense of rotation of the various gears may now be examined. Looking again at Fig. 68, it is observed that for two spur wheels (which have one contact) the sense is reversed, where there are two contacts, as between 1 and c the sense remains unchanged, and with three contacts such as between 1 and 2 the sense is reversed and the rule for determining the relative sense of rotation of the first and last wheels may be stated thus: **In any spur-wheel train, if the number of contacts between the first and last gears are even then both turn in the same sense, and if the number of contacts is odd, then the first and last wheels turn in the opposite sense.** Should the train contain annular gears, the same rule will apply if it is remembered that any contact with an annular gear has the same effect as two contacts between spur gear. The same rule also applies in case belts are used, an open belt corresponding to an annular gear and a crossed one to spur gears.

The rules both for ratio and sense of rotation are the same for bevel gears as for spur gears.

130. Idlers.—It not infrequently happens that in a compound train the two gears on an intermediate axle are made of the same size and combined into one; thus r_a may be made equal to r_b or $t_a = t_b$. This single intermediate wheel, then, has no effect on the velocity ratio R, as an inspection of the formula for R will show, and is therefore called an **idler.** The sole purpose in using such wheels is either to change the sense of rotation or else to increase the distance between the centers of other wheels without increasing their diameters.

131. Examples.—The application of the formula may be best explained by some examples which will now be given:

1. A wheel of 144 teeth drives one of 12 teeth on a shaft which makes one revolution in 12 sec., while a second one driven by it turns once in 5 sec. On the latter shaft is a 40-in. pulley connected by a crossed belt to a 12-in. pulley; this latter pulley turns twice while one geared with it turns three times. Show that the ratio is 144 and that the first and last wheels turn in the same sense.

2. It is required to arrange a train of gearing having a ratio of $\frac{250}{13}$.

It is possible to solve this problem by using two gears having 250 teeth and 13 teeth respectively, but in general the larger wheel will be too big and it will be well to make up a train of four or six gears. Break the ratio up into factors, thus: $R = \frac{250}{13} = \frac{50}{13} \times \frac{60}{12}$ and referring to the formula for the ratio of a train it is evident that one could be made up of four gears having 50 teeth, 13 teeth, 60 teeth and 12 teeth, and these would be arranged with the first wheel on the first axle, the gears of 60 teeth and 13 teeth would be keyed together and turn on the intermediate axle and the 12-tooth gear would be on the last axle and the contacts would be the 50 to the 13 and the 60 to the 12-tooth wheel.

Evidently, the data given allow of a great many solutions for this problem, another with six wheels being,

$$R = \frac{250}{13} = \frac{5}{1} \times \frac{5}{4} \times \frac{40}{13} = \frac{60}{12} \times \frac{15}{12} \times \frac{40}{13}.$$

This would give a train similar to Fig. 68 in which the gears are $t_1 = 60$, $t_a = 12$, $t_b = 15$, $t_c = 12$, $t_e = 40$ and $t_2 = 13$ teeth.

3. To design a train of wheels suitable for connecting the second hand of a watch to the hour hand. Here the ratio is $R = 720$ and the first and last wheels must turn in the same sense, and as annular wheels are not used for this purpose, the number of contacts must be even. The following two solutions would be satisfactory for eight wheels:

$$R = 720 = \frac{4 \times 4 \times 5 \times 9}{1} = \frac{56}{14} \times \frac{48}{12} \times \frac{50}{10} \times \frac{108}{12}$$

or

$$R = 720 = \frac{6 \times 6 \times 4 \times 5}{1} = \frac{72}{12} \times \frac{60}{10} \times \frac{52}{13} \times \frac{60}{12}.$$

Attention is called to the statement in Sec. 89 that it is unusual to have wheels of less than 12 teeth.

4. Required the train of gears suitable for connecting the minute and hour hands of a clock.

Here $R = 12$ and the train must be reverted; further, since both hands turn in the same sense there must be an even number of contacts, and four wheels will be selected. In addition to obtaining the correct velocity ratio it is necessary that $r_1 + r_a = r_b + r_2$, and if all the wheels have the same pitch $t_1 + t_a = t_b + t_2$. The following train will evidently produce the correct result:

$$R = 12 = \frac{4 \times 3}{1} = \frac{48}{12} \times \frac{45}{15}$$

or $t_1 = 48$, $t_a = 12$, $t_b = 45$ and $t_2 = 15$ teeth, the hour hand carrying the 48 teeth and the minute hand the 15 teeth.

132. Automobile Gear Box.—Very many applications of trains of gearing have been made to automobiles and a drawing of a variable-speed transmission is shown in Fig. 69. The drawing shows an arrangement for three forward speeds (one without using the gears) and one reverse. The shaft *E* is the crankshaft and to it is secured a gear *A* having also a part of a jaw clutch *B* on its right-hand side. Gear *A* meshes with another *G* on a countershaft *M*, which carries also the gears *H*, *J* and *K* keyed to it, and whenever the engine shaft *E* operates the gears *G*, *H*, *J* and *K* are running. On the right is shown the power shaft *P* which extends back to the rear or driving axle; this shaft is central with *E* and carries the gears *D* and *F* and also the inner part *C* of the jaw clutch for *B*.

The gears *D* and *F* are forced to rotate along with *P* by means of keys at *T*, but both gears may be slid along the shaft by means of the collars at *N* and *R* respectively. In addition to the gears already mentioned there is another, *L*, which always meshes with *K* and runs on a bearing behind the gear *K*.

Assuming the driver wishes to operate the car at maximum speed he throws *F* into the position shown and pushes *D* to the left so that the clutch piece *C* engages with *B*, in which case *P* runs at the same speed as the engine shaft *E*. Second highest speed is obtained by slipping *D* to the right until it comes into contact with H, the ratio of gears then being A to G and H to D; F remains as shown. For lowest speed D is placed as shown in the figure and F slid into contact with J; the shaft P and the car are reversed by moving F to the right until it meshes with L, the gear ratio being A to G and K to L to F.

Builders of automobiles so design the operating levers that it is possible to have only one set of gears in operation at once.

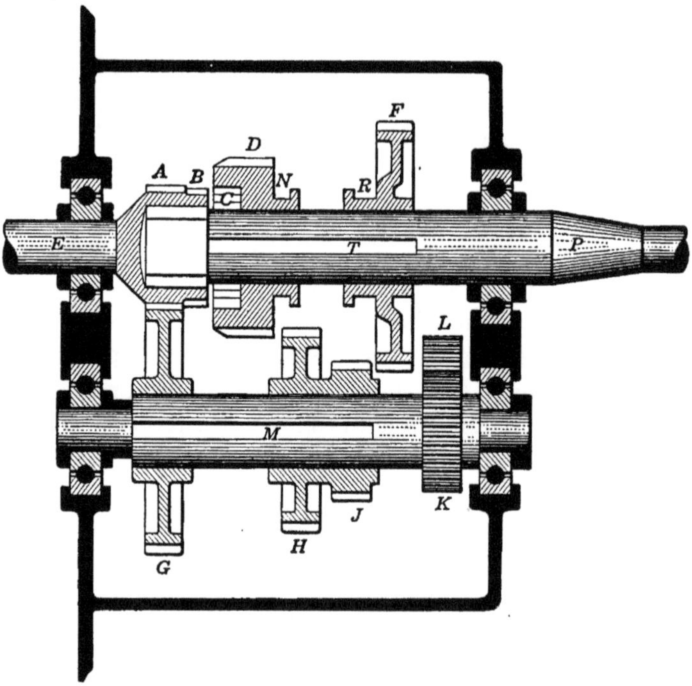

Fig. 69.—Automobile gear box.

133. The Screw-cutting Lathe.—Most lathes are arranged for cutting threads on a piece of work, and as this forms a very interesting application of the principles already described, it will be used as an illustration.

The general arrangement of the headstock of a lathe is shown in Fig. 70, and in this case in order to make the present discussion as simple as possible, it is assumed that the back gear is not in use. The cone C is connected by belt to the countershaft which supplies the power, the four pulleys permitting the operation of the lathe at four different speeds. This cone is secured to the spindle S, which carries the chuck K to which the work is attached and by which it is driven at the same rate as the cone C. On the other end of S is a gear e, which drives the gear h through one idler g or two idlers f and g. The shaft which carries h also has a gear 1 which is keyed to it, and must turn with the shaft at the same speed as h. The gear 1 meshes with a pinion a on a separate shaft, this pinion being also rigidly connected to and revolving with gear b, which latter gear meshes with a wheel 2 keyed to the leading screw L. Thus the spindle S is geared to the leading screw L through the wheels e, f, g, h, 1, a, b, 2 of which the first four are permanent, while the latter four may be changed to suit conditions, and are called **change gears.**

The work is attached to the chuck K on S and is supported by the center on the tail stock so that it rotates with K. The lead-

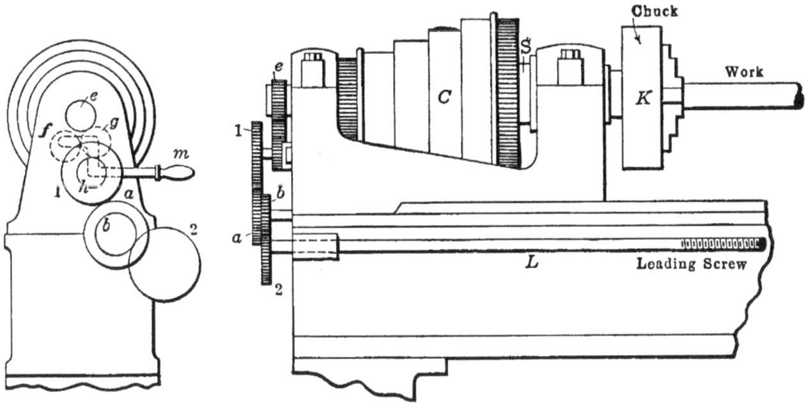

FIG. 70.—Lathe head stock.

ing screw L passes through a nut in the carriage carrying the cutting tool, and it will be evident that for given gears on 1, a, b, 2 a definite number of turns of S correspond to a definite number of turns of L, and hence to a certain horizontal travel of the carriage and cutting tool. Suppose that it is desired to cut a screw on the work having s threads per inch, the number of threads per inch l on the leading screw being given. This requires that while the tool travels 1 in. horizontally, corresponding to l turns of the leading screw L, the work must revolve s times, or if n_1 represents the revolutions per minute of the work, and n_2 those of the leading screw, then

$$= \frac{n_2}{n_1} = \frac{l}{s} = \frac{t_e}{t_h} \times \frac{t_1}{t_a} \times \frac{t_b}{t_2}$$

where t_e, t_h, etc., represent the numbers of teeth in the gears. Evidently f and g are idlers and have no effect on the ratio.

In many lathes the gears e and h arc made the same size so that gear 1 turns at the same speed as gear h.

Then $$R = \frac{t_1}{t_a} \times \frac{t_b}{t_2}.$$

This ratio is used in the example here. For many purposes also $t_a = t_b$.

Further, if L and S turn in the same sense, and if the leading screw has a right-hand thread, as is usual, then the thread cut on the work will also be right-hand. The idlers f and g arc provided to facilitate this matter, and if a right-hand thread is to be cut, the handle m carrying the axes of f and g is moved so that g alone connects e and h, while, if a left-hand thread is to be cut the handle is depressed so that f meshes with e and g with h. The figure shows the setting for a right-hand thread.

An illustration will show the method of setting the gears to do a given piece of work. Suppose that a lathe has a leading screw cut with 4 threads per inch, and the change gears have respectively 20, 40, 45, 50, 55, 60, 65, 70, 75, 80 and 115 teeth. Assume $t_e = t_h$.

1. It is required to cut a right-hand screw with 20 threads per inch. Then $\frac{l}{s} = \frac{t_1}{t_a} \times \frac{t_b}{t_2}$ where $l = 4$ and s is to be 20.

Thus $$\frac{t_1}{t_a} \times \frac{t_b}{t_2} = \frac{4}{20} = \frac{1}{5}.$$

This ratio may be satisfied by using the following gears $t_1 = 20$. $t_a = 50$, $t_b = 40$ and $t_2 = 80$. Only the one idler g would be used to give the right-hand thread.

2. To cut a standard thread on a 2-in. gas pipe in the lathe. The proper number of threads here would be $11\frac{1}{2}$ per inch and hence $l = 4$, $s = 11\frac{1}{2}$ and $\frac{t_1}{t_a} \times \frac{t_b}{t_2} = \frac{4}{11\frac{1}{2}} = \frac{8}{23}$. This could be done by making $t_1 = 40$ and $t_2 = 115$, and $t_b = t_a$ both acting as one idler.

3. If it were required to cut 100 threads per inch then $l = 4$, $s = 100$ and $\frac{t_1}{t_a} \times \frac{t_b}{t_2} = \frac{4}{100} = \frac{1}{25}$, which may be divided into two parts, thus $\frac{1}{25} = \frac{1}{4} \times \frac{1}{6\frac{1}{4}}$, so that making $t_1 = 20$, $t_a = 80$,

$t_2 = 75$, would require an extra gear of 12 teeth to take the place of b, as $t_b = 12$.

The axle holding the gears a and b may be changed in position so as to make these gears fit in all cases between 1 and 2. The details of the method of doing this are omitted in the drawing.

In order to show how much gearing has been used in the modern lathe, the details of the gearing for the headstock of the Hendey-Norton lathe are given in Figs. 71, 72 and 73, from figures made up from drawings kindly supplied by the Hendey Machine Co., Torrington, Conn.

FIG. 71.—Hendey-Norton lathe.

A general view of the Hendey-Norton lathe is shown in Fig. 71, and a detailed drawing in Fig. 72 in the latter of which is shown a belt cone with four pulleys, P, running freely on the live spindle S. Keyed to the same spindle is the gear shown at Q, and secured to the cone P is the pinion T, and Q and T mesh, when required, with corresponding gears on the back gearshaft R. When the cone is driving the spindle directly the pin W, shown in the gear Q, is left in the position shown in the drawing, thus forcing P to drive the spindle through Q, but when the back gear is to be used, the pin W is drawn back out of contact with the cone pulley, the shaft R is revolved by means of the handle A so as to throw the gears on it into mesh with T and Q and

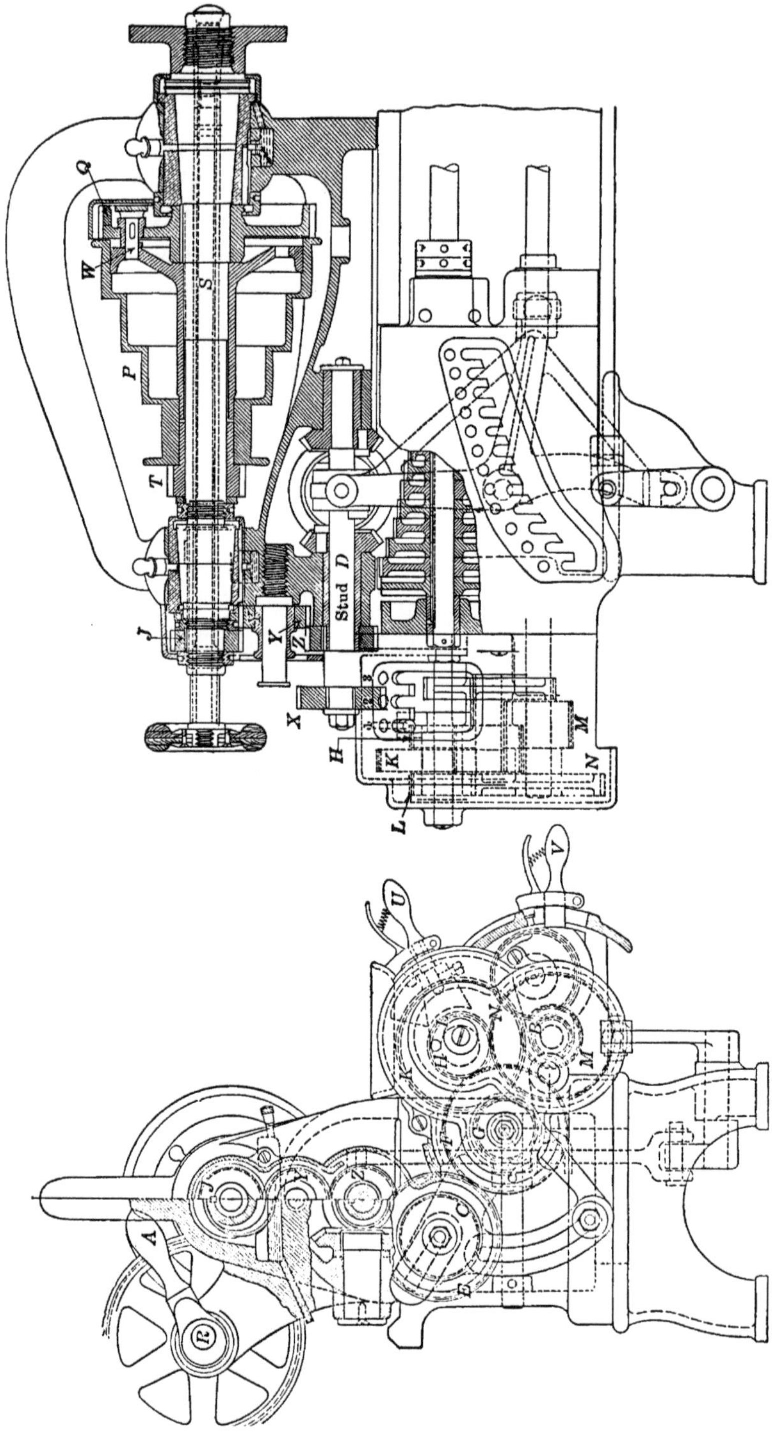

FIG. 72.—Headstock of Hendey-Norton lathe.

then the spindle is driven from the cone, through T and the back gears, and back through gear Q which is keyed to the spindle. The use of the back gears enables the spindle to be run at a much slower speed than the cone pulley.

For screw cutting the back gear is not used, but a train of gears, J, Y, Z, X, C, E, F, G, H, K, L, M, N, and B, and an idler (or tumble gear as it is called) delivers motion from the live spindle to the set of gears A on the lead screw. These gears are partly shown on Fig. 72 and partly on Fig. 73 which latter is diagrammatic. Of the gears mentioned J is keyed to the live

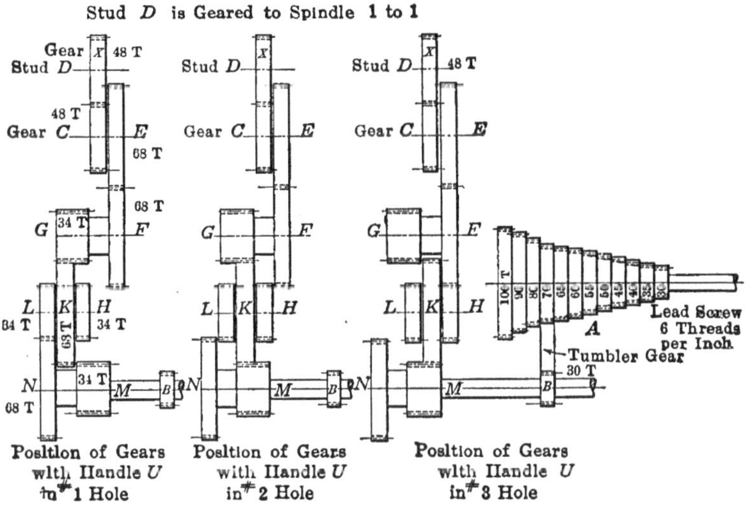

FIG. 73.—Gears on Hendey-Norton lathe.

spindle, the idler Y is slipped over into gear when a screw is to be cut and causes the gear Z to turn at the same speed as the spindle. The gear Z gives motion to the stud D by means of the bevel gears and jaw clutch shown below gear T, and the sense of motion of D will depend on whether the jaw clutch is put into contact with the right- or left-hand bevel gear, these bevel gears being to adapt the lathe to the cutting of a right- or left-hand thread. Gear X, keyed to stud D, revolves at the same speed as Z and therefore as the spindle S and in the same or opposite sense to it according to the position of the clutch; with the clutch to the left both would turn in the same sense. The remainder of the train of gearing is indicated clearly on Fig. 73, and it is to be noted that where several of these gears are on the same shaft they are fastened together, such as C and E, G and F,

L, *K*, *H*, etc. The numbers of teeth shown in the various gears correspond to those used in the 16-in. and 18-in. lathes.

The handles shown control the gear ratios; thus *U* controls the positions of the gears *L*, *K*, and *H* and the figure shows the three possible positions provided by the maker and corresponding to the three holes 1, 2 and 3 in Fig. 72. The gear *B* is provided with a feather running in a long key seat cut in the shaft shown, and the handle *V* is arranged so as to control the horizontal position of the gear *B* and its tumbler gear; that is the handle *V* enables the operator to bring *B* into gear with any of the 12 gears on the lead screw. The lead screw has 6 threads per inch.

With the handle *U* in No. 3 hole and the handle *V* in the fourth hole as shown in the right-hand diagram of Fig. 73 the ratio is

$$R = \frac{n_2}{n_1} = \frac{l}{s} = \frac{t_J}{t_Z} \times \frac{t_X}{t_C} \times \frac{t_E}{t_F} \times \frac{t_L}{t_H} \times \frac{t_K}{t_M} \times \frac{t_B}{70}$$

$$= \frac{1}{1} \times \frac{48}{48} \times \frac{68}{68} \times \frac{68}{34} \times \frac{68}{34} \times \frac{30}{70} = \frac{12}{7}.$$

$$\therefore\ s = \frac{7}{12} \times 6 = \frac{7}{12} \times 6 = 3\tfrac{1}{2}$$

or the lathe would be set to cut 3½ threads per inch.

134. Cutting Special Threads, Etc.—When odd numbers of threads are to be cut, various artifices are resorted to to get the required gearing, sometimes approximations only being employed. Thus if it were required to cut threads on a 2-in. gas pipe, which has properly 11½ threads per inch, and the lathe had not gears for the purpose, it might be possible to cut 11¼ threads per inch or 11¾ threads per inch, either of which would serve such a purpose quite well. There are cases, however, where exact threads of odd pitches must be cut and an example will show one method of getting at the proper gears.

Let it be required to cut a screw with an exact pitch of 1 mm. (0.0393708 in.) with a lathe having 8 threads per inch on the leading screw, and assume $t_e = t_h$.

A convenient means of working out this problem is the method of continued fractions.

The exact value of the ratio $\frac{1}{R}$ is

$$\frac{1}{R} = \frac{\tfrac{1}{8}}{0.0393708} = \frac{0.125}{0.0393708}.$$

The first approximation is

$$\frac{1}{R} = 3, \text{ the real value being } \frac{1}{R} = 3 + \frac{68{,}876}{393{,}708}.$$

The second is

$$3 + \frac{1}{5}, \text{ the real value being } 3 + \cfrac{1}{5 + \cfrac{49{,}328}{68{,}876}}$$

and in this way the third, fourth, fifth, etc., approximations are readily found. The sixth is

$$\frac{1}{R} = 3 + \cfrac{1}{5 + \cfrac{1}{1 + \cfrac{1}{2 + \cfrac{1}{1 + 1}}}}$$

$$= 3\frac{7}{40} = \frac{127}{40}.$$

Thus with a gear of 40 teeth at 1 or $t_1 = 40$ and $t_2 = 127$ on the leading screw, and an idler in place of *a* and *b* the thread could be cut.

(Note that $\frac{0.125}{0.0393708} = 3.17494$ while $\frac{127}{40} = 3.175$, so that the arrangement of gears would give the result with great accuracy.)

Problems of this nature frequently lend themselves to this method of solution, but other methods are sometimes more convenient and the ingenuity of the designer will lead him to devise other means.

135. Hunting Tooth Gears.—These are not much used now but were formerly employed a good deal by millwrights who thought that greater evenness of wear on the teeth would result when a given pair of teeth in two gears came into contact the least number of times. To illustrate this, suppose a pair of gears had 80 teeth each, the velocity ratio between them thus being unity; then a given tooth of one gear would come into contact with a given tooth of the other gear at each revolution of each gear, but if the number of teeth in one gear were increased to 81 then the velocity ratio is nearly the same as before and yet a given pair of teeth would come into contact only after 80 revolutions of one of the gears and 81 revolutions of the other. The odd tooth is called a **hunting tooth.** Compare the case where the gears have 40 teeth and 41 teeth with the case cited.

EPICYCLIC GEARING, ALSO CALLED PLANETARY GEARING

136. Epicyclic Gearing.—An epicyclic train has been defined at the beginning of the chapter as one in which one of the gears in the train is held stationary or is prevented from turning, while all the other gears revolve relative to it. The frame carrying the revolving gear or gears must also revolve. The train is called epicyclic because a point on the revolving gear describes epicyclic curves on the fixed one, and the term planetary gearing appears to be due to the use of such a train by Watt in his "sun and planet" motion between the crankshaft and connecting rod of his early engines.

An epicyclic train of gears is made up in exactly the same way as an ordinary train already examined, the only difference

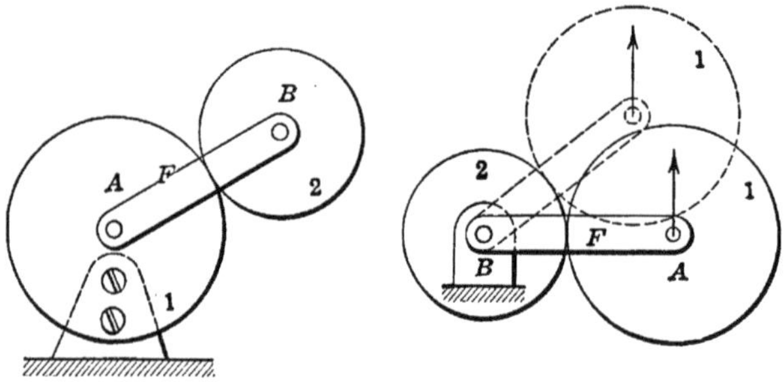

FIG. 74.—Epicyclic trains.

between the two is in the part of the combination that is fixed; in the ordinary train the axles on which the gears revolve are fixed in space, that is, the frame is fixed and all the gears revolve, whereas in the epicyclic train one of the gears is prevented from turning and all of the other gears and the frame revolve. This is another example of the inversion of the chain explained in Chapter *I*.

The general purpose of the train is to obtain a very **low** velocity ratio without the use of a large number of gears; thus a ratio of $1/10,000$ may very simply be obtained with four gears, the largest of which contains 101 teeth. *It* also has other applications. Any number of wheels may be used, although it is unusual to employ over four.

In discussing the train, the term "**first wheel**" will correspond with wheel 1 and "**last wheel**" with wheel 2 in the train shown in Fig. 68, and it will always be the first wheel which is prevented from turning. The ratio of the train is the number of turns of the last wheel for each revolution of the frame.

In Fig. 74 two forms of the train, each containing two wheels, are shown. In the left-hand figure, wheel 1 is fixed in space and the frame F and wheel 2 revolve, whereas in the right-hand figure the wheel 1 is fixed only in direction, being connected to links in such a way that the arrow shown on it always remains vertical (a construction easily effected in practice), that is wheel 1 does not revolve on its axis, and the frame F and wheel 2 both revolve about the center *B*. The following discussion applies to either case.

137. Ratio of Epicyclic Gearing.—*L*et the gears 1 and 2 contain t_1 and t_2 teeth respectively; then as a simple train the ratio is $R = \frac{t_1}{t_2}$ and is negative, since the first and last wheels turn in the opposite sense. The method of obtaining the velocity ratio of the corresponding epicyclic train may now be explained. Assume first that the frame and both wheels are fastened together as one body and the whole given one revolution in space; then frame F turns one revolution, and also the gears 1 and 2 each turn one revolution on their axes (not axles). But in the epicyclic train the gear 1 must not turn at all, hence it must be turned back one revolution to bring it back to its original state, and this will cause the wheel 2 to make *R* revolutions in the same sense as before, since the ratio *R* is negative. During the whole operation gear 1 has not moved, the frame F has made 1 revolution and the last wheel $1 + R$ revolutions in the same sense, hence the ratio of the train is

$$E = \frac{1 + R}{1} = \frac{\text{Revolutions made by the last wheel.}}{\text{Revolutions made by the frame.}}$$

A study of the problem will show that if *R* were positive then $E = 1 - R$ and in fact the correct algebraic formula is

$$E = 1 - R$$

and in substituting in this formula care must be taken to attach to *R* the correct sign which belongs to it in connection with an ordinary train.

Owing to the difficulty presented by this matter the following method of arriving at the result may be helpful, and in this case a train will be considered where *R* is positive, *i.e.*, there are an even number of contacts.

1. Assume the frame fixed and first wheel revolved once; then:
Frame makes 0 revolutions.
First wheel makes + 1 revolutions.
Last wheel makes + R revolutions.

But the epicyclic train is one in which the first wheel does not revolve, and therefore to bring it back to rest let all the parts be turned one revolution in opposite sense to the former motion.

2. After all parts have been turned backward one revolution the total net result of both operations is:

Frame has made $0 - 1$ revolutions.
First wheel has made $+ 1 - 1$ revolutions.
Last wheel has made $+ R - 1$ revolutions,

which has brought the wheel 1 to rest; hence

$$E = \frac{R - 1}{0 - 1} = 1 - R \quad \text{as before.}$$

138. Examples.—The following examples will illustrate the meaning of the formula and the application of the train in practice.

1. Let wheel 1 have 60 teeth and wheel 2 have 59 teeth; then $t_1 = 60$, $t_2 = 59$ and therefore $R = -\frac{60}{59}$.

Hence, $E = 1 - R = 1 - \left(-\frac{60}{59}\right) = 1 + \frac{60}{59} = \frac{119}{59}$ or the last wheel turns in the same sense as the frame and at about double its speed.

2. Suppose now that an idler is inserted between 1 and 2, keeping $R = \frac{60}{59}$ still, but making it positive.

Then $E = 1 - R = 1 - \left(+\frac{60}{59}\right) = 1 - \frac{60}{59} = -\frac{1}{59}$.

Thus, the wheel 2 turns in opposite sense to the frame and at $\frac{1}{59}$ its speed.

3. If in example (2) wheels 1 and 2 are interchanged, then $R = \frac{59}{60}$ and is positive, so that

$$E = 1 - \frac{59}{60} = +\frac{1}{60}$$

in which case the last wheel will turn in the same sense as the frame and at $\frac{1}{60}$ of its velocity.

4. To design a train having a positive ratio of $\frac{1}{10,000}$, that is,

one in which the last wheel turns in the same sense as the frame and at $\frac{1}{10{,}000}$ the speed.

Here
$$E = 1 - R = \frac{1}{10{,}000}$$

or
$$R = 1 - \frac{1}{10{,}000} = \left(1 - \frac{1}{100}\right)\left(1 + \frac{1}{100}\right) = \frac{99}{100} \times \frac{101}{100}.$$

The train thus consists of four gears 1, *a*, *b* and 2 and the numbers of teeth are $t_1 = 99$, $t_a = 100$, $t_b = 101$, $t_2 = 100$.

In practice such a train could easily be reverted, although the numbers of teeth are not exactly suited to it, and would work quite smoothly. The train is frequently made up in the form

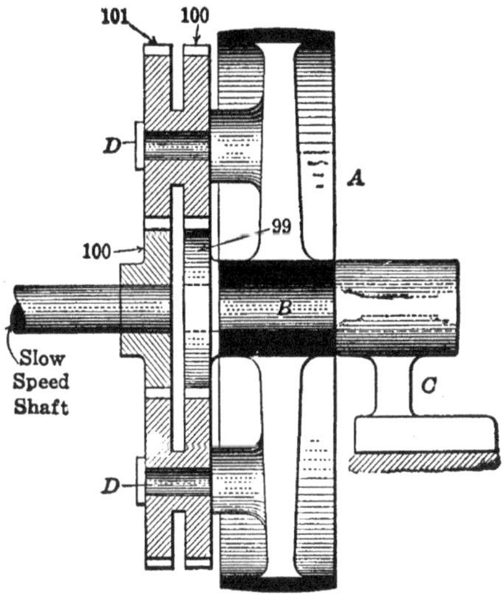

FIG. 75.

shown in Fig. 75 where the frame takes the form of a loose pulley *A*, carrying axles *D* on which the intermediate gears run and the 99-toothed gear is keyed to *B*, which is also keyed to the frame *C*. The pulley will turn 10,000 times for each revolution that the shaft makes. Should the gears be changed around, so that the 100-tooth wheel is fixed, while the 99-tooth wheel is on the shaft and gears with the 100-tooth wheel on *D*, then

$$E = 1 - R = 1 - \frac{100}{101} \times \frac{100}{99} = -\frac{1}{9{,}999},$$

or the shaft will turn slowly in opposite sense to the wheel *A*

The arrangement sketched in Fig. 75, in a slightly modified form, is used in screw-cutting machines, but with a much larger value of E. In this case there are two pulleys, one as shown at A and one keyed to the shaft, while the gears on D are usually replaced by a broad idler. When the die is running up on the stock the operation is slow and the belt is on the pulley A, but for other operations the speed is much increased by pushing the belt over to the pulley keyed to the shaft, the gears then running idly.

5. **Watt's Sun and Planet Motion.**—In this case gear 2 was the same size as 1 and was keyed to the crankshaft, while the gear 1 was secured to the end of the connecting rod and a link kept the two gears at the proper distance apart, as in the right-hand diagram of Fig. 74. There was, of course, no crank.

Here $\qquad R = -1$ and $E = 1 - R = 2.$

Therefore, the crankshaft made two revolutions for each two strokes of the piston.

139. Machines Using Epicyclic Gearing.—There are a great many illustrations of this interesting arrangement and space permits the introduction of only a very few of these.

(*a*) **The Weston Triplex Pulley Block.**—A form of this block, which contains an epicyclic train of gearing, is shown in Fig. 76. The frame D contains bearings which carry the hoisting sprocket F, and on the casting carrying the hoisting sprocket are axles each carrying a pair of compound gears BC, the smaller one C gearing with an annular gear made in the frame D, while the other and larger gear B of the pair meshes with a pinion A on the end of the shaft S to which the hand sprocket wheel H is attached. When a workman pulls on the hand sprocket chain he revolves H and with it the pinion A on the other end of the shaft, which in turn sets the compound gears BC in motion. As one of these gears meshes with the fixed annular gear on the frame D the only motion possible is for the axles carrying the compound gears to revolve and thus carry with them the hoisting sprocket F.

In the one ton Weston block the annular wheel has 49 teeth, the gear B has 31 teeth, C has 12 teeth and the pinion A has 13 teeth. For the train, then, evidently R is negative since one wheel is annular and

$$R = -\frac{49}{12} \times \frac{31}{13} = -9.73.$$

Hence

$$E = 1 - R = 1 - (-9.73) = 10.73.$$

So that there must be 10.73 turns of the hand wheel to cause one turn of the hoisting wheel F. As these wheels are respectively $9\frac{3}{4}$ in. and $3\frac{1}{8}$ in. diameter, the hand chain must be moved $\frac{9\frac{3}{4}}{3\frac{1}{8}} \times 10.73 = 33.2$ ft. to cause the hoisting chain to

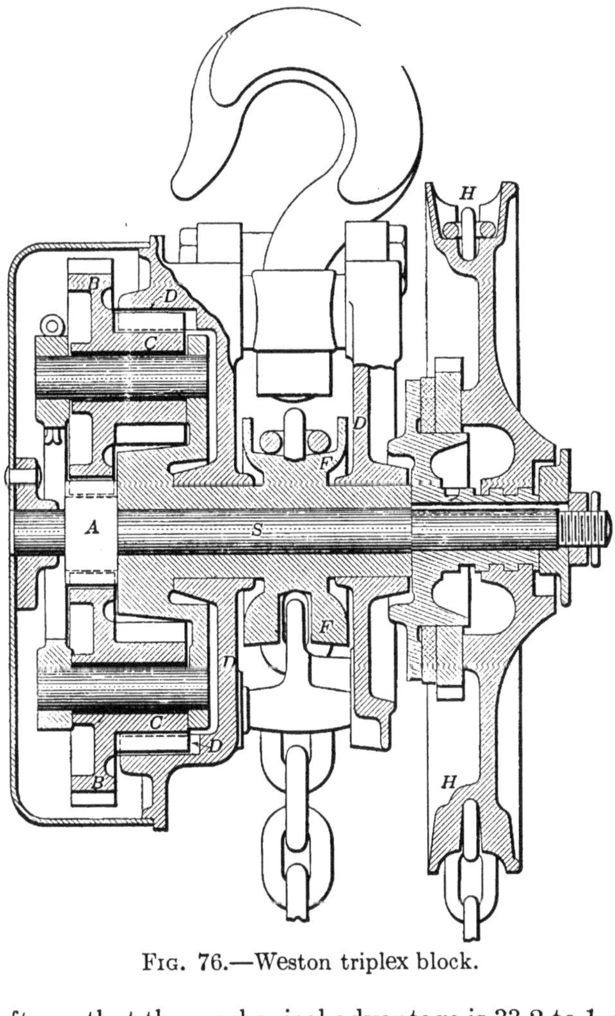

FIG. 76.—Weston triplex block.

move 1 ft., so that the mechanical advantage is 33.2 to 1, neglecting friction.

(*b*) **Motor-driven Portable Drill.**—A form of air-driven drill is shown at Fig. 77 in which epicyclic gearing is used. This drill is made by the Cleveland Drill Co., Cleveland, and the figure shows the general construction of the drill while a detail is also given of the train of gearing employed.

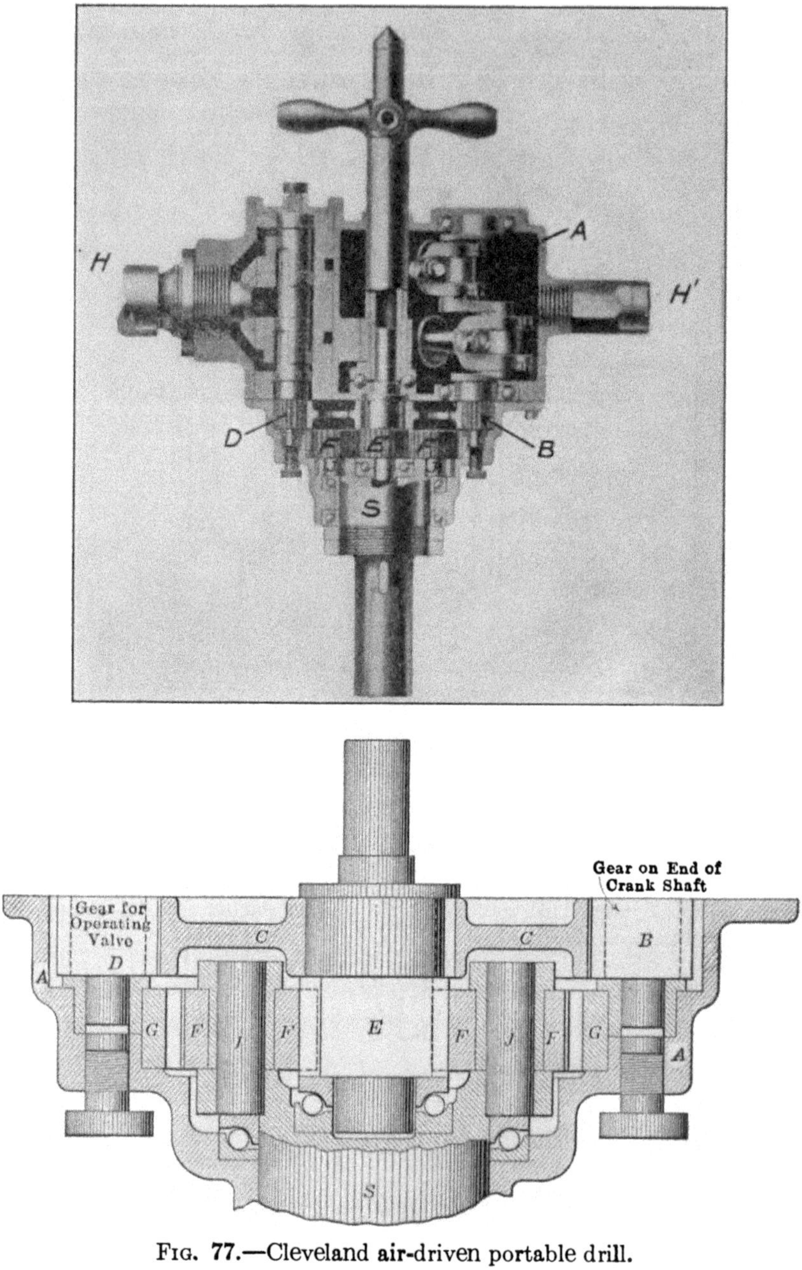

FIG. 77.—Cleveland air-driven portable drill.

The outer casing A of the drill is held from revolving by means of the handles H and H', air to drive the motors passing in through H. Fastened to the bottom of the crankshaft of the air motors is a pinion B which drives a gear C and through it a second pinion D, which latter revolves at the same speed as B and operates the valve for the motors. The gear C is keyed to a shaft which has another gear E also secured to it, the latter meshing with pinions F which in turn mesh with the internal gear G secured to the frame A. The gears F run freely on shafts J which are in turn secured in a flange on the socket S, which carries the drill.

As the motors operate on the crankshaft causing it to revolve, the pinion B turns with it and also the gear C and with it the gear E. As E revolves it sets the gears F in motion and as these mesh with the fixed gear G the only thing possible is for the spindles J carrying F to revolve in a circle about the center of E and as these revolve they carry with them the drill socket S. In one of these drills the motor runs at 1,275 revolutions per minute and the numbers of teeth in the gears are $t_B = 14$, $t_C = 70$, $t_E = 15$, $t_F = 15$ and $t_G = 45$. The speed of the spindle S will then be $1{,}275 \div \left\{1 - \frac{45}{15} \times \frac{70}{14}\right\} = 91$ revolutions per minute.

(*c*) **Automobile Transmission.**—Epicyclic gearing is now commonly used in automobiles and two examples are given here, in concluding the chapter. The mechanism shown in Fig. 78[1] was used in Ford cars for variable speed and reversing. The engine flywheel A carries three axles X uniformly spaced around a circle and each carrying loosely three gears H, G and K, the three gears being fastened together and rotating as one solid body. Each of these gears meshes with another one which is connected by a sleeve to a drum; thus H gears with B and through it to the drum C which is keyed to the shaft P passing back to the rear axle. Gear G meshes with the gear F. which is attached to the drum E, while K meshes with the gear J on the drum I. The disc M is keyed to an extension of the crankshaft as shown and carries one part of a disc clutch, the other part of which is carried on casting C. The mechanism for operating this clutch is not shown completely.

The driver of the car has pedals and a lever under his control

[1] A drawing was kindly furnished by the Ford Motor Co., Ford, Ontario, for the purpose of this cut.

and it is beyond the present purpose to discuss the action of these in detail, but it may be explained that these control band brakes, one about the drum I, another about the drum E and a third about the drum C, and in addition the pedals and lever control the disc clutch between C and M.

In this mechanism the gears have the following numbers of teeth: $t_H = 27$ teeth, $t_G = 33$ teeth, $t_K = 24$ teeth, $t_B = 27$ teeth, $t_F = 21$ teeth and $t_J = 30$ teeth.

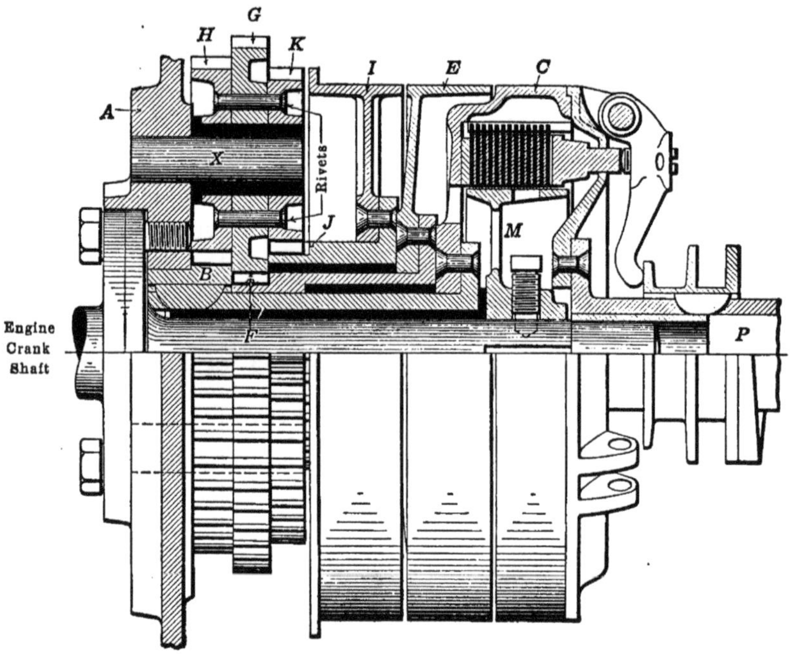

FIG. 78.—Old model Ford transmission.

Should the driver wish the car to travel at maximum speed he throws the disc clutch into action which connects M and C and thus causes the power shaft P to run at the same speed as that of the engine crankshaft. If he wishes to run at slow speed he operates the pedal which applies the band brake to the drum E, causing the latter to come to rest. The gears F, G, H and B then form an epicyclic train and for this

$$R = \frac{21}{33} \times \frac{27}{27} = 0.636 \text{ and } E = 1 - R = 0.36.$$

So that the power shaft P will turn forward, making 36 revolutions for each 100 made by the crank.

If the car is to be reversed, drum I is brought to rest and the train consists of gears J, K, H and B.

Then $$R = \frac{30}{24} \times \frac{27}{27} = \frac{5}{4} \text{ and } E' = I - R = -\frac{1}{4}$$

or the power shaft P will turn in opposite sense to the crank and at one-fourth its speed.

The brake about C is for applying the brakes to the car.

(*d*) **Automobile Differential Gear.**—The final illustration is the differential used on the rear axle of Packard cars. This is

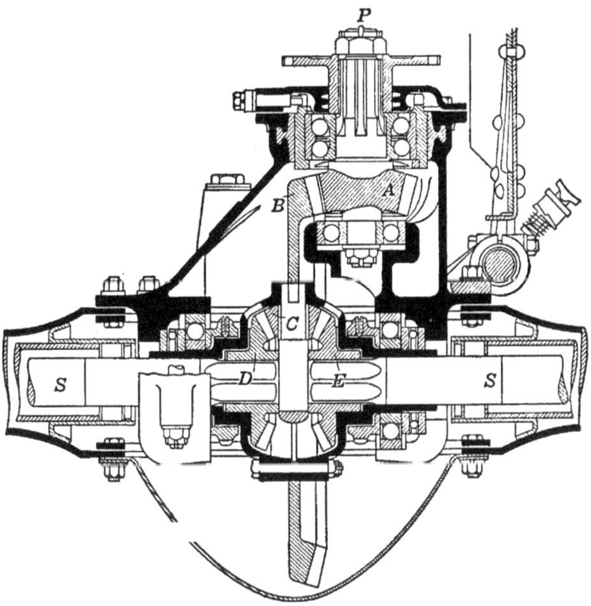

FIG. 79.—Automobile differential gear.

shown in Fig. 79 which is from a Packard pamphlet. The power shaft P attached to the bevel pinion A drives the bevel gear B which has its axis at the rear axle but is not directly connected thereto. The wheel B carries in its web bevel pinions C, the axles of which are mounted radially in B, and the pinions C may rotate freely on these axles.

The rear axle S is divided where it passes B and on one part of the axle there is a bevel gear D and on the other one the bevel gear E of the same size as D. When the car is running on a straight smooth road the two wheels and therefore the two parts

S of the rear axle run at the same speed and then the power is transmitted from P through A and B just as if the gears C, D and E formed one solid body.

In turning a corner, however, the rear wheel on the outer part of the curve runs faster than the inner one, that is D and E run at different speeds and gear C rotates slowly on its axle. When the one wheel spins in the mud, and the other one remains stationary, as not infrequently happens when a car becomes stalled, the arrangement acts as an epicyclic train purely.

QUESTIONS ON CHAPTER VII

1. Find the velocity ratio for a train of gears as follows: A gear of 30 teeth drives one of 24 teeth, which is on the same shaft with one of 48 teeth; this last wheel gears with a pinion of 16 teeth.

2. The handle of a winch carries two pinions, one of 24 teeth, the other of 15 teeth. The former may mesh with a 60-tooth gear on the rope drum or, if desired, the 15-tooth gear may mesh with one of 56 teeth on the same shaft with one of 14 teeth, this latter gear also meshing with the gear of 60 teeth on the drum. Find the ratio in each case.

3. Design a reverted train for a ratio 4 to 1, the largest gear to be not over 9 in. diameter, 6 pitch.

4. A gear a of 40 teeth is driven from a pinion c of 15 teeth, through an idler b of 90 teeth. Retaining c as before, also the positions of the centers of a and c, it is required to drive a 60 per cent. faster, how may it be done?

5. A car is to be driven at 15 miles per hour by a motor running at 1,200 revolutions per minute. The car wheels are 12 in. diameter and the motor pinion has 20 teeth, driving through a compound train to the axle; design the train.

6. In a simple geared lathe the lead screw has 5 threads per inch, gear $e = 21$ teeth, $h = 42$ teeth, $1 = 60$ teeth and $2 = 72$ teeth; find the thread cut on the work.

7. It is desired to cut a worm of 0.194 in. pitch with a lathe as shown at Fig. 70, using these change gears; find the gears necessary.

8. Make out a table of the threads that can be cut with the lathe in Fig. 70 with different gears.

9. Make a similar table to the above for the Hendey-Norton lathe illustrated.

10. Design an auto change-gear box of the selective type, with three speeds and reverse, ratios 1.8 and 3.2 with $\frac{6}{8}$ pitch stub gears, shaft centers not over 10 in.

11. A motor car is to have a speed of 45 miles per hour maximum with an engine speed of 1,400 revolutions per minute. What reduction will be required at the rear axle bevel gears, 36-in. tires? At the same engine speed find the road speed at reductions of 4 and 2 respectively.

12. Design the gear box for the above car with $\frac{6}{8}$ stub-tooth gears, shafts 9 in. centers.

13. Prove that the velocity ratio of an epicyclic train is $E = 1 - R$.

14. Design a reverted epicyclic train for a ratio of 1 to 2,500.

15. In a train of gears *a* has 24 teeth, and meshes with a 12-tooth pinion *b* which revolves bodily about *a*, and *b* also meshes with an internal gear *c* of 48 teeth. Find the ratio with *a* fixed and also with *c* fixed.

CHAPTER *VIII*

CAMS

140. Purpose of Cams.—In many classes of machinery certain parts have to move in a non-uniform and more or less irregular way. For example, the belt shifter of a planer moves in an irregular way, during the greater part of the motion of the planer table it remains at rest, the open and crossed belts driving their respective pulleys, but at the end of the stroke of the table the belts must be shifted and then the shifter must operate quickly, moving the belts, after which the shifter comes again to rest and remains thus until the planer table has completed its next stroke, when the shifter operates again. The valves of a gas engine afford another illustration, for these must be quickly opened at the proper time, held open and then again quickly closed. The operation of the needle bar of a sewing machine is well known and the irregular way in which it moves is familiar to everyone.

In the machines just described, and indeed in almost all machines in which this class of motion occurs, the part which moves irregularly must derive its motion from some other part of the machine which moves regularly and uniformly. Thus, in the planer all the motions of the machine are derived from the belts which always run at steady velocity; further, the shaft operating the valves of a gas engine runs at speed proportional to the crankshaft while the needle bar of a sewing machine is operated from a shaft turning uniformly.

The problem which presents itself then is to obtain a non-uniform motion in one part of a machine from another part which has a uniform motion, and it is evident that at least one of the links connecting these two parts must be unsymmetrical in shape, and the whole irregularity is usually confined to one part which is called a **cam. Thus a cam may be defined as a link of a machine, which has generally an irregular form and by means of which the uniform motion of one part of the machine may be made to impart a desired kind of non-uniform motion to another part.**

Cams are of many different forms and designs depending upon the conditions to be fulfilled. Thus in the sewing machine the cam is usually a slot in a flat plate attached to the needle bar, in the gas engine the cam is generally a non-circular disc secured to a shaft, whereas in screw-cutting machines it often takes the form of a slot running across the face of a cylinder, and many other cases might be cited, the variations in its form being very great. Some forms of cams are shown in Fig. 80.

Several problems connected with the use of cams will explain their application and method of design.

141 Stamp-mill Cam.—The first illustration will be that of the stamp mill used in mining districts for crushing ores, and a general view of such a mill is shown in Fig. 81. Such a mill consists essentially of a number of stamps A, which are merely

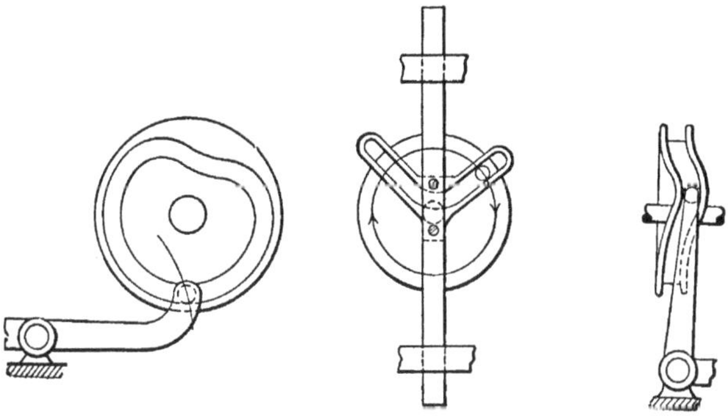

FIG. 80.—Forms of cams.

heavy pieces of metal, and during the operation of the mill these stamps are lifted by a cam to a desired height, and then suddenly allowed to drop so as to crush the ore below them. The power to lift the stamps is supplied through a shaft B which is driven at constant speed by a belt, and as no work is done by the stamps as they are raised, the problem is to design a cam which will lift them with the least power at shaft B, and after they have been lifted the cam passes out of gear and the weights drop by gravity alone.

Now, it may be readily shown that the force required to move the stamp at any time will depend upon its acceleration, being least when the acceleration is zero, because then the only force necessary is that which must overcome the weight of the stamp alone, no force being required to accelerate it. Thus, for the minimum expenditure of energy, the stamp must be lifted at a uniform velocity, and the problem, therefore, resolves itself into that of designing a cam which will lift the stamp A at uniform velocity.

The general disposition of the parts involved, is shown in Fig. 82, where B represents the end of the shaft B shown in Fig. 81,

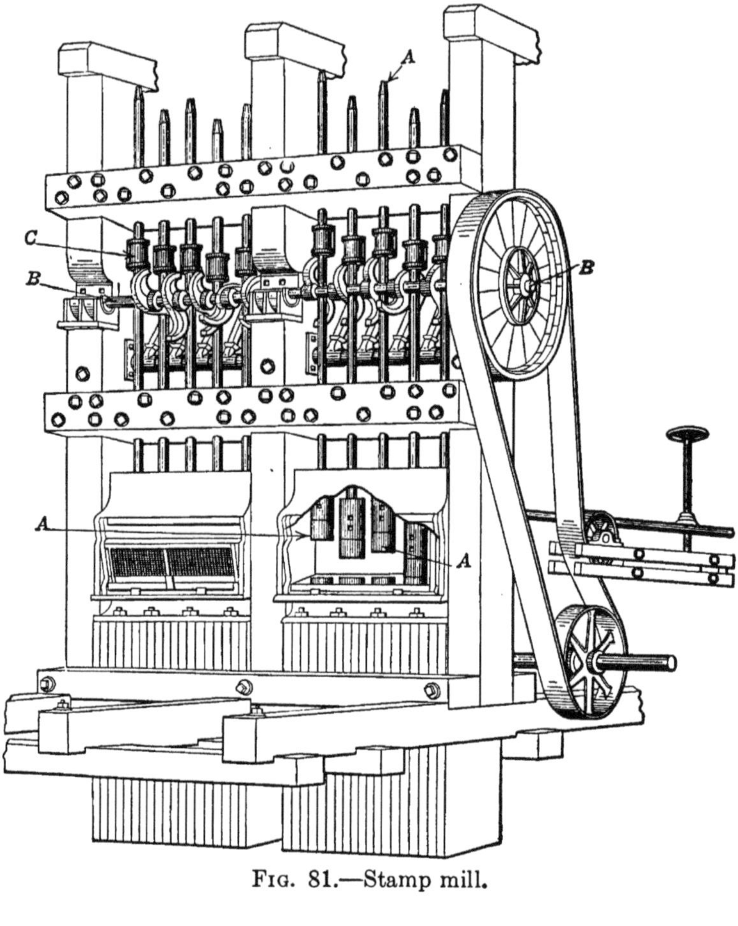

FIG. 81.—Stamp mill.

and YF represents the center line of the stamp A, which does not pass through the center of B. Let the vertical shank of the stamp have a collar C attached to it, which collar comes into direct contact with the cam on B; then the part C is usually called the **follower**, being the part of the machine directly actuated by the cam.

It will be further assumed that the stamp is to be raised twice for each revolution of the shaft B, and as some time will be taken by the stamp in falling, the latter must be raised its full distance while the shaft B turns through less than 180°. Let the total lift occur while B turns through 102°.

Further, let the total lift of the cam be h ft., that is, let the distance 0 – 6, Fig. 82, through which the bottom of the follower C rises, be h ft.

The construction of the cam may now be begun. Draw BF perpendicular to YF and lay off the angle FBE equal to 102°. Next divide the distance $0 - 6 = h$, and also the angle FBE, into

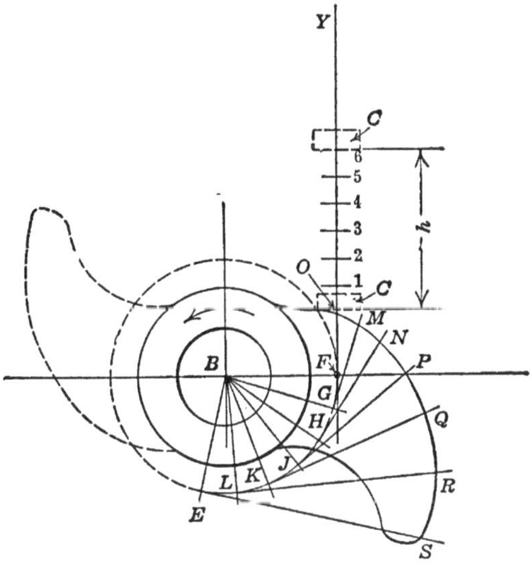

FIG. 82.—Stamp mill cam.

any convenient number of equal parts, the same number being used in each case; six parts have been used in the drawing.

Now a little consideration will show that since the stamp A and also the shaft B are to move at uniform speed, the distances 0 – 1, 1 – 2, 2 – 3, etc. and also the angles FBG, GBH, HBJ, etc., must each be passed through in the same intervals of time and all these intervals of time must be equal. With center B and radius BF draw a circle $FGH \ldots E$ tangent to the line 0 – 6 and draw GM, HN, etc., tangent to this circle at G, H, etc. Now while the follower is being lifted from 0 to 1 the shaft B is revolved through the angle FBG, and then the line GM will be vertical and must be long enough to reach from F to 1 or GM should equal FI. The construction is completed by making HN = F – 2, JP

= F − 3, etc., and in this way locating the points *O*, *M*, *N*, *P*, *Q*, *R* and *S* and a smooth curve through these points gives the face of the cam. As a guide in drawing the curve it is to be remembered that *MG*, *NH*, etc., are normals to it.

A hub of suitable size is now drawn on the shaft, the dimensions of the hub being determined from the principles of machine design, and curves drawn from *S* and *O* down to the hub complete the design; the curve from *S* must be so drawn that the follower will not strike the cam while falling.

The curve *OMN* . . . *S* is clearly an involute having a base circle of radius *BF*, or the curve of the cam is that which would be described by a pencil attached to a cord on a drum of radius

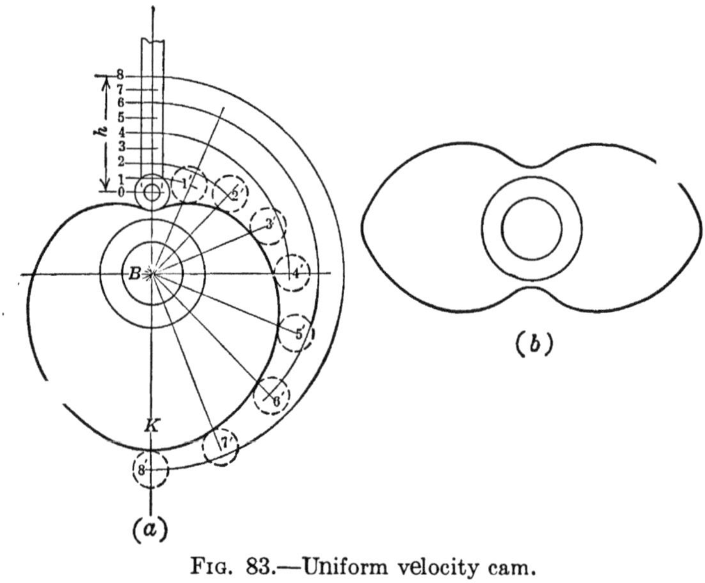

FIG. 83.—Uniform velocity cam.

BF, the cord being unwound and kept taut. The dotted line shows the other half of the cam.

In this case there is line contact between the cam and its follower, that is, it is a case of higher pairing, as is frequently, though not always, the case with cams.

142. Uniform Velocity Cam.—As a second illustration, take a problem similar to the latter, except that the follower is to have a uniform velocity on the up and down stroke and its line of motion is to pass through the shaft *B*. It will be further assumed that a complete revolution of the shaft will be necessary for the up and down motion of the follower.

Let 0 – 8, Fig. 83 (*a*) represent the travel of the follower, the latter being on a vertical shaft, with a roller where it comes into contact with the cam. Divide 0 – 8 into, say, eight equal parts as shown, further, divide the angle OBK (= 180°) into the same number of equal parts, giving the angles $OB1'$, $1'B2'$, etc. Now since the shaft B turns at uniform speed the center of the follower is at 1 when B1′ is vertical and at 2 when $B2'$ is vertical, etc., hence it is only necessary to revolve the lengths $B1$, $B2$, etc., about B till they coincide with the lines $B1'$, $B2'$, etc., respectively. The points 1′, 2′, 3′, will be obtained on the radial lines $B1'$, $B2'$, etc., as the distances from B which the center of the follower must have when the corresponding line is vertical. With centers 1′, 2′, 3′, etc., draw circles to represent the roller and the heavy line shown tangent to these will be the proper outline for one-half of the cam, the other half being exactly the same as this about the vertical center line. Here again there is higher pairing and some external force is supposed to keep the follower always in contact with the cam.

A double cam corresponding to the one above described is shown at Fig. 83 (*b*), this double cam making the follower perform two double strokes at uniform speed for each revolution of the camshaft.

143. Cam for a Shear.—The problem may appear in many different forms and the case now under consideration assumes somewhat different data from the former two, and the shear shown in Fig. 84 may serve as a good illustration. Suppose it is required to design a cam for this shear; it would usually be desirable to have the shear remain wide open during about one-half the time of rotation of the cam, after which the jaw should begin to move uniformly down in cutting the plate or bar, and then again drop quickly back to the wide-open position. With the shear wide open, let the arm be in the position A_1B_1 where it is to remain during nearly one-half the revolution of the cam; then let it be required to move uniformly from A_1B_1 to A_2B_2 while the cam turns through 120°, after which it must drop back again very quickly to A_1B_1.

An enlarged drawing of the right-hand end of the machine is shown at Fig. 85, the same letters being used as in Fig. 84, the lines A_1B_1 and A_2B_2 representing the extreme positions of the arm AB. Draw the vertical line QB_1B_2 and lay off the angle $B_1QB'_2$ equal to 120°; this then is the angle through which the camshaft must turn while the arm is moving over its range from A_1B_1 to A_2B_2. Now divide the angle B_1OB_2, Fig. 84, into any number of equal parts, say four, by the lines OC, OD, and OE; these lines are shown on Fig. 85. Next, divide the angle

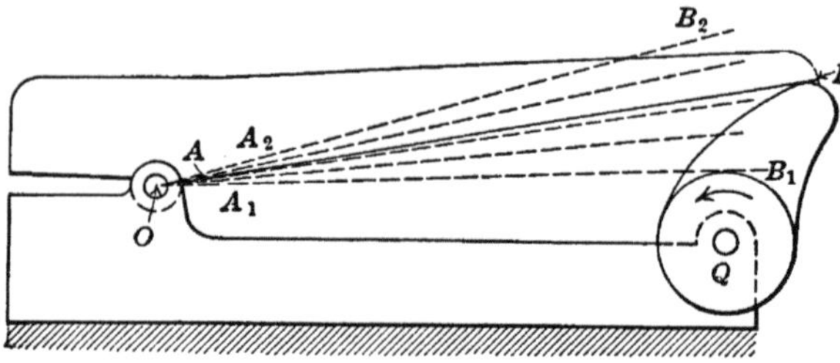

FIG. 84.

B_1OB_2' into the same number of parts as B_1OB_2, that is four, by the lines QC', QD' and QE'.

Now, when the line QB_1 is vertical as shown, the cam must be tangent to A_1B_1. Next, when the cam turns so that QC_1 becomes

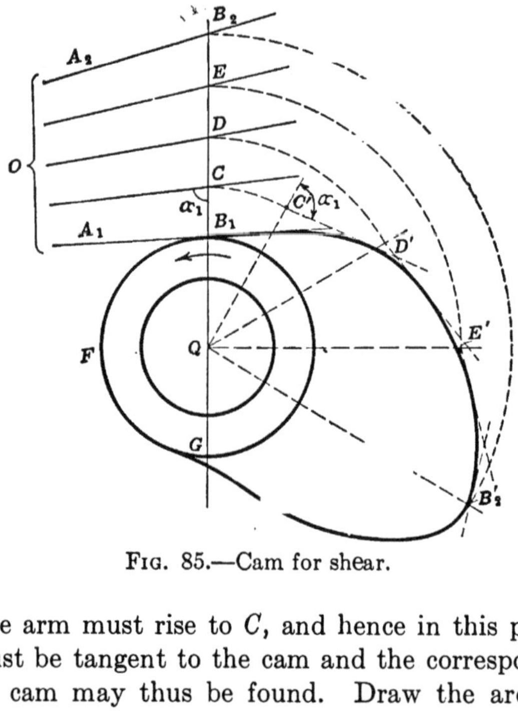

FIG. 85.—Cam for shear.

vertical, the arm must rise to C, and hence in this position the line OC must be tangent to the cam and the corresponding outline of the cam may thus be found. Draw the arc CC' with center Q, and through C' draw a line making the same angle

α_1 with QC' that OC does with QC. The line through C' is a tangent to the cam. Similarly, tangents to the cam through D', E' and B'_2 may be drawn and a smooth curve drawn in tangent to these lines, as shown in Fig. 85.

The details of design for the part B'_2 G may be worked out if proper data are given, and evidently the part GFB is circular and corresponds with the wide-open position of the shear.

144. Gas-engine Cam.—It not infrequently happens that the follower has not a straight-line motion but is pivoted at some point and moves in the arc of a circle. This is the case with some gas engines and an outline of the exhaust cam, camshaft

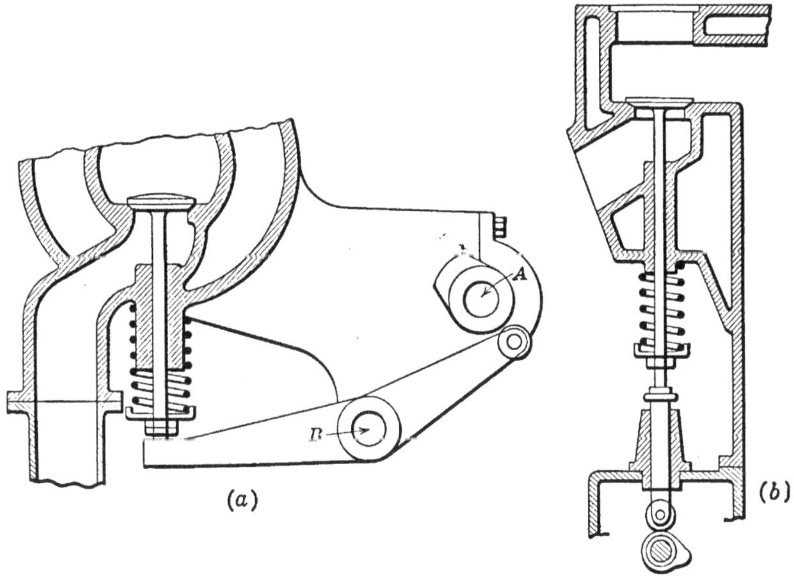

FIG. 86.—Gas-engine cams.

lever and exhaust valve for such an engine is shown at Fig. 86 (*a*), where A is the camshaft and B is the pin about which the follower swings. This presents no difficulties not already discussed but in executing such a design care must be taken to allow for the deviation of the follower from a radial line, and if this is not done the cam will not do the work for which it was intended.

As this problem occurs commonly in practice, it may be as well to work out the proper form of cam. The real difficulty is not in making the design of the cam, but in choosing the correct data and in determining the conditions which it is desired to have the cam fulfil. A very great deal of discussion has taken place on this point, and as the matter depends primarily on the conditions set in the engine, it is out of place here to enter into it at any length. Such a cam should open and close the valve at the right instants and should push it open far enough, but in addition to these requirements it is necessary that the valve should come back to its seat quietly, and that in moving it should always remain in contact with the cam-actuated operating lever. Further, there should be no undue strain at any part of the motion, or the pressure of the valve on the lever should be as nearly uniform and as low as possible, during its entire motion.

The total force required to move the valve at any instant is that necessary to overcome the gas pressure on top of it, plus that necessary to overcome the spring, plus that necessary to lift and accelerate the valve if it has not uniform velocity. The gas pressure is great just at the moment the valve is opened (the exhaust valve is here spoken of) and immediately falls almost to that of the atmosphere, while the spring force is least when the valve is closed and most when the valve is wide open. The weight of the valve is constant and its acceleration is entirely under the control of the designer of the cam. Under the above circumstances it would seem that the acceleration should be low at the moments the valve is opened and closed, and that it might be increased as the valve is raised, although the increasing spring pressure would prevent undue increase in acceleration.

Again, the velocity of the valve at the moment it returns to its seat must be low or there will be a good deal of noise, and the cam should be so designed that the valve can fall rapidly enough to keep the follower in contact with the cam, or the noise will be objectionable. The general conditions should then be that the follower should start with a small acceleration which may be increased as the valve opens more, and that it must finish its stroke at comparatively low velocity.

In lieu of more complete data, let it be assumed that the valve is to remain open for 120° of rotation of the cam, and is to close at low velocity. The travel of the valve is also given and it is to remain nearly wide open during 20° of rotation. It will first be assumed that the follower moves on a radial line as at Fig. 86 (*b*) and correction made later for the deviation due to the arc.

From the data assumed the valve is to move upward for 50° of rotation of the cam and downward during the same interval, and as the camshaft turns at constant speed each degree of rotation represents the same interval of time. Let the acceleration be as shown on the diagram Fig. 87 (*a*) on a base representing degrees of camshaft rotation, which is also a time base; then the assumed form of acceleration curve will mean that at first the acceleration is zero but that this rapidly increases during the first 5° of rotation to its maximum value at which it remains for

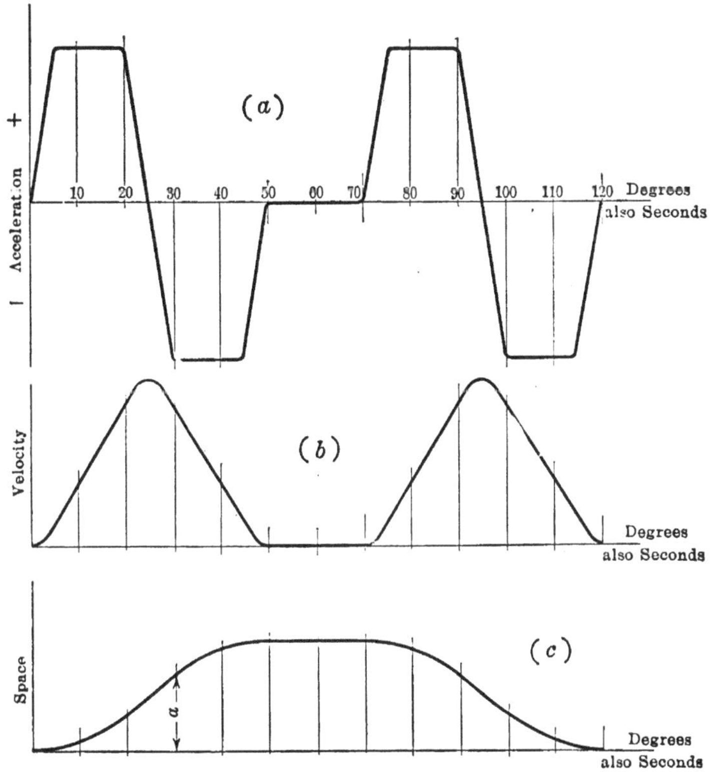

Fig. 87.

the next 15°. It then drops rapidly to the greatest negative value where it also remains constant for a short interval and then rapidly returns to zero at which it remains for 20°, after which the process is repeated. Such a curve means a rapidly increasing velocity of the valve to its maximum value, followed by a rapid decrease to zero velocity corresponding to the full opening of the valve and in which position the valve remains at rest for 20°. The valve then drops rapidly, reaching its seat at the end of 120° at zero velocity.

By integrating the acceleration curve the velocity curve is found as shown at (*b*) Fig. 87, and making a second integration gives the space curve shown at (*c*), the maximum height of the space curve representing the assumed lift of the cam. These curves show that the valve starts from rest, rises and finally comes to rest at maximum opening; it then comes down with rapid acceleration near the middle of its stroke and comes back on its seat again with zero velocity and acceleration and therefore without noise.

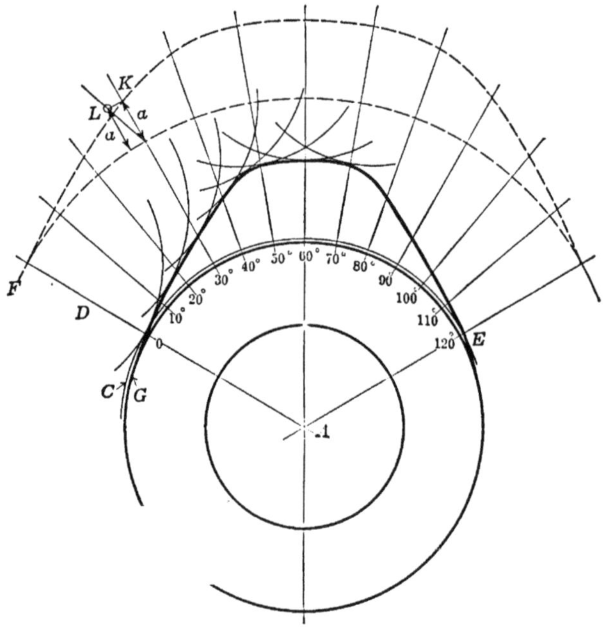

Fig. 88.—Gas-engine cam.

Having now obtained the space curve the design of the cam is made as follows:

In Fig. 88, let A represent the camshaft and the circle G represent the end of the hub of the cam, the diameter of which is determined by considerations of strength. There is always a slight clearance left between the hub and follower so that the valve may be sure to seat properly and this clearance circle is indicated in light lines by C. Lay off radii (say) 10° apart as shown; then AD and AE, 120° apart, represent the angle of action of the cam. Draw a circle F with center A and at distance from C equal to the radius of the roller; then this circle F is the base circle from which the displacements shown in (*c*), Fig. 87, are to be laid off, and this is now done, one case being shown to indicate the exact method. The result is the curve shown in dotted lines which begins and ends on the circle F. A pair of compasses are now set with radius equal to the radius of the roller of the follower and a series of arcs drawn, as shown, all having centers on the dotted curve. The solid curve drawn tangent to these arcs is the outline of the cam which would fulfil the desired conditions provided the follower moved in and out along the radial line from the center *A* as shown at Fig. 86 (*b*).

Should the follower move in the arc of a circle as is the case in Fig. 86 (*a*), where the follower moves in the arc of a circle described about *B*, then a slight modification must be made in laying out the cam although the curves shown at (*a*), (*b*) and (*c*), Fig. 87, would not be altered. The method of laying out the cam from Fig. 87 (*c*) may be explained as follows: From center *A* draw a circle *H* (not shown on the drawings) of radius *AB* equal to the distance from the center of camshaft to the center of the fulcrum for the lever. Then set a pair of compasses with a radius equal to the distance from *B* to the center of the follower, and with centers on *H* draw arcs of circles outward from the points where the radial lines *AD*, etc., intersect the circle F; one of these is shown in Fig. 88. All distances such as *a* are then laid off radially from F but so that their termini will be on the arcs just described; thus the point *K* will be moved over to *L*, and so with other points. The rest of the procedure is as in the former case. For ordinary proportions the two cams will be nearly alike.

Should the follower have a flat end without a roller, as is often the case, then the circle F is not used at all and all distances such as *a* are laid off on radial lines from the circle *C* and on each radius a line is drawn at right angles to such radius and of length to represent the face of the follower. The outline of the cam is then made tangent to these latter lines.

Lack of space prevents further discussion of this very interesting machine part, which enters so commonly and in such a great variety of forms into modern machinery. No discussion has been given of cams having reciprocating motion, nor of those used very commonly in screw machines, in which bars of various shapes are secured to the face of a drum and form a cam which may be easily altered to suit the work to be done by simply removing one bar and putting another of different shape in its place.

After a careful study of the cases worked out, however, there should be no great difficulty in designing a cam to suit almost any required set of conditions. The real difficulty, in most cases, is in selecting the conditions which the cam should fulfil, but once these are selected the solution may be made as explained.

QUESTIONS ON CHAPTER VIII

1. Design a disk cam for a stamp mill, for a flat-faced follower, the line of the stamp being 4 in. from the camshaft. The stamp is to be lifted 9 in. at a uniform rate.

2. Design a disk cam with roller follower to give a uniform rate of rise and fall of 3 in. per revolution to a spindle the axis of which passes through the center line of the shaft.

3. Taking the proportions of the parts from Fig. 84, design a suitable cam for the shear.

4. A cam is required for a 1 in. shaft to give motion to a roller follower ¾ in. diameter, and placed on an arm pivoted 6 in. to the left and 2 in. above the camshaft. The roller (center) is to remain 2 in. above the camshaft center for 200° of camshaft rotation, to rise ½ in. at uniform rate during 65°, to remain stationary during the next 30°, and then to fall uniformly to its original position during the next 65°. Design the cam.

5. Design a cam similar to Fig. 88 to give a lift of 0.375 in. during 45°, a full open period of valve of 25° and a closing period of 45°. Base radius of cam to be 0.625 in. and roller 1 in. diameter.

CHAPTER IX

FORCES ACTING IN MACHINES

145. External Forces.—When a machine is performing any useful work, or even when it is at rest there are certain forces acting on it from without, such as the steam pressure on an engine piston, the belt pull on the driving pulley, the force of gravity due to the weight of the part, the pressure of the water on a pump plunger, the pressure produced by the stone which is being crushed in a stone crusher, etc. These forces are called **external** because they are not due to the motion of the machine, but to outside influence, and these external forces are transmitted from link to link, producing pressures at the bearings and stresses in the links themselves. In problems in machine design it is necessary to know the effect of the external forces in producing stresses in the links, and further what the stresses are, and what forces or pressures are produced at the bearings, for the dimensions of the bearings and sliding blocks depend to a very large extent upon the pressures they have to bear, and the shape and dimensions of the links are determined by these stresses.

The matter of determining the sizes of the bearings or links does not belong to this treatise, but it is in place here to determine the stresses produced and leave to the machine designer the work of making the links, etc., of proper strength.

In most machines one part usually travels with nearly uniform motion, such as an engine crankshaft, or the belt wheel of a shaper or planer, many of the other parts moving at variable rates from moment to moment. If the links move with variable speed then they must have acceleration and a force must be exerted upon the link to produce this. This is a very important matter, as the forces required to accelerate the parts of a machine are often very great, but the consideration of this question is left to a later chapter, and for the present the acceleration of the parts will be neglected and a mechanism consisting of light, strong parts, which require no force to accelerate them, will be assumed in place of the actual one.

146. Machine is Assumed to be in Equilibrium.—It will be further assumed that at any instant under consideration, the machine is in equilibrium, that is, no matter what the forces acting are, that they are balanced among themselves, or the whole machine is not being accelerated. Thus, in case of a shaper, certain of the parts are undergoing acceleration at various times during the motion, but as the belt wheel makes a constant number of revolutions per minute there must be a balance between the resistance due to the cutting and friction on the one hand and the power brought in by the belt on the other. In the case of a train which is just starting up, the speed is steadily increasing and the train is being accelerated, which simply means that more energy is being supplied through the steam than is being used up by the train, the balance of the power being free to produce the acceleration, and the forces acting are not balanced. When, however, the train is up to speed and running at a uniform rate the input and output must be equal, or the locomotive is in equilibrium, the forces acting upon it being balanced.

147. Nature of Problems Presented.—The most general form of problem of this kind which comes up in practice is such as this: Given the force required to crush a piece of rock, what belt pull in a crusher will be required for the purpose? or: What turning moment will be required on the driving pulley of a punch to punch a given hole in a given thickness of plate? or: Given an indicator diagram for a steam engine, what is the resulting turning moment produced on a crankshaft?, etc. Such problems may be solved in two ways: (*a*) by the use of the virtual center; (*b*) by the use of the phorograph, and as both methods are instructive each will be discussed briefly.

148. Solutions by Use of Virtual Centers.—This method depends upon the fundamental principles of statics and the general knowledge of the virtual center discussed in Chapter II. The essential principles may be summed up in the following three statements:

If a set of forces act on any link of a machine then there will be equilibrium, provided:

1. That the resultant of the forces is zero.

2. That if the resultant is a single force it passes through a point on the link which is at the instant at rest. Such a point

may, of course, be permanently fixed or at rest, or only temporarily so.

3. That if the resultant is a couple the link has, at the instant, a motion of translation.

The first statement expresses a well-known fact and requires no explanation. The second statement is rather less known but it simply means that the forces will be in equilibrium if their resultant passes through a point which is at rest relative to the fixed frame of the machine. No force acting on the frame of the machine can disturb its equilibrium, for the reason that the frame is assumed fixed and if the frame should move in any case where it was supposed to remain fixed, it would simply mean that the machine had been damaged. Further, a force passing through a point at rest is incapable of producing motion.

The third statement is a necessary consequence of the second and corresponds to it. If the resultant is a couple, or two parallel forces, then both forces must pass through a point at rest, which is only possible if the point is at an infinite distance, or the link has a motion about a point infinitely distantly attached to the frame, that is the link has a motion of translation.

Let a set of forces act on any link *b* of a mechanism in which the fixed link is *d*; then the only point on *b* even temporarily at rest is the virtual center *bd*, which may possibly be a permanent center. Then the forces acting can be in equilibrium only if their resultant passes through *bd*, and if the resultant is a couple both forces must pass through *bd*, which must therefore be at an infinite distance, or *b* must at the instant, have a motion of translation. These points may be best explained by some examples.

149. Examples.—1. Three forces P_1, P_2 and P_3, Fig. 89, act on the link *b*; under what conditions will there be equilibrium? In the first place the three forces must all pass through the same point *A* on the link, and treating P_2 as the force balancing P_1 and P_3, then in addition to P_2 passing through *A* it must also pass through a point on the link *b* which is at rest, that is the point *bd*. This fixes the direction of P_2, by fixing two points on it, and thus the directions of the three forces P_1, P_2 and P_3 are fixed and their magnitudes may be found from the vector triangle to the right of the figure.

2. To find the force P_2 acting at the crankpin, in the direction

of the connecting rod, Fig. 90, which will balance the pressure P_1 on the piston. In this case P_1 and P_2 may both be regarded as forces acting on the two ends of the connecting rod and the problem is thus similar to the last one. P_1 and P_2 intersect at bc; hence their resultant P must pass through bc and also through the only point on b at rest, that is bd, which fixes the position and

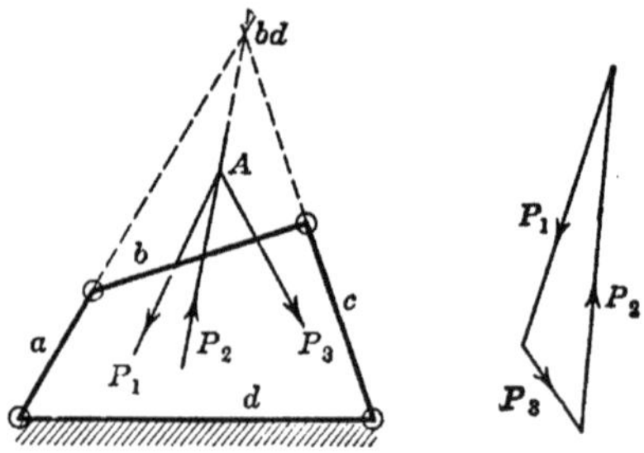

FIG. 89.

direction of P and hence the relation between the forces may be determined from the vector triangle. This enables P_2 to be found as in the upper right-hand figure.

The moment of P_2 on the crankshaft is $P_2 \times OD$, which may readily be shown by geometry to be equal to $P_1 \times O - ac$ since $\frac{P_1}{P_2} = \frac{OD}{O - ac}$, that is, the turning effect on the crankshaft

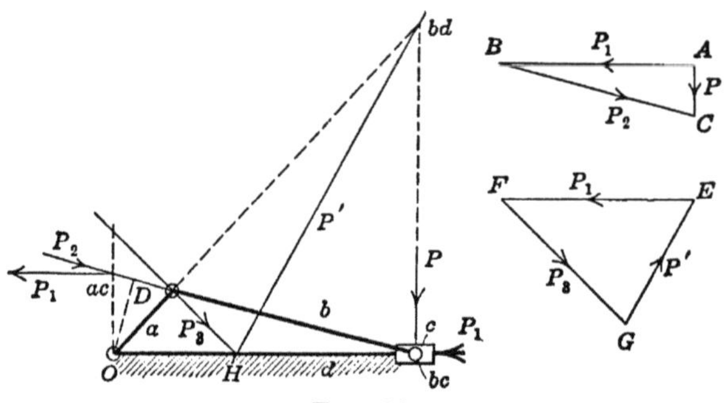

FIG. 90.

due to the piston pressure P_1 is the same as if P_1 was transferred to the point ac on the crankshaft.

Let P_3, acting normal to the crank a through the crankpin, be the force which just balances P_1; it is required to find P_3. Now P_3 and P_1 intersect at H, and their balancing force P' must pass through H and through bd which gives the direction and position of P' and the vector triangle EFG gives P_3 corresponding to a known value of P_1. The force P_3 is called the **crank effort** and may be defined as the force, passing through the crankpin and normal to the crank, which would produce the same turning moment on the crank that the piston pressure does. More will be said about this in the next chapter.

3. Forces P_1 and P_2 act on a pair of gear wheels, the pitch circles of which are shown in Fig. 91; it is required to find the relation between them, friction of the teeth being neglected. Since friction is not considered, the direction of pressure between

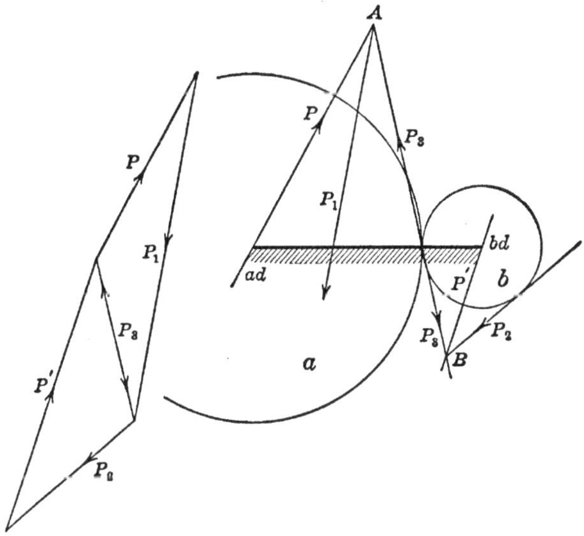

FIG. 91.

the teeth must be normal to them at their point of contact, and is shown at P_3 in the figure, this force always passing through the point of contact of the teeth and always through the pitch point or point of tangency of the pitch circles.[1] For the involute system of teeth P_3 is fixed in direction and coincides with the line of obliquity, but with the cycloidal system P_3 becomes more and more nearly vertical as the point of contact approaches the pitch point. Knowing the direction of P_3 from these considerations, let it intersect P_1 and P_2 at A and B respectively. On the wheel a there are the forces P_1, P_3 and P, the latter acting through A and ad, and their values are obtained from the vector triangle; and on b the forces P_3, P_2 and P', the latter acting through B

[1] For a complete discussion on these points see Chapter V.

and bd, and representing the bearing pressure, are found in the same way, the vector polygon on the left giving the values of the several forces concerned and hence P_2 if P_1 is known.

4. The last example taken here is the beam engine illustrated in outline in Fig. 92, and the problem is to find the turning moment produced on the crankshaft due to a given pressure P_1 acting on the walking beam from the piston. Two convenient methods of solution are available, the first being to take moments about cd and in this way to find the force P_2 acting through bc which is the equivalent of the force P_1 at C, the remainder of the problem there being solved as in Example 2.

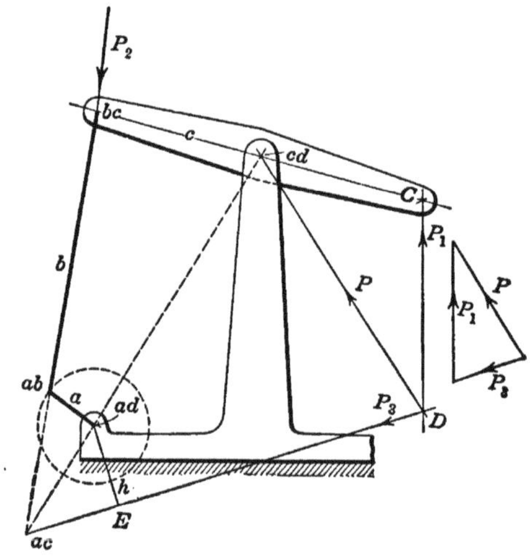

FIG. 92.

It is more general, however, and usually simpler to determine the equivalent force on the crankshaft directly. Select any point D on P_1 and resolve P_1 into two components, one P passing through the only point on the beam c at rest, that is cd, the other, P_3, passing through the common point ac of a and c. The positions of P and P_3 and their directions are known, since both pass through D and also through cd and ac respectively; hence the vector triangle on the right gives the forces P_3 and P. But P_3 acts through ac on a, and if $ad - E = h$, be drawn from ad normal to P_3, the moment of P_3 about the crankshaft is P_3h, which therefore balances the moment produced on the crankshaft by the pressure P_1 on the walking beam. The magnitude of this moment is, of course, independent of the position of the point D.

150. General Formula.—The general formula for the solution of all such problems by use of virtual centers is as follows: A force P_1 acts through any point B on a link b; it is required to find the magnitude of a force P_2, of known direction and position, acting on a link e which will exactly balance P_1, d being the fixed link. Find the centers bd, be and ed. Join B to be and bd and resolve P_1 into P_3 in the direction $B - be$ and P_4 in the direction $B - bd$; then the moment of P_3 about de must be the same as the moment of P_2 about the same point and thus P_2 is known.

151. Solution of Such Problems by the Use of the Phorograph.—In solving such problems as are now under consideration by the use of the phorograph the matter is approached from a somewhat different standpoint, and as there is frequent occasion to use the method it will be explained in some detail.

It has already been pointed out that the present investigation deals only with the case where the machine is in equilibrium, or where it is not, on the whole, being accelerated. This is always the case where the energy put into the machine per second by the source of energy is equal to that delivered by the machine, for example, where the energy per second delivered by a gas engine to a generator is equal to the energy delivered to the piston by the explosion of the gaseous mixture, friction being neglected.

Suppose now that on any mechanism there is a set of forces P_1, P_2, P_3, etc., acting on various links, and that these forces are acting through points having the respective velocities v_1, v_2, v_3, etc., feet per second in the directions of P_1, P_2, P_3. The energy which any force will impart to the mechanism per second is proportional to the magnitude of the force and the velocity with which it moves in its own direction; thus if a force of 20 pd. acts at a point moving at 4 feet per second in the direction of the force, the energy imparted by the latter will be 80 ft.-pd. per second, and this will be positive or negative according to whether the sense of force and velocity are the same or different.

The above forces will then impart respectively P_1v_1, P_2v_2, P_3v_3, etc., ft.-pd. per second of energy, some of the terms being negative frequently and the direction of action of the various forces are usually different. The total energy given to the machine per second is $P_1v_1 + P_2v_2 + P_3v_3 +$ etc., ft.-pd. and if this total sum is zero there will be equilibrium, since the net energy delivered to the machine is zero. This leads to the important statement that if in the machine any two points in the same or different links have identical motions, then, as far as the equilibrium of the machine is concerned, a given force may be applied at either of the points as desired, or if at the two points forces of equal magnitude and in the same direction but opposite in sense are applied then the equilibrium of the machine will be unaffected by these two forces, for the product Pv will be the same in each case, but opposite in sense, and the sum of the products Pv will be zero.

To illustrate these points further let any two points B and B' in the same or different links in the mechanism have the same motion, and let any force P act through B, then the previous paragraph asserts that without affecting the conditions of equilibrium in any way, the force P may be transferred from B to B', that is to say that if a force P act through a point B in any link, and there is found in any other link in the mechanism a point B' with the same motion as B, the force P will produce the same effect as far as the equilibrium of the mechanism is concerned, whether it acts at B or B'.

It has been shown in Chapter IV that for each point in a mechanism there may be found a point called its image on a selected link, which point has the same motion as the point under discussion, and thus it is possible to find on a single link a collection of points having the same motions as the various points of application of the acting forces. Without affecting the conditions of equilibrium, any force may be moved from its actual point of application to the image of this point, and thus the whole problem be reduced to the condition of equilibrium of a set of forces acting on a single link. There will be equilibrium provided the sum of the moments of the forces about the center of rotation relative to the frame is zero.

152. Examples Using the Phorograph.—As this matter is somewhat difficult to understand it may best be explained by a few practical examples in which the application is given and in the solution of the problems it will be found that the only difficulty offered is in the finding of the phorograph of the mechanism, so that Chapter IV must be carefully mastered and understood.

1. To find the turning effect produced on the crankshaft of an engine due to the weight of the connecting rod. Let Fig. 93 represent the engine mechanism, with connecting rod AB

having a weight W lb. and with its center of gravity at G; the weight W then acts vertically downward through G. Find A', B' and G', the images of A, B and G on the crank OA; then since, by the principle of the phorograph, the motion of G' is identical with that of G, it follows that G' must have exactly the same velocity as G, that is to say energy will be imparted to the mechanism at the same rate per second by the force W acting at G' as it will by the same force acting at G, so that the force W may be moved to G' without affecting the conditions of equilibrium, and this has been done in the figures. It must not be supposed that W acts both at G and G' at the same time; it is simply transferred from G to G'.

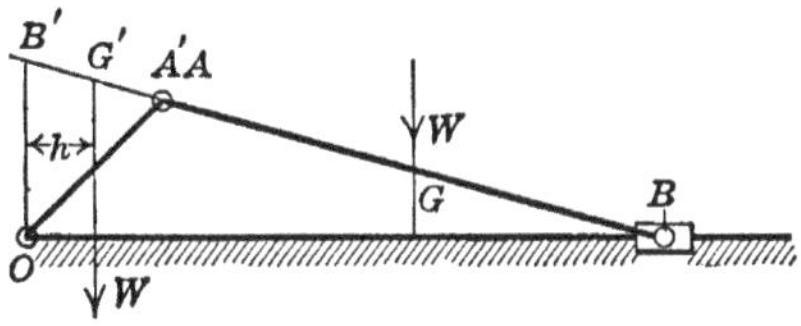

FIG. 93.

Since G' is a point on the crankshaft, the moment due to the weight of the rod is Wh ft.-pd., where h is the shortest distance, in feet, from O to the direction of the force W.

2. A shear shown in Fig. 94 is operated by a cam a attached to the main shaft O, the shaft being driven at constant speed by a belt pulley. Knowing the force F necessary to shear the bar

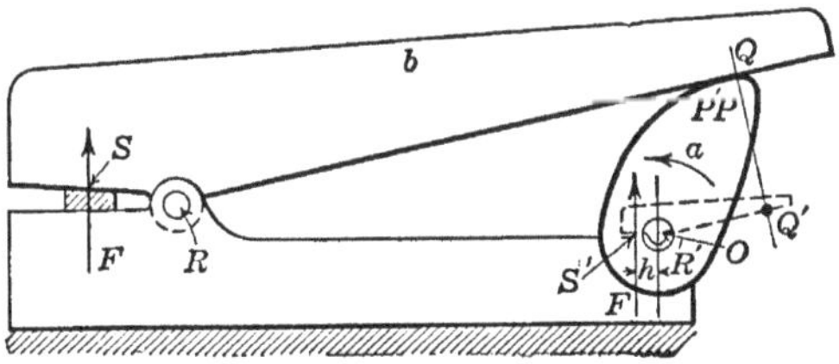

FIG. 94.—Shear.

at S, the turning moment which must be applied at the camshaft O is required. Let P be the point on the cam a where it touches the arm b at Q, then the motion of P with regard to Q is one of sliding along the common tangent at P. Choosing a as the link of reference, P' will lie at P, R' at O, $R'Q'$ will be parallel to RQ and Q' will lie in $P'Q'$ the common normal to the surfaces at P, this locates Q'. Having now two points on b', viz., R' and Q', complete the figure by drawing from Q' the line $Q'S'$ parallel to QS, also drawing $R'S'$ parallel to RS and thus locating S'. The construction lines have not been drawn on the diagram. The figure shows the whole jaw dotted in, although it is quite unnecessary. Having now found S' a point on a with the same velocity as S on b, the force F may be transferred to S' and the moment $\mathrm{F} \times h$ of F about O is the moment which must be produced on the shaft in the opposite sense. By finding the

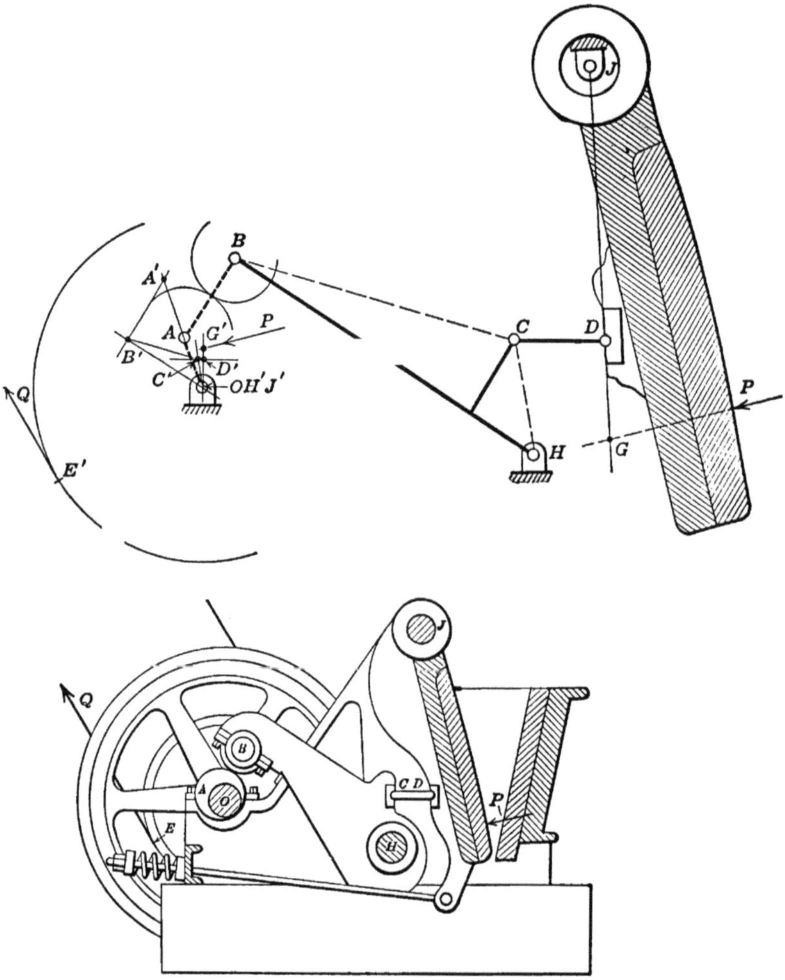

FIG. 95.—Rock crusher.

moment in a number of positions it is quite easy to find the necessary power to be delivered by the belt for the complete shearing operation.

3. A somewhat more complicated machine is shown in Fig. 95, which represents a belt-driven rock crusher built by the Fair-

banks-Morse Co., the lower figure having been redrawn from their catalogue, and the upper figure shows the mechanism on a larger scale.

On a belt wheel shown, the belt exerts a net pull Q which causes the shaft O, having the eccentric OA attached to it, to revolve. The shaft H carried by the frame has the arm HB attached to it, to the left-hand end of which is a roller resting on the eccentric OA. The crusher jaw is pivoted on the frame at J and a strong link CD keeps the jaw and the arm HB a fixed distance apart. As shaft O turns, the eccentric imparts a motion to the arm HB which in turn causes the jaw to have a pendulum motion about J and to exert a pressure P on a stone to be crushed. It is required to find the relation between belt pull Q and the pressure P.

Select OA as the link of reference and make the phorograph of double scale as in Fig. 39, making $OA' = 2\ OA$. As the device simply employs two chains similar to Fig. 32, viz., $OABCH$ and $JDCH$, the images of all the points may readily be found and these are shown on the figure. Then P is transferred from G to G' and Q from E to E' and then P and Q both act on the one link and hence their moments must be equal, or $Q \times OE'$ = moment of P about O, from which P is readily found for a given value of Q.

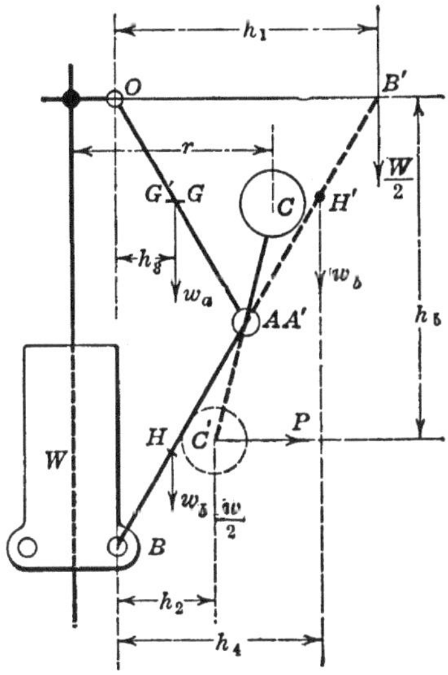

FIG. 96.—Proell governor.

4. The application to a governor[1] is shown in Fig. 96 which represents one-half of a Proell governor, and it is required to find the speed of the vertical spindle which will hold the parts in equilibrium in the position shown. In the sketch the arm OA is pivoted to the spindle at O and to the arm BA at A, the latter arm carrying the ball C on an extension of it and being attached to the central weight W at B. The weight of each revolving ball at C is $\frac{w}{2}$ lb. and of the central weight is W lb.

[1] A complete discussion of governors is given in Chapter XII.

Treating OA as the link of reference and G as the center of it, find the images of A' at A and also B' and C', then transfer $\frac{W}{2}$ (one-half the central weight acts on each side) to B' and $\frac{w}{2}$ to C', and if it is desired to allow for the weights w_a and w_b of the arms OA and AB the centers of gravity G and H of the latter are found and also their images G' and H', then w_b is transferred to H', but as G' is at G, w_a is not moved. If the balls revolve with linear velocity v ft. per second in a circle of radius r ft., then the centrifugal force acting on each ball will be $P = \frac{w}{2g} \times \frac{v^2}{r}$. pds. in the horizontal direction, and this force P is transferred to C'. Let the shortest distances from the vertical line through O to B', C', G' and H' be h_1, h_2, h_3 and h_4 respectively, and let the vertical distance from C' to OB' be h_5, then for equilibrium of the parts (neglecting friction), taking moments about O.

$$\frac{W}{2} \cdot h_1 + \frac{w}{2} \cdot h_2 + w_a h_3 + w_b h_4 = \frac{w}{2g} \times \frac{v^2}{r} \times h_5$$

which enables the velocity v necessary to hold the governor in equilibrium in any given position to be found, and from this the speed of the spindle may be computed.

5. The chapter will be concluded by showing two very interesting applications to riveters of toggle-joint construction. The first one is shown in Fig. 97, the drawing on the left showing the construction of the machine, while on the right is shown the mechanism involved and the solution for finding the pressure P at the piston necessary to exert a desired rivet pressure R. The frame d carries the cylinder g, with piston f, which is connected to the rod e by the pin C. At the other end of e is a pin A which connects e with two links a and b, the former of which is pivoted to the frame at O. The link b is pivoted at B to the slide c which produces the pressure on the rivet.

Select a as the primary link because it is the only one having a fixed point; then A' is at A, and since B has vertical motion, therefore B' will lie on a horizontal line through O and also on a line through A' in the direction of b, that is, on b produced so that B' is found. C' lies on a line through O normal to the direction of motion of C, that is to the axis of the cylinder g, and since it also lies on a line through A' parallel to e, therefore C' is found.

By transferring P to C' and R to B' as shown dotted, the relations between the forces P and R are easily found, since their moments about O must be equal, that is, $P \times OC' = R \times OB'$.

By comparing the first and later positions in this and the following figures the rapid increase in the mechanical advantage of the mechanism, as the piston advances, will be quite evident.

6. Another form of riveter is shown at Fig. 98 and the solution for finding the rivet pressure R corresponding to a given piston pressure P is shown along with the mechanism on the right in

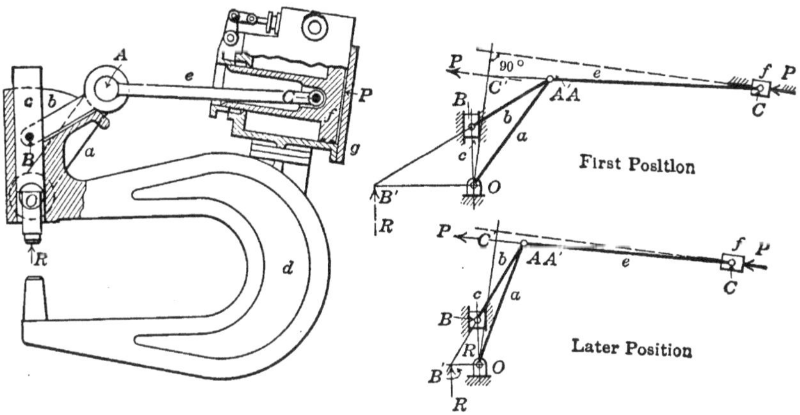

Fig. 97.—Riveter.

two positions. The proportions in the mechanism have been altered to make the illustration more clear. The loose link b contains four pivots, C, B, A, F; C being jointed to the frame at D by the link e; B having a connection to the link c, which link is also connected at E to the sliding block e acting directly on the rivet. A is connected to the frame at O by means of the link a, and F is connected to the piston g at G by means of the link f.

Either links a or e may be used as the link of reference, as each has a fixed center, the link a having been chosen. The images are found in the following order: C' is on $A'C'$ and on $D'C'$ parallel to DC; B' is next found by proportion, as is also F$'$ and thus the image of the whole link b. Next E' is on $B'E'$, parallel to BE and on $O'E'$ drawn perpendicular to the motion of the slide e, while G' is on a line through O perpendicular to the motion of the piston g and is also on the line $F'G'$ parallel to FG.

Transfer the force P from G to G', and the force R from E to E', and then the moment about O of R through E' must equal the moment of P through G', that is, $R \times OE' = P \times OG'$ from which the relation between R and P is computed and this may be done for all the different positions of the piston g.

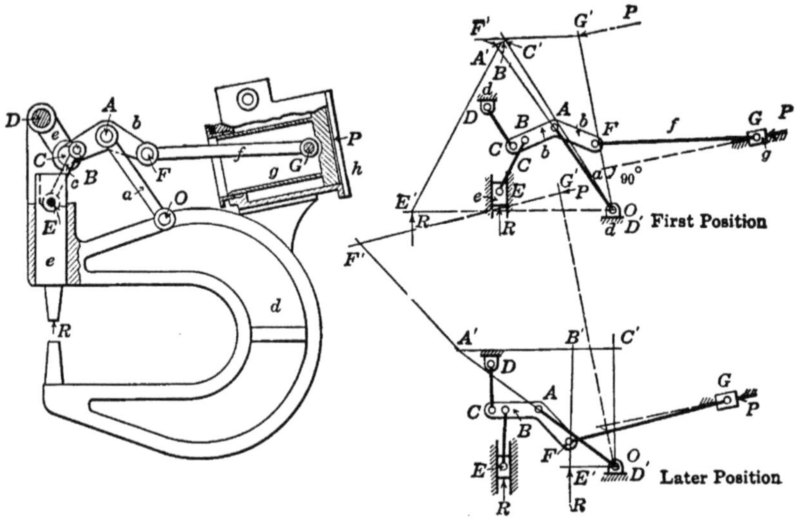

FIG. 98.—Riveter.

QUESTIONS ON CHAPTER IX

1. Why are external forces so named? What effects do they produce in the machine?

Solve the following by virtual centers:

2. Determine the crank éffort and torque when the crank angle is 45° in an 8 in. by 10-in. engine with rod 20 in. long, the steam pressure being 40 pds. per square inch.

3. In a pair of gears 15 in. and 12 in. diameter respectively the direction of pressure between the teeth is at 75½° to the line of centers, which is horizontal. On the large gear there is a pressure of 200 pds. sloping upward at 8° and its line of action is 3 in. from the gear center. On the smaller gear is a force P acting downward at 10° and to the left, its line of action being 4 in. from the gear center. Find P.

4. In a mechanism like Fig. 37 a = 15 in., b = 24 in., d = 4 in. and e = 60 in. and the link a is driven by a belt on a 10-in. pulley sloping upward at 60°. Find the relation between the net belt pull and the pressure on f when a is at 45°.

5. The connecting rod of a 10 in. by 12-in. engine is 30 in. long and weighs 30 lb., its center of gravity being 12 in. from the crankpin. What turning effect does the rod's weight produce for a 30° crank angle?

Solve the following by the phorograph:

6. In a crusher like Fig. 95, using the same proportions as are there given, find the ratio of the belt pull to the jaw pressure and plot this ratio for the complete revolution of the belt-wheel.

7. In the Gnome motor, Fig. 178, with a fixed link 2 in. long, and the others in proportion from the figure, find the turning moment on the cylinder due to a given cylinder pressure.

CHAPTER X

CRANK-EFFORT AND TURNING-MOMENT DIAGRAMS

153. Variations in Available Energy.—In the case of all engines, whether driven by steam, gas or liquid, the working fluid delivers its energy to one part of the machine, conveniently called the piston, and it is the purpose of the machine to convert the energy so received into some useful form and deliver it at the shaft to some external machine. Pumps and compressors work in exactly the opposite way, the energy being delivered to them through the crankshaft, and it is their function to transfer the energy so received to the water or gas and to deliver the fluid in some desired state.

In the case of machines having pure rotary motion, such as steam and water turbines, turbine pumps, turbo-compressors, etc., there is always an exact balance between the energy supplied and that delivered, and the input to and output from the machine is constant from instant to instant. Where reciprocating machines, having pistons, are used the case becomes somewhat different, and in general, the energy going to or from the piston at one instant differs from that at the next instant and so on. This of necessity causes the energy available at the crankshaft to vary from time to time and it is essential that this latter energy be known for any machine under working conditions.

These facts are comparatively well known among engineers. Steam turbines are never made with flywheels because of the steadiness of motion resulting from the manner of transforming the energy received from the steam. On the other hand, reciprocating steam engines are always constructed with a flywheel, or what corresponds to one, which will produce a steadying effect and the size of the wheel depends on the type of engine very largely. Thus, a single-cylinder engine would have a heavy wheel, a tandem compound engine would also have a heavy one, while for a cross-compound engine for the same purpose the flywheel could be much smaller and lighter.

Again, a single-cylinder, four-cycle, single-acting gas engine would have a much larger wheel than any form of steam engine, and the flywheel size would diminish as the number of cylinders increased, or as the engine was made double-acting or made to run on the two-cycle principle, simply because the input to the pistons becomes more constant from instant to instant, and the energy delivered by the fluid becomes more steady.

In order that the engineer may understand the causes of these differences, and may know how the machines can best be designed, the matter will here be dealt with in detail and the first case examined will be the steam engine.

154. Torque.—An outline of a steam engine is shown in Fig. 99, and at the instant that the machine is in this position let

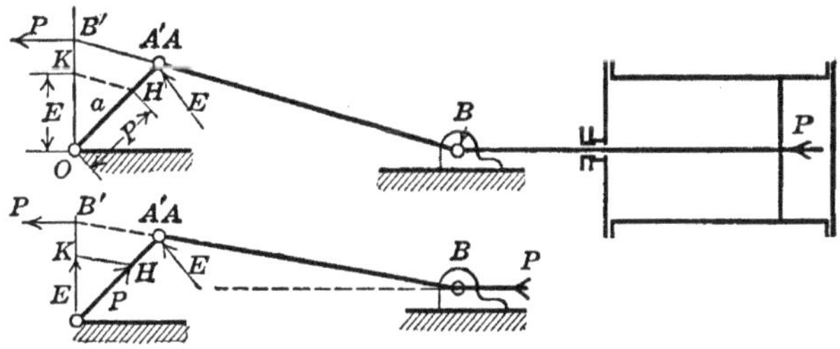

FIG. 99.

the steam produce a pressure P on the piston as indicated (the method of arriving at P will be explained later), then it is required to find the turning moment produced by this pressure on the crankshaft. It is assumed that the force P acts through the center of the wristpin B.

Construct the phorograph of the machine and find the image B' of B by the principles laid down in Chapter IV. Now in Chapter IX it is shown that, for the purposes of determining the equilibrium of a machine, any acting force may be transferred from its actual point of application to the image of its point of application. Hence, the force P acting through B will produce the same effect as if this force were transferred to B' on the crankdisc, so that the turning moment produced on the crankdisc and shaft is $P \times OB'$ ft.-pds., and this turning moment will be called the **torque** T.

Thus $T = P \times OB'$ ft.-pds. where OB' is measured in feet.

155. Crank Effort.—Now let the torque T be divided by the length a of the crank in feet, then since a is constant for all crank positions, the quantity so obtained is a force which is proportional to the torque T produced by the steam on the crankshaft. This force is usually termed the **crank effort** and may be defined as the force which if acting through the crankpin at right angles to the crank would produce the same turning effect that the actual steam pressure does (see Sec. 149 (2)).

Let E denote the crank effort; then

$$E \times a = T = P \times OB' \text{ ft.-pds.}$$

or

$$E = P \cdot \frac{OB'}{a} \text{pds.}$$

It is evident that the turning moment produced on the crankshaft by the steam may be represented by either the torque T ft.-pds. or by the crank effort E pds., since these two always bear a constant relation to one another. For this reason, crank efforts and torques are very frequently confused, but it must be remembered that they are different and measured in different units, and the one always bears a definite relation to the other.

The graphical solution for finding the effort E corresponding to the pressure P is shown in Fig. 99. It is only necessary to lay off OH along a to represent P on any convenient scale, and to draw HK parallel to $A'B'$, and then the length OK will represent E on the same scale that OH represents P. The proof is simple. Since the triangles $OB'A'$ and OKH are similar, it is evident that:

$$\frac{OK}{OH} = \frac{OB'}{OA} = \frac{OB'}{a} = \frac{E}{P} \text{ since } E \times a = P \times OB'.$$

156. Crank Effort and Torque Diagrams.—Having now shown how to obtain the crank effort and torque, it will be well to plot a diagram showing the value of these for each position of the crank during its revolution. Such a diagram is called a **crank-effort diagram** or a **torque diagram.** In drawing these diagrams the usual method is to use a straight base for crank positions, the length of the base being equal to that of the circumference of the crankpin circle.

157. Example.—Steam Engine.—The method of plotting such curve from the indicator diagrams of a steam engine is given in detail so that it may be quite clear.

Let the indicator diagrams be drawn as shown in Fig. 100, an outline of the engine being shown in the same figure, and the crank efforts and torques will be plotted for 24 equidistant positions of the crankpin, that is for each 15° of crank angle. The straight line OX in Fig. 101 is to be used as the base of the new diagram, and is made equal in length to the crankpin circle, being divided into 24 equal parts. The corresponding numbers in the two figures refer to the same positions.

The vertical line OL through O will serve as the axis for torques and crank efforts, but, of course, the scale for crank efforts must be different from that for torques.

Let A_1 and A_2 represent respectively the areas in square inches of the head and crank ends of the piston, the difference between the two being due to the area of the piston rod; the stroke of the piston is L ft.

Suppose the indicator diagrams to be drawn to scale s, by which is meant that such a spring was used in the indicator that

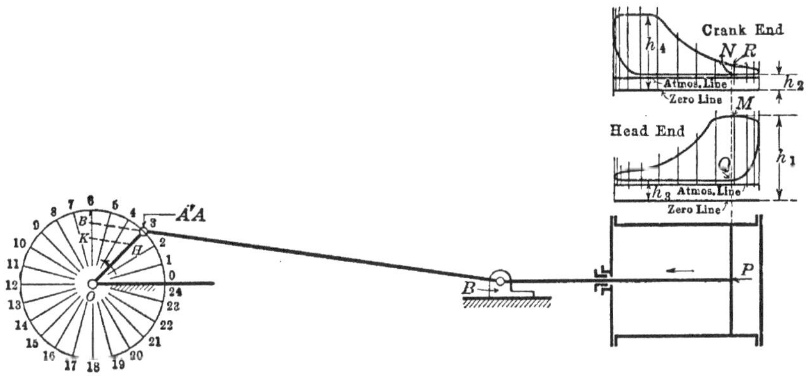

FIG. 100.

1 in. in height on the diagram represents s pds. per square inch pressure on the engine piston; thus if $s = 60$ then each inch in height on the diagram represents a pressure of 60 pds.per square inch on the piston. The lengths of the head- and crank-end diagrams are assumed as l_1 and l_2 in. (usually $l_1 = l_2$) and these lengths rarely exceed 4 in. irrespective of the size of the engine.

Now place the diagrams above the cylinder as in Fig. 100 with the atmospheric lines parallel to the line of motion of the piston. The two diagrams have been separated here for the sake of clearness, although often they are superimposed with the atmospheric lines coinciding. Further, the indicator diagram lengths have been adjusted to suit the length representing the travel of the piston. While this is not necessary, it will frequently be found convenient, but all that is really required is to draw on the diagrams a series of vertical lines showing the points on the diagrams corresponding to each of the 24 crank positions; these lines are shown very light on the diagrams. Next, draw on the diagrams the lines of zero pressure which are parallel to the atmospheric lines and at distances below them equal to the atmospheric pressure on scale s.

Having done this preliminary work, it is next necessary to find the image $A'B'$ of b for each of the 24 crank positions, one of the images being shown on the figure For the crank position 3 shown, it will be observed that the engine is taking steam on the head end and exhausting on the crank end, since the piston is moving to the left, and hence at this instant the indicator pencils would be at M and N on the head- and crank-end dia-

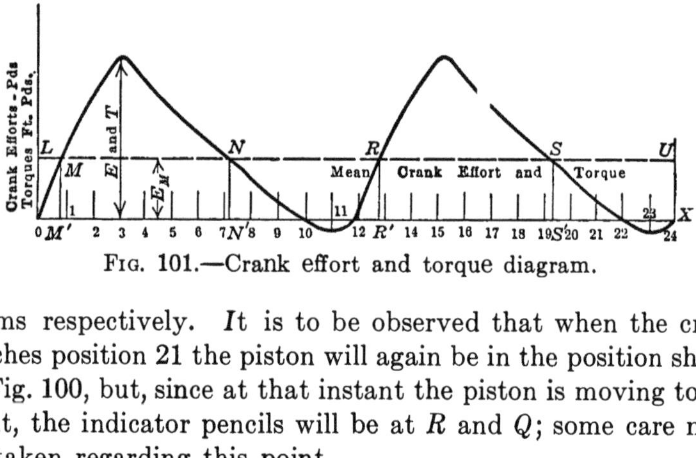

FIG. 101.—Crank effort and torque diagram.

grams respectively. It is to be observed that when the crank reaches position 21 the piston will again be in the position shown in Fig. 100, but, since at that instant the piston is moving to the right, the indicator pencils will be at R and Q; some care must be taken regarding this point.

Now let h_1 in. represent the height of M above the zero line and h_2 in. the height of N; then the force urging the piston forward is $h_1 \times s \times A_1$, while that opposing it is $h_2 \times s \times A_2$ and hence the piston is moving forward under a positive net force of

$$P = h_1 \times s \times A_1 - h_2 \times s \times A_2 \text{ pds.}$$

While it is clear that P is positive in this position, and as a matter of fact is positive for most crank positions, yet there are some in which it is negative, the meaning of which is that in these positions the mechanism has to force the piston to move **against** an opposing steam pressure; the mechanism is able to do this to a limited extent by means of the energy stored up in its parts.

From the value of P thus found, the crank effort E is determined by the method already explained and the process repeated for each of the 24 crank positions, obtaining in this way 24 values of E. These will be found to vary within fairly wide limits. Then, using the axis of Fig. 101, having a base OX equal the circumference of the crankpin circle, plot the values of E thus found at each of the 24 positions marked and in this way the crank-effort diagram $OMNRSX$ is found, vertical heights on the diagram representing crank efforts for the corresponding crankpin positions, and these heights may also be taken to represent the torques on a proper scale determined from the crank-effort scale.

158. Relation between Crank-effort and Indicator Diagrams. —From its construction, horizontal distances on the crank-effort diagram represent space in feet travelled by the pin, while vertical distances represent forces in pounds, in the direction of motion of the crankpin, and therefore the area under this curve represents the work done on the crankshaft in foot-pounds. Since the areas of the indicator diagrams represent foot-pounds of work delivered to the piston, and from it to the crank, therefore the work represented by the indicator diagrams must be exactly equal to that represented by the crank-effort diagram. The stroke of the piston has been taken as L ft. and hence the length of the base OX will represent $\pi \times L$ ft., while the length of each indicator diagram will represent L ft. Calling p_m the mean pressure corresponding to the two diagrams and E_M the mean crank effort, then $2L \times p_m = \pi \times L \times E_M$ ft.-pds. or $E_M = \frac{2}{\pi} p_m$ pds., that is, the mean height of the crank-effort diagram in pounds is $\frac{2}{\pi}$ times the mean indicated pressure as shown by the indicator diagrams. In this way the mean crank-effort line LU may be located, and this location may be checked by finding the area under the crank-effort diagram in foot-pounds, by planimeter and then dividing this by OX will give E_M.

The crank-effort diagram may also be taken to represent torques. Thus, if the diagram is drawn on a vertical scale of E pds. equal 1 in., and if the crank radius is a ft. then torques may be scaled from the diagram using a scale of $E \times a$ ft.-pds. equal to 1 in.

The investigation above takes no account of the effect of inertia of the parts as this matter is treated extensively in Chapter XV under accelerations in machinery.

159. Various Types of Steam Engines.—An examination of Fig. 101 shows that the turning moment on the crankshaft, in the engine discussed, is very variable indeed and this would cause certain variations in the operation of the engine which will be discussed later. In the meantime it may be stated that designers try to arrange the machinery as far as possible to produce uniform effort and torque.

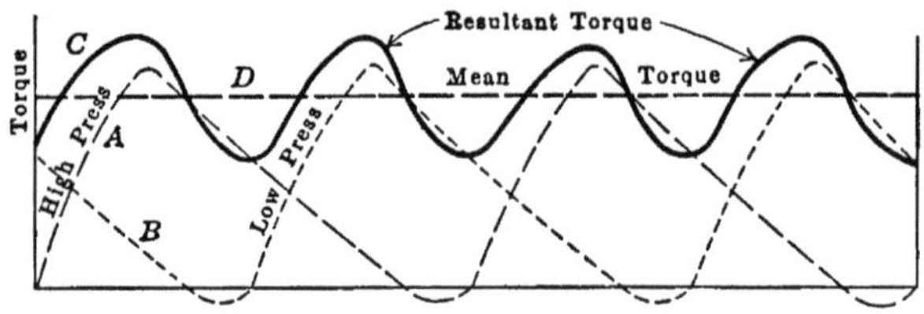

FIG 102.—Torque diagrams for cross-compound engine.

Steam engines are frequently designed with more than one cylinder, sometimes as compound engines and sometimes as twin arrangements, as in the locomotive and in many rolling-mill engines. Compound engines may have two or three and sometimes four expansions, requiring at least two, three or four cylinders, respectively. Engines having two expansions are arranged either with the cylinders tandem and having both pistons

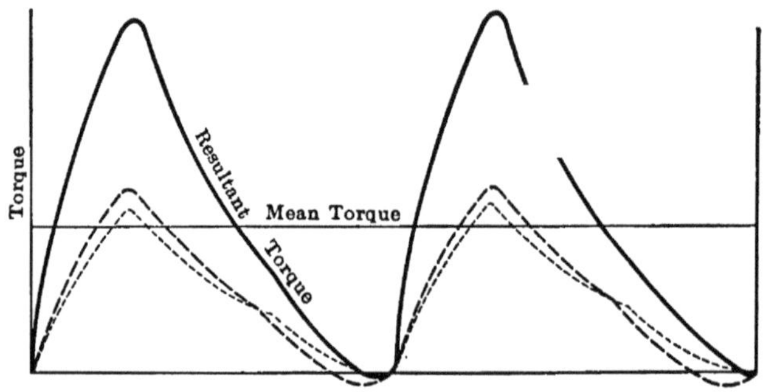

FIG. 103.—Torque diagrams for tandem engine.

connected to the same crosshead, or as cross-compound engines with the cylinders placed side by side and each connected through its own crosshead and connecting rod to the one crankshaft, the cranks being usually of the same radius and being set at 90° to one another. In Fig. 102 are shown torque diagrams for twin engines as used in the locomotive or for a cross-compound engine with cranks at 90°, the curve A showing the torque corresponding to the high-pressure cylinder with leading crank and *B* that for the low-pressure cylinder, while the curve *C* in plain lines gives the resultant torque on the crankshaft, and the horizontal dotted line *D* shows the corresponding mean torque. The very great improvement in the torque diagram resulting from this arrangement of the engine is evident, for the torque diagram *C* varies very little from the mean line *D* and is never negative as it was with the single-cylinder engine.

On the other hand, the tandem engine shows no improvement in this respect over the single-cylinder machine as is shown by the torque diagram corresponding to it shown in Fig. 103, the dotted curves corresponding to the separate cylinders and the plain curve being the resultant torque on the shaft.

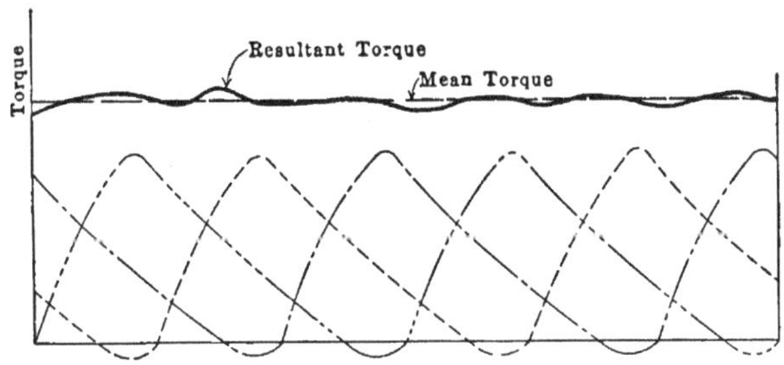

Fig. 104.—Torque diagram for triple-expansion engine.

Increasing the number of cylinders and cranks usually smooths out the torque curve and Fig. 104 gives the results obtained from a triple-expansion engine with cranks set at 120°, in which it is seen that the curve of mean torque differs very little from the actual torque produced by the cylinders.

160. Internal-combustion Engines.—It will be well in connection with this question to examine its bearing on internal-combustion engines, now so largely used on self-propelled vehicles of all kinds. Internal-combustion engines are of two general classes, two-cycle and four-cycle, and almost all machines of this class are single-acting, and only such machines are discussed here as the treatment of the double-acting engine offers no difficulties not encountered in the present case.

In the case of four-cycle engines the first outward stroke of the piston draws in the explosive mixture which is compressed in the return stroke. At the end of this stroke the charge is ignited and the pressure rises sufficiently to drive the piston forward on the third or power stroke, on the completion of which the exhaust valve opens and the burnt products of combustion are driven out by the next instroke of the piston. Thus, there is only one power stroke (the third) for each four strokes of the piston, or for each two revolutions. An indicator diagram for this type of engine is shown in (*a*) Fig. 105, and the first and fourth strokes are represented by straight lines a little below and a little above the atmospheric line respectively.

The indicator diagram from a two-cycle engine is also shown in (*b*) Fig. 105 and differs very little from the four-cycle card

Fig. 105.—Gas-engine diagrams.

except that the first and fourth strokes are omitted. The action of this type may be readily explained. Imagine the piston at its outer end and the cylinder containing an explosive mixture, then as the piston moves in the charge is compressed, ignited near the inner dead point, and this forces the piston out on the next or power stroke. Near the end of this stroke the exhaust is opened and the burnt gases are displaced and driven out by a fresh charge of combustible gas which is forced in under slight pressure; this charge is then compressed on the next instroke. In this cycle there is one power stroke to each two strokes of the piston or to each revolution, and thus the machine gets the same number of power strokes as a single-acting steam engine.

The torque diagram for a four-cycle engine is shown in Fig. 106 and its appearance is very striking as compared with those for the steam engine, for evidently the torque is negative for three out of the four strokes, that is to say, there has to be sufficient energy in the machine parts to move the piston during these strokes, and all the energy is supplied by the gas through the one power or expansion stroke. The torque has evidently very large variations and the total resultant mean torque is very small indeed.

For the two-cycle engine the torque diagram will be similar to the part of the curve shown in Fig. 106 and included in the compression and expansion strokes, the suction and exhaust strokes being omitted. Evidently also the mean torque line will be much higher than for the four-cycle curve.

Returning now to the four-cycle engine it is seen that the turning moment is very irregular and if such an engine were used with a small flywheel in driving a motorcycle or dynamo, the motion would be very unsteady indeed, and would give so much trouble that some special means must be used to control it. Various methods are taken of doing this, one of the most common

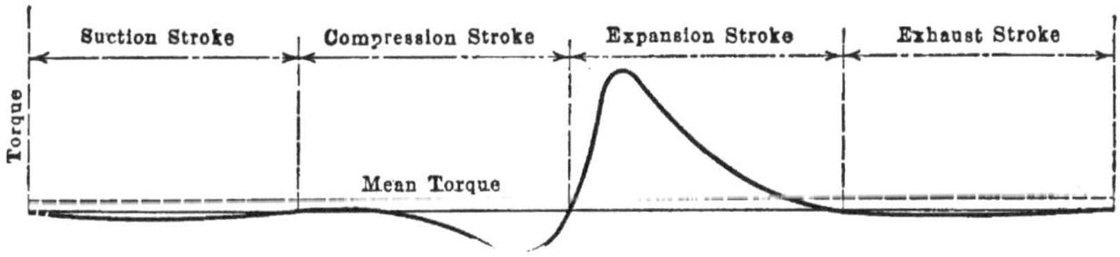

FIG. 106.—Torque diagram for four-cycle gas engine.

in automobiles, etc., being to increase the number of cylinders. Torque diagrams from two of the more common arrangements are shown in Fig. 107. The diagram marked (*a*) gives the results for a two-cylinder engine where these are either opposed or are placed side by side and the cranks are at 180°. Diagram (*b*) gives the results from a four-cylinder engine and corresponds

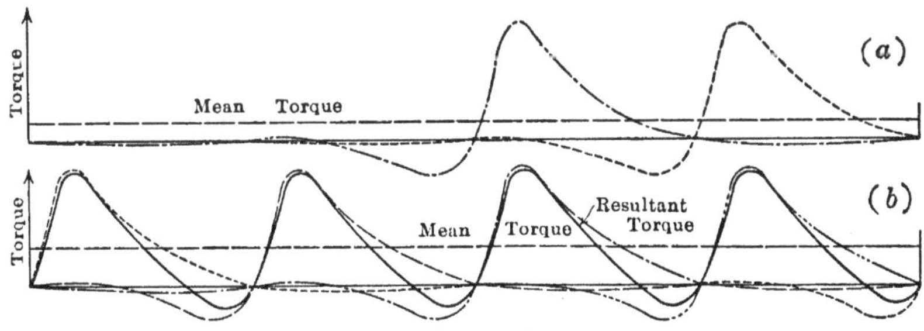

FIG. 107.—Torque diagrams for multicylinder gasoline engines.

to the use of two opposed engines on the same shaft or of four cylinders side by side, each crank being 180° from the one next it.

All of the arrangements shown clearly raise the line of mean torque and thus make the irregularities in the turning moment very much less, and by sufficiently increasing the number of cylinders this moment may be made very regular. Some automobiles now have twelve-cylinder engines resulting in very uniform turning moment and much steadiness, and the arrangements made of cylinders in aeroplanes are particularly satisfactory. Space does not permit the discussion of the matter in any further detail as the subject is one which might profitably form a subject for a special treatise.

161. General Discussion on Torque Diagrams.—The unsteadiness resulting from the variable nature of the torque has been referred to already and may now be discussed more in detail, although a more complete treatment of the subject will be found in Chapter *XIII* under the heading of "Speed Fluctuations in Machinery."

For the purpose of the discussion it is necessary to assume the kind of load which the engine is driving, and this affects what is to be said. Air compressors and reciprocating pumps produce variable resisting torques, the diagram representing the torque required to run them being somewhat similar to that of the engine, as shown in Fig. 101. It will be assumed here, however, that the engine is driving a dynamo or generator, or turbine pump, or automobile or some machine of this nature which requires a constant torque to keep it moving; then the torque required for the load will be that represented by the mean torque line in the various figures, and this mean torque is therefore also what might be called the load curve for the engine.

Consider Fig. 101; it is clear that at the beginning of the revolution the load is greater than the torque available, whereas between M and N the torque produced by the engine is in excess of the load and the same thing is true from R to S, whilc NR and SU on the other hand represent times when the load is in excess. Further, the area between the torque curve and MN plus the corresponding area above RS represents the total work which the engine is able to do, during these periods, in excess of the load, and must be equal to the sum of the areas between LM, NR and SU and the torque curve.

Now during MN the excess energy must be used up in some way and evidently the only way is to store up energy in the parts of the machine during this period, which energy will be restored by the parts during the period NR and so on. The net result is that the engine is always varying in speed, reaching a maximum at N and S and minimum values at M and R, and the amount of these speed variations will depend upon the mass of the moving parts. It is always the purpose of the designer to limit these variations to the least practical amounts and the torque curves show one means of doing this. Thus, the tandem compound engine has a very decided disadvantage relative to the cross-compound engine in this respect.

Internal-combustion engines with one cylinder are also very deficient because Fig. 106 shows that the mean torque is only a small fraction of the maximum and further that enough energy must be stored up in the moving parts during the expansion stroke to carry the engine over the next three strokes. When such engines are used with a single cylinder they are always constructed with very heavy flywheels in order that the parts may be able to store up a large amount of energy without too great variation in speed. In automobiles large heavy wheels are not possible and so the makers of these machines always use a number of cylinders, and in this way stamp out very largely the cause of the difficulty. This is very well shown in the figures representing the torques from multicylinder engines, and in such engines it is well known that the action is very smooth and even and yet all parts of the machine including the flywheel are quite light. It has not yet been possible, however, to leave off the flywheel from these machines.

QUESTIONS ON CHAPTER X

1. Using the data given in the engine of Chapter XIII and the indicator diagrams there given, plot the crank effort and torque diagrams. For what crank angle are these a maximum?

2. What would be the torque curve for two engines similar to that in question 1, with cranks coupled at 90°? At what crank angle would these give the maximum torque?

3. Compare the last results with two cranks at 180° and three cranks at 120°.

4. Using the diagram Fig. 161 and the data connected therewith, plot the crank-effort curve.

5. In an automobile motor 3½ in. bore and 5 in. stroke, the rod is 12 in. long; assuming that the diagram is similar to Fig. 161, but the pressures are only two-thirds as great, draw the crank-effort curve and torque diagram. Draw the resulting curve for two cylinders, cranks at 180°; four cylinders, cranks at 90°; six cylinders, cranks at 60° and at 120° respectively. Try the effect of different sequence of firing.

CHAPTER XI

THE EFFICIENCY OF MACHINES

162. Input and Output.—The accurate determination of the efficiency of machines and the loss by friction is extremely complicated and difficult, and it is doubtful whether it is possible to deal with the matter except through fairly close approximations. All machines are constructed for the purpose of doing some specific form of work, the machine receiving energy in one form and delivering this energy, or so much of it as is not wasted, in some other form; thus, the water turbine receives energy from the water and transforms the energy thus received into electrical energy by means of a dynamo, or a motor receives energy from the electric circuit, and changes this energy into that necessary to drive an automobile, and so for any machine. For convenience, the energy received by the machine will be referred to as the input and the energy delivered by the machine as the output.

Now a machine cannot create energy of itself, but is only used to change the form of the available energy into some other which is desired, so that for a complete cycle of the machine (*e.g.*, one revolution of a steam engine, or two revolutions of a four-cycle gas engine or the forward and return stroke of a shaper) there must be some relation between the input and the output. If no energy were lost during the transformation, the input and output would be equal and the machine would be perfect, as it would change the form of the energy and lose none. However, if the input per cycle were twice the output then the machine would be imperfect, for there would be a loss of one-half of the energy available during the transformation. The output can, of course, never exceed the input. It is then the province of the designer to make a machine so that the output will be as nearly equal to the input as possible and the more nearly these are to being equal the more perfect will the machine be.

163. Efficiency.—In dealing with machinery it is customary to use the term **mechanical efficiency** or **efficiency** to denote the ratio of the output per cycle to the input, or the efficiency $\eta = \frac{\text{output per cycle}}{\text{input per cycle}}$. The maximum value of the efficiency is unity, which corresponds to the perfect machine, and the minimum value is zero which means that the machine is of no value in transmitting energy; the efficiency of the ordinary machine lies between these two limits, electric motors having an efficiency of 0.92 or over, turbine pumps usually not over 0.80, large steam pumping engines over 0.90, etc., while in the case where the clutch is disconnected in an automobile engine the efficiency of the latter is zero, all the input being used up in friction.

The quantity $1 - \eta$ represents the proportion of the input which is lost in the bearings of the machine and in various other ways; thus in the turbine pump above mentioned, $\eta = 0.80$ and $1 - \eta = 0.20$, or 20 per cent. of the energy is wasted in this case in the bearings and the friction of the water in the pump. The amount of energy lost in the machine, and which helps to heat up the bearings, etc., will depend on such items as the nature of lubricant used, the nature of the metals at the bearings and other considerations to be discussed later.

Suppose now that on a given machine there is at any instant a force P acting at a certain point on one of the links which point is moving at velocity v_1 in the direction and sense of P; then the energy put into the machine will be at the rate of Pv_1 ft.-pds. per second. At the same instant let there be a resisting force Q acting on some part of the machine and let the point of application of Q have a velocity with resolved part v_2 in the direction of Q so that the energy output is at the rate of Qv_2 ft.-pds. per second. The force P may for example be the pressure acting on an engine piston or the difference between the tensions on the tight and slack sides of a belt driving a lathe, while Q may represent the resistance offered by the main belt on an engine or by the metal being cut off in a lathe. Now from what has been already stated the efficiency at the instant is $\eta = \frac{\text{output}}{\text{input}} = \frac{Q.v_2}{P.v_1}$ and if no losses occurred this ratio would be unity, but is always less than unity in the actual case. Now, as in practice Qv_2 is always less than Pv_1, choose a force P_0 acting in the direction of, and through the point of application of P such that $P_0v_1 =$

Qv_2, then clearly P_0 is the force which, if applied to a frictionless machine of the given type, would just balance the resistance Q, and

$$\eta = \frac{Qv_2}{Pv_1} = \frac{P_0v_1}{Pv_1} = \frac{P_0}{P}$$

so that evidently the efficiency will be $\frac{P_0}{P}$ at the instant, and P_0 will always be less than P.

The efficiency may also be expressed in a different form. Thus, let Q_0 be the force which could be overcome by the force P if there were no friction in the machine; then $Pv_1 = Q_0v_2$ and therefore

$$\eta = \frac{Qv_2}{Pv_1} = \frac{Qv_2}{Q_0v_2} = \frac{Q}{Q_0} \text{ and } Q_0 \text{ is always greater than } Q.$$

164. Friction.—Whenever two bodies touch each other there is always some resistance to their relative motion, this resistance being called friction. Suppose a pulley to be suitably mounted in a frame attached to a beam and that a rope is over this pulley, each end of the rope holding up a weight w lb. Now, since each of these weights is the same they will be in equilibrium and it would be expected that if the slightest amount were added to either weight the latter would descend. Such is, however, not the case, and it is found by experiment that one weight may be considerably increased without disturbing the conditions of rest.

It will also be found that the amount it is possible to add to one weight without producing motion will depend upon such quantities as the amount of the original weight w, being greater as w increases, the kind and amount of lubricant used in the bearing of the pulley, the stiffness of the rope, the materials used in the bearing and the nature of the mechanical work done on it, and upon very many other considerations which the reader will readily think of for himself.

One more illustration might be given of this point. Suppose a block of iron weighing 10 lb. is placed upon a horizontal table and that there is a wire attached to this block of iron so that a force may be produced on it parallel to the table. If now a tension is put on the wire and there is no loss the block of iron should move even with the slightest tension, because no change is being made in the potential energy of the block by moving it from place to place on the table, as no alteration is taking place in its height.

It will be found, however, that the block will not begin to move until considerable force is produced in the wire, the force possibly running as high as 1.5 pds. The magnitude of the force necessary will, as before, depend upon the material of the table, the nature of the surface of the table, the area of the face of the block of iron touching the table, etc.

These two examples serve to illustrate a very important matter connected with machinery. Taking the case of the pulley, it is found that a very small additional weight will not cause motion, and since there must always be equilibrium, there must be some resisting force coming into play which is exactly equal to that produced by the additional weight. As the additional weight increases, the resisting force must increase by the same amount, but as the additional weight is increased more and more the resisting force finally reaches a maximum amount, after which it is no longer able to counteract the additional weight and then motion of the weights begins. There is a peculiarity about this resisting force then, it begins at zero where the weights are equal and increases with the inequality of the weights but finally reaches a maximum value for a certain difference between them, and if the difference is increased beyond this amount the weights move with acceleration.

In the case of the block of iron on the table something of the same nature occurs. At first there is no tension in the wire and therefore no resisting force is necessary, but as the tension increases the resisting force must also increase, finally reaching a maximum value, after which it is no longer able to resist the tension produced in the wire and the block moves, and the motion of the block will be accelerated if the tension is still further increased. This resisting force must be in the direction of the force in the wire but opposite in sense, so that it must act parallel to the table, that is, to the relative direction of sliding, and increases from zero to a limiting value.

The resisting force referred to above always acts in a way to oppose motion of the parts and also always acts tangent to the surfaces in contact, and to this resisting force the name of **friction** has been applied. Much discussion has taken place as to the nature of the force, or whether it is a force at all, but for the present discussion this idea will be adopted and this method of treatment will give a satisfactory solution of all problems connected with machinery.

Wherever motion exists friction is always acting in a sense opposed to the motion, although in many cases its very presence is essential to motion taking place. Thus it would be quite impossible to walk were it not for the friction between one's feet and the earth, a train could not run were there no friction between the wheels and rails, and a belt would be of no use in transmitting power if there were no friction between the belt and pulley. Friction, therefore, acts as a resistance to motion and yet without it many motions would be impossible.

165. Laws of Friction.—A great many experiments have been made for the purpose of finding the relation between the friction and other forces acting between two surfaces in contact. Morin stated that the frictional resistance to the sliding of one body upon another depended upon the normal pressure between the surfaces and not upon the areas in contact nor upon the velocity of slipping, and further that if F is the frictional resistance to slipping and N the pressure between the surfaces, then $F = \mu N$ where μ is the **coefficient of friction** and depends upon the nature of the surfaces in contact as well as the materials composing these surfaces.

A discussion of this subject would be too lengthy to place here and the student is referred to the numerous experiments and discussions in the current engineering periodicals and in books on mechanics, such as Kennedy's "Mechanics of Machinery," and Unwin's "Machine Design." It may only be stated that Morin's statements are known to be quite untrue in the case of machines where the pressures are great, the velocities of sliding high and the methods of lubrication very variable, and special laws must be formulated in such cases. In machinery the nature of the rubbing surfaces, the intensity of the pressures, the velocity of slipping, methods of lubrication, etc., vary within very wide limits and it has been found quite impossible to devise any formula that would include all of the cases occurring, or even any great number of them, when conditions are so variable. The only practical method seems to be to draw up formulas for each particular class of machinery and method of lubrication. Thus, before it is possible to tell what friction there will be in the main bearing of a steam engine, it is necessary to know by experiment what laws exist for the friction in case of a similar engine having similar materials in the shaft and bearing and oiled in the same way, and if the machine is a horizontal Corliss engine the laws would not be the same as with a vertical high-speed engine; again the laws will depend upon whether the lubrication is forced or gravity and on a great many other things. For each type of bearing and lubrication there will be a law for determining the frictional loss and these laws must in each case be determined by careful experiment.

166. Friction Factor.—Following the method of Kennedy and other writers, the formula used in all cases will be $F = fN$ for determining the frictional force F corresponding to a normal pressure N between the rubbing surfaces, where f is called the **friction factor** and differs from the coefficient of friction of Morin in that it depends upon a greater number of elements, and the law for f must be known for each class of surfaces, method of lubrication, etc., from a series of experiments performed on similarly constructed and operated surfaces.

In dealing with machines it has been shown that they are made up of parts united usually by sliding or turning pairs, so that it will be well at first to study the friction in these pairs separately.

FRICTION IN SLIDING PAIRS

167. Friction in Sliding Pairs.—Consider a pair of sliding elements as shown in Fig. 108 and let the normal component of the pressure between these two elements be N, and let R be the resultant external force acting upon the upper element which is moving, the lower one, for the present being considered stationary. Let the force R act parallel to the surfaces in the sense shown, the tendency for the body is then to move to the right. Now, from the previous discussion, there is a certain resistance to the motion of a the amount of which is fN, where f is the friction factor, and this force must in the very nature of the case act tangent to the surfaces in contact (Sec. 164); thus, from the way in which R is chosen, the friction force $F = fN$ and R are parallel.

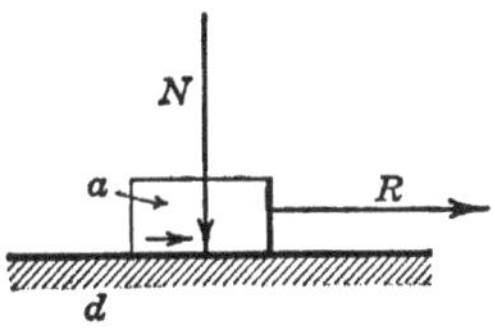

FIG. 108.

Now if R is small, there is no motion, as is well known, for the maximum value of F due to the normal pressure N is greater than R; this corresponds to a sleigh stalled on a level road, the horses being unable to move it. If, however, R be increased steadily it reaches a point where it is equal to the maximum value of F and then the body will begin to move, and so long as

R and F are equal, will continue to move at uniform speed because the force R is just balanced by the resistance to motion; this corresponds to the case where the sleigh is drawn along a level road at uniform speed by a team of horses. Should R be still further increased, then since the frictional resistance F will be less than R, the body will move with increasing speed, the acceleration it has depending upon the excess of R over F; this corresponds to horses drawing a sleigh on a level road at an increasing speed, and just here it may be pointed out that the friction factor must depend upon the speed in some way because the horses soon reach a speed beyond which they cannot go.

These results may be summarized as follows:

1. If R is less than F, that is $R < fN$, there is no relative motion.
2. If R is equal to F, that is $R = fN$, the relative motion of the bodies will be at uniform velocity.
3. If R is greater than F, that is $R > fN$, there will be accelerated motion, relatively, between the bodies. R is the resultant external force acting on the body and is parallel to the surfaces in contact.

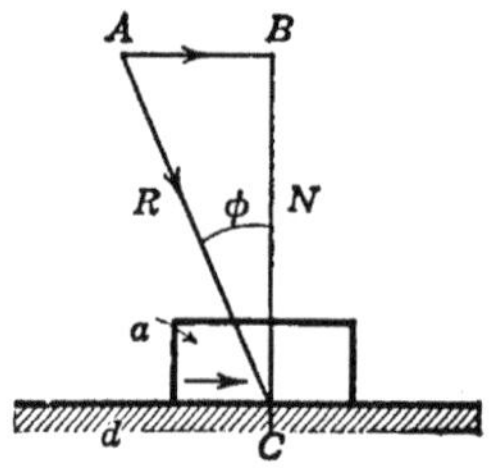

FIG. 109.

Consider next the case shown in Fig. 109, where the resultant external force R acts at an angle ϕ to the normal to the surfaces in contact, and let it be assumed that the motion of a relative to d is to the right as shown by the arrow. The bodies are taken to be in equilibrium, that is, the velocity of slipping is uniform and without acceleration. Resolve R into two components AB and BC, parallel and normal respectively to the surfaces of contact, then since $BC = N$ is the normal pressure between the surfaces, the frictional resistance to slipping will be $F = fN$, from Sec. 166, where f is the friction factor, and since there is equilibrium, the velocity being uniform, the value of F must be exactly equal and opposite to AB, these two forces being in the same direction. Should AB exceed $F = fN$ there would be acceleration, and should it be less than fN there would be no motion.

Now from Fig. 109, $AB = R \sin \phi$ and also $AB = BC \tan \phi = N \tan \phi$. Hence, since $AB = fN$, there results the relation $fN = N \tan \phi$ or $f = \tan \phi$; this is to say, **in order that two bodies may have relative motion at uniform velocity, the resultant force must act at an angle ϕ to the normal to the rubbing surfaces, and on such a side of the normal as to have a resolved** part in the direction of motion. The angle ϕ is fixed by the fact that its tangent is the friction factor f.

168. Angle of Friction.—The angle ϕ may be conveniently called the angle of friction and wherever the symbol ϕ occurs in the rest of this chapter it stands for the angle of friction and is such that its tangent is the friction factor f. The angle ϕ is, of course, the limiting inclination of the resultant to the normal and if the resultant act at any other angle less than ϕ to the normal, motion will not occur; whereas if it should act at an angle greater than ϕ there will be accelerated motion, for the simple reason that in the latter case, the resolved part of the resultant parallel to the surfaces would exceed the frictional resistance, and there would then be an unbalanced force to cause acceleration.

169. Examples.—A few examples should make the principles clear, and in those first given all friction is neglected except that

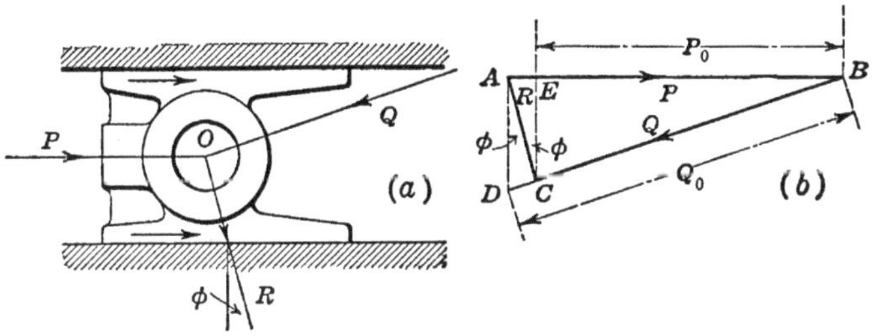

FIG. 110.—Crosshead.

in the sliding pair. The friction in other parts will be considered later.

1. As an illustration, take an engine crosshead moving to the right under the steam pressure P acting on the piston, Fig. 110. The forces acting on the crosshead are the steam pressure P, the thrust Q due to the connecting rod and the resultant R of these two which also represents the pressure of the crosshead on the guide. Now from the principles of statics, P, Q and R must all intersect at one point, in this case the center of the wristpin O, since P and Q pass through O, and further the resultant R must be inclined at an angle ϕ to the normal to the surfaces in contact, (Sec. 167); thus R has the direction shown. Note that the side of the normal on which R lies must be so chosen that R has a component in the direction of motion. Now draw $AB = P$ the steam pressure, and draw AC and BC parallel respectively to R and Q, then $BC = Q$ the thrust of the rod and $AC = R$ the resultant pressure on the crosshead shoe.

If there were no friction in the sliding pair R would be normal to the surface and in the triangle ABD the angle BAD would be 90°; BD is the force in the connecting rod and AD is the pressure on the shoe. The efficiency in this position will thus be $\eta = \frac{BC}{BD} = \frac{Q}{Q_0}$. Or it is just as direct to find P_0 the force necessary to overcome Q if there were no friction by drawing CE normal to AB then $P_0 = BE$ and $\eta = \frac{P_0}{P} = \frac{BE}{BA}$.

2. A cotter is to be designed to connect two rods, Fig. 111;

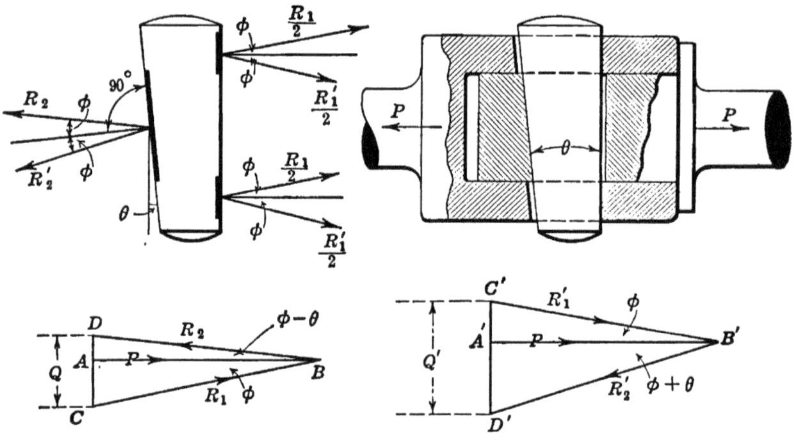

FIG. 111.—Cotter pin.

it is required to find the limiting taper of the cotter to prevent it slipping out when the rod is in tension. It will be assumed that both parts of the joint have the same friction factor f, and hence the same friction angle ϕ, and that the cotter tapers only on one side with an angle θ. The sides of the cotter on which the pressure comes are marked in heavy lines and on the right-hand side the total pressure R_1 is divided into two parts by the shape of the outer piece of the connection. Both the forces R_1 and R_2 act at angle ϕ to the normal to their surfaces and, from what has already been said, it will be understood that when the cotter just begins to slide out they act on the side of the normal shown, so that by drawing the vector triangle on the left of height $AB = P$ and having CB and BD respectively parallel to R_1 and R_2, the force Q necessary to force the cotter out is given by the side CD.

In the figure the angle $ABC = \phi$ and $ABD = \phi - \theta$. Therefore

$$Q = P\,[\tan\phi + \tan(\phi - \theta)]$$

The cotter will slip out of itself when $Q = 0$, that is

$$\tan\phi + \tan(\phi - \theta) = 0,$$

or

$$\theta = 2\phi$$

This angle θ is evidently independent of P except in so far as ϕ is affected by the tension P in the rod.

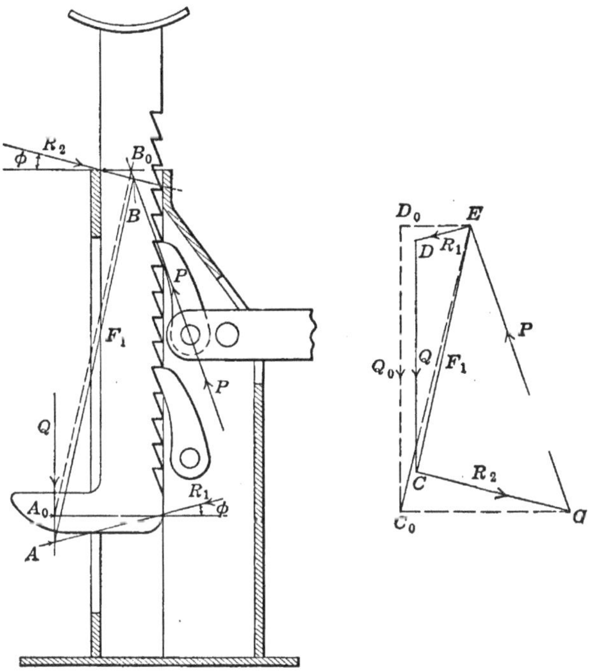

FIG. 112.—Lifting jack.

If the cotter is being driven in then the sense of the relative motion of the parts is reversed and hence the forces R_1 and R_2 take the directions R_1' and R_2' and the vector diagram for this case is also shown on the right in the figure. The force $Q' = C'D'$ necessary to drive the cotter in is $Q' = P[\tan\phi + \tan(\theta + \phi)]$ and Q' increases with θ. Small values of θ make the cotter easy to drive in and harder to drive out.

3. An interesting example of the friction in sliding connections is given in Fig. 112, which shows a jack commonly used in lifting automobiles, etc.; the outlines of the jack only are shown, and

no details shown of the arrangement for lowering the load. In the figure the force P applied to lifting the load Q on the jack is assumed to act in the direction of the pawl on the end of the handle, and this would represent its direction closely although the direction of P will vary with each position of the handle. The load Q is assumed applied to the toe of the lifting piece, and when the load is being raised the heel of the moving part presses against the body of the jack with a force R_1 in the direction shown and the top pressure between the parts is R_2, both R_1 and R_2 being inclined to the normals at angle ϕ.

At the base of the jack are the forces Q and R_1, the resultant of which must pass through A, while at the top are the forces R_2 and P, the resultant of which must pass through B; and if there is equilibrium the resultant F_1 of Q and R_1 must balance the resultant F_1 of R_2 and P, which can only be the case if F_1 passes through A and B; thus the direction of F_1 is known.

Now draw the vector triangle ECG with sides parallel to F_1, R_2 and P, and for a given value of P, so that $F_1 = EC$ and $R_2 = CG$. Next through E draw ED parallel to R_1 and through C draw CD parallel to Q from which $Q = CD$ is found. If there were no friction the reactions between the jack and the frame would be normal to the surfaces at the points of contact, thus A would move up to A_0 and B to B_0 and the vector diagram would take the form ED_0C_0G where $EG = P$ as before and $D_0C_0 = Q_0$ so that Q_0 is found.

The efficiency of the device in this position is evidently $\eta = \dfrac{Q}{Q_0}$.
It is evident that with the load on the toe, the efficiency is a maximum when the jack is at its lowest position because AB is then most nearly vertical, while for the very highest positions the efficiency will be low.

4. One more example of this kind will suffice to illustrate the principles. Fig. 113 shows in a very elementary form a quick-return motion used on shapers and machine tools, and illustrated at Fig. 12. Let Q be the resistance offered to the cutting tool which is moving to the right and let P be the net force applied by the belt to the circumference of the belt pulley. For the present problem only the friction losses in the sliding elements will be considered leaving the other parts till later. Here the tool holder g presses on the upper guide and the pressure on this guide is R_1, the force in the rod e is denoted by F_1. Further the

pressure of b on c is to the right and as the former is moving downward for this position of the machine, the direction of pressure between the two is R_2 through the center of the pin.

Now on the driving link a the forces acting are P and R_2, the resultant F_2 of which must pass through O and A. In the vector diagram draw BC equal and parallel to P, then CD and BD parallel respectively to F_2 and R_2 will represent these two forces

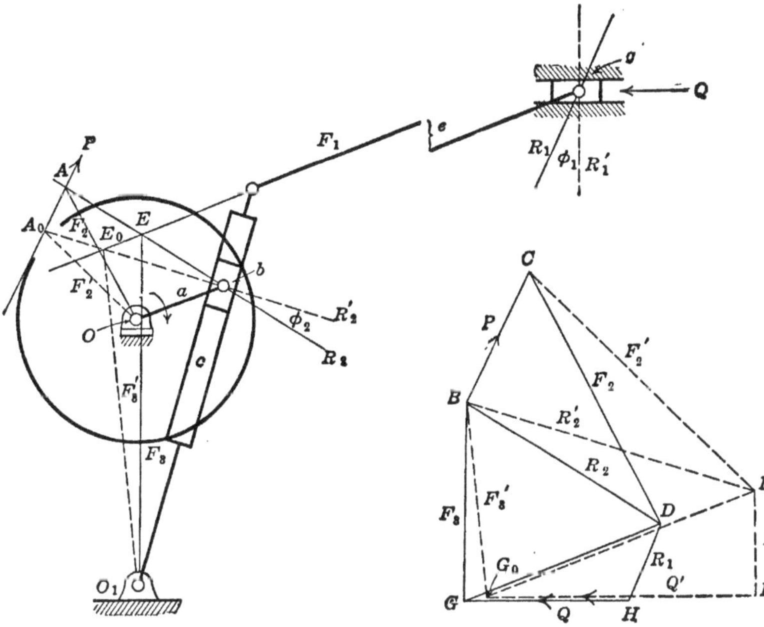

FIG. 113.—Quick-return motion.

so that R_2 is determined. Again on c the forces acting are R_2 and F_1, and their resultant passes through O_1 and also through E, the intersection of F_1 and R_2, so that drawing BG and DG in the vector diagram parallel respectively to F_3 and F_1 gives the force $F_1 = DG$ in the rod e. Acting on the tool holder g are the forces F_1, Q and R_1 and the directions of them are known and also the magnitude of F_1, hence complete the triangle GHD with sides parallel to the forces concerned and then $GH = Q$ and $HD = R_1$ which gives at once the resistance Q which can be overcome at the tool by a given net force P applied by the belt.

If there were no friction in these sliding pairs then the forces R_1 and R_2 would act normal to the sliding surfaces instead of at angles ϕ_1 and ϕ_2 to the normals so that A moves to A_0 and

E to E_0 and the construction is shown by the dotted lines, from which the value of Q_0 is obtained. The efficiency for this position of the machine is $\eta = \frac{Q}{Q_0}$. The value of η should be found for a number of other positions of the machine, and, if desirable, a curve may be plotted so that the effect of friction may be properly studied.

Before passing on to the case of turning pairs the attention of the reader is called to the fact that the greater part of the problem is the determination of the condition of static equilibrium as described in Chapter IX, the method of solution being by means of the virtual center, in these cases the permanent center being used. The only difficulty here is in the determination of the direction of the pressures R between the sliding surfaces, and the following suggestions may be found helpful in this regard.

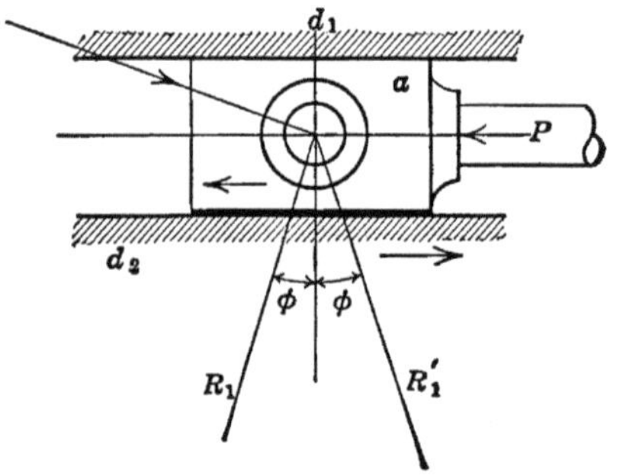

FIG. 114.

Let a crosshead a, Fig. 114, slide between the two guides d_1 and d_2, first find out, by inspection generally, from the forces acting whether the pressure is on the guide d_1 or d_2. Thus if the connecting rod and piston rod are in compression the pressure is on d_2, if both are in tension it is on d_1, etc., suppose for this case that both are in compression, the heavy line showing the surface bearing the pressure.

Next find the **relative** direction of sliding. It does not matter whether both surfaces are moving or not, only the relative direction is required it is assumed in the sense shown, *i.e.*, the sense of motion of a relative to d_2 is to the left (and, of course, the sense of motion of d_2 relative to a is to the right). Now the resultant pressure between the surfaces is inclined at angle ϕ to the normal where $\phi = \tan^{-1} f$, f being the friction factor, so that the resultant must be either in the direction of R_1 or R'_1.

Now R_1 the pressure of a upon d_2 acts downward, and in order that it may have a resolved part in the direction of motion, then R_1 and not R_1' is the correct direction. If R_1 is treated as the pressure of d_2 upon a then R_1 acts upward, but the sense of motion of d_2 relative to a is the opposite of that of a relative to d_2, and hence from this point of view also R_1 is correct.

It is easy to find the direction of R_1 by the following simple rule: *Imagine either of the sliding pieces to be an ordinary carpenter's wood plane, the other sliding piece being the wood to be dressed, then the force will have the same direction as the tongue of the plane when the latter is being pushed in the given direction on the cutting stroke, the angle to the normal to the surfaces being* ϕ.

170. Turning Pairs.—In dealing with turning pairs the same principles are adopted as are used with the sliding pairs and should not cause any difficulty. Let a, Fig. 115, represent the outer

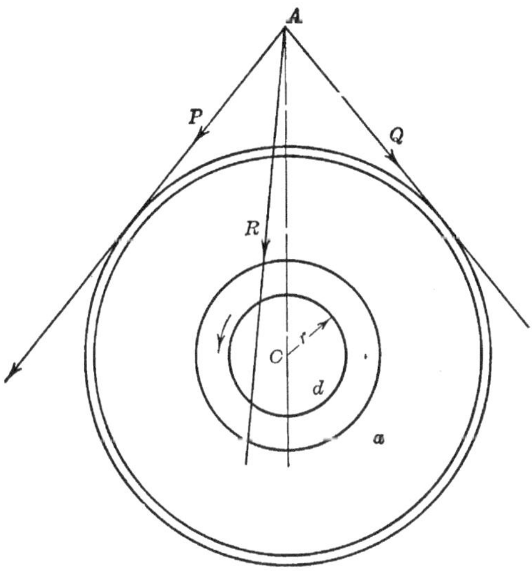

Fig. 115.

element of a turning pair, such as a loose pulley turning in the sense shown upon the fixed shaft d of radius r, and let the forces P and Q act upon the outer element. It must be explained that the arrow shows the sense in which the pulley turns relatively to the shaft and this is to be understood as the meaning of the arrow in the rest of the present discussion. It may be that both elements are turning in a given case, and the two elements may also turn in the same or in opposite sense, but the arrow indicates the relative sense of motion and the forces P and Q are assumed to act upon the link on which they are drawn, that is upon a in Fig. 115.

If there were no friction then the resultant of P and Q would pass through the intersection A of these forces and also through the center O of the bearing, so that under these circumstances it would be possible to find Q for a given value of P by drawing the vector triangle.

There is, however, frictional resistance offered to motion at the surface of contact, hence if the resultant R of P and Q acted through O, there could be no motion. In order that motion may exist it is necessary that the resultant produce a turning moment about the center of the bearing equal and opposite to the resistance offered by the friction between the surfaces. It is known already that the frictional resistance is of such a nature as to oppose motion, and hence the resultant force must act in such a way as to produce a turning moment in the sense of motion equal to the moment offered by friction in the opposite sense. Thus in the case shown in the figure the resultant must pass through A and lie to the left of O.

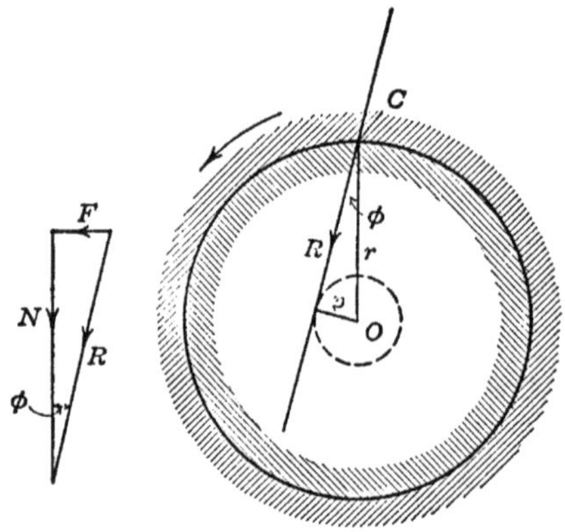

FIG. 116.

In Fig. 116, which shows an enlarged view of the bearing, let p be the perpendicular distance from O to R, so that the moment of R about O is $R \times p$. The point C may be conveniently called the center of pressure, being the point of intersection of R and the surfaces under pressure. Join CO. Now resolve R into two components, the first, F, tangent to the surfaces at C, and the second, N, normal to the surfaces at the same point. Following the method employed with sliding pairs, N is the normal pressure between the surfaces and the frictional resistance to motion will be fN, where f is the friction factor (Sec. 166), and since the parts are assumed to be in equilibrium, there must be no unbalanced force, so that the resolved part F of the resultant R must be equal in magnitude to the frictional resistance, or $F = fN$. But $f = \tan \phi$, where ϕ is the friction angle, so that

$$\tan \phi = f = \frac{F}{N},$$

from which it follows that the angle between N and R must be ϕ, and hence the resultant R must make an angle ϕ with the radius r at the center of pressure C.

171. Friction Circle.—With center O draw a circle tangent to R as shown dotted; then this circle is the one to which the resultant R must be tangent to maintain uniform relative motion, and the circle may be designated as the **friction circle.** The radius p of the friction circle is $p = r \sin \phi$, where r is the radius of the journal, and this circle is concentric with the journal and much smaller than the latter, since ϕ is always a small angle in practice. Thus, in turning pairs the resultant must always

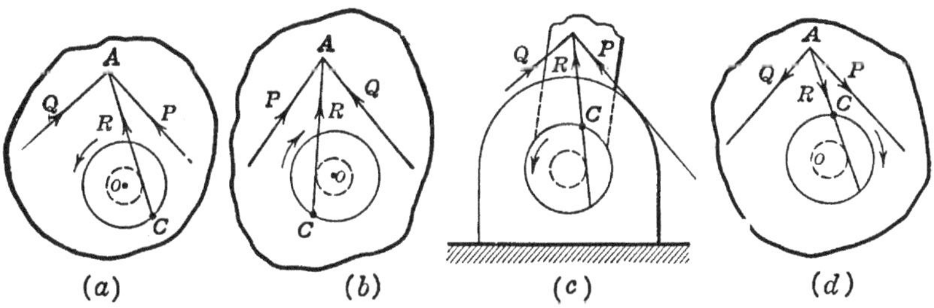

FIG. 117.

be at an angle ϕ to the normal to the surfaces, and this is most easily accomplished by drawing the resultant tangent to the friction circle, and on such a side of it that it produces a turning moment in the sense of the relative motion of the parts. Since f is always small in actual bearings, ϕ is also small, and hence $\tan \phi = \sin \phi$ nearly, so that approximately $p = r \sin \phi = r \tan \phi = rf$.

Four different arrangements of the forces on a turning pair are shown at Fig. 117, similar letters being used to Fig. 115. At (a) P and Q act on the outer element but their resultant R acts in opposite sense to the former case and hence on opposite side of the friction circle, since the relative sense of rotation is the same. In case (c) P and Q act on the inner element and the relative sense of rotation is reversed from (a), hence R passes on the right of the friction circle; at (b) conditions are the same as (a) except for the relative sense of motion which also changes the position of R; at (d) the forces act on the outer element and the sense of rotation and position of R are both as indicated.

172. Examples.—The construction already shown will be applied in a few practical cases.

1. The first case considered will be an ordinary bell-crank lever, Fig. 118, on which the force P is assumed to act horizontally and Q vertically on the links a and c respectively, the whole lever b turning in the clockwise sense. An examination of the figure shows that the sense of motion of a relative to b is counter-clockwise as is also the motion of c relative to b, therefore P will be tangent to the lower side of the friction circle at bearing 1, and Q will be tangent to the left-hand side of the friction circle

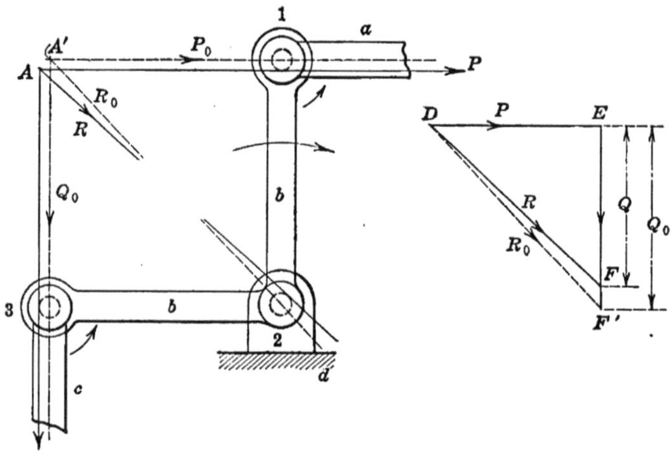

FIG. 118.

at bearing 3, and the resultant of P and Q must pass through A and must be tangent to the upper side of the friction circle on the pair 2, so that the direction of R becomes fixed. Now draw DE in the direction of P to represent this force and then draw EF and DF parallel respectively to Q and R and intersecting at F, then $EF = Q$ and $DF = R$.

In case there was no friction and assuming the **directions** of P and Q to remain unchanged (this would be unusual in practice), then P, Q and their resultant, would act through the centers of the joints 1, 3 and 2 respectively. Assuming the magnitude of P to be unchanged, then the vector triangle DEF' has its sides EF' and DF' parallel respectively to the resistance Q_0 and the resultant R_0 so that there is at once obtained the force $Q_0 = EF'$.

Then the efficiency of the lever in this position is $\eta = \dfrac{Q}{Q_0}$ and for any other position may be similarly found.

The friction circles are not drawn to scale but are made larger than they should be in order to make the drawing clear.

2. Let it be required to find the line of action of the force in the connecting rod of a steam engine taking into account friction at the crank- and wristpins. To avoid confusion the details of the rod are omitted and it is represented by a line, the friction circles being to a very much exaggerated scale. Let Fig. 119(*a*) represent the rod in the position under consideration, the direc-

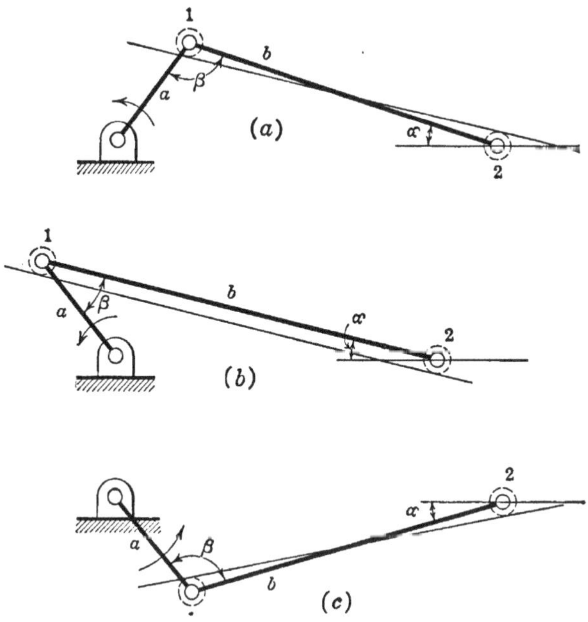

FIG. 119.—Steam-engine mechanism.

tion of the crank is also shown and the piston rod is assumed to be in compression, this being the usual condition for this position of the crank. Inspection of the figure shows that the angle α is increasing and the angle β is decreasing, so that the line of action of the force in the connecting rod must be tangent to the top of the friction circle at 2 and also to the bottom of the friction circle at 1, hence it takes the position shown in the light line and crosses the line of the rod. This position of the line of action of the force is seen on examination to be correct, because in both cases the force acts on such a side of the center of the bearing as to produce a turning moment in the direction of relative motion.

Two other positions of the engine are shown in Fig. 119 at (*b*) and (*c*), the direction of revolution being the same as before and the line of action of the force in the rod is in light lines. In the case (*b*), the rod is assumed in compression and evidently both the angles α and β are decreasing so that the line of action of the force lies below the axis of the rod; while in the position shown in (*c*), the connecting rod is assumed in tension, α is decreasing, and β is increasing so that the line of the force intersects the rod. In all cases the determining factor is that the force must lie on such a side of the center of the pin as to produce a turning moment in the direction of relative motion.

173. Governor—Turning Pairs Only.—A complete device in which turning pairs alone occur is shown at Fig. 120, which

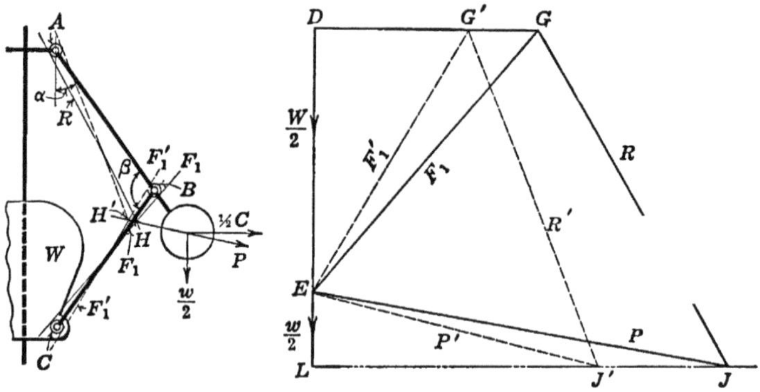

FIG. 120.—Governor.

represents one of the governors discussed fully in the following chapter, except for the effect of friction. The governor herewith is shown also at Fig. 125 and only one-half of it has been drawn in, the total weight of the two rotating balls is w lb. while that of the central weight including the pull of the valve gear is taken as W lb. In Chapter XII no account has been taken of friction or pin pressures while these are essential to the present purpose. There will be no frictional resistance between the central weight and the spindle and the friction circles at A, B and C are drawn exaggerated in order to make the construction more clear.

It is assumed that the balls are moving slowly outward and that when passing through the position illustrated the spindle rotates at n revolutions per minute or at ω radians per second; it is required to find n and also the speed n' of the spindle as the balls pass through this same position when travelling inward. The difference between these two speeds indicates to some extent the quality of the governor, as it shows what change must be made before the balls will reverse their motion.

On one ball there is a centrifugal force $\frac{C}{2}$ pds., where $\frac{C}{2} = \frac{w}{2g} r\omega^2$, r being the radius of rotation of the balls in feet, also $\frac{C}{2}$ acts horizontally while the weight of the ball $\frac{w}{2}$ lb. acts vertically, and their resultant is a force P inclined as shown in the left-hand figure. The arms AB and BC are both in tension evidently, and as the balls are moving outward, α is increasing and β is decreasing (see Fig. 120); hence the direction of the force in the arm BC crosses the axis of the latter as shown, F_1 representing the force.

Now the direction of the force P is unknown and it cannot be determined without first finding $\frac{C}{2}$ which, however, depends upon n, the quantity sought. An approximation to the slope of P may be found by neglecting friction and with this approximate value the first trial may be made. With the assumed direction of P the point H, where P intersects F_1, is determined and then the resultant R of F_1 and P must pass through H and also be tangent to the friction circle at A. (If there were no friction, R would pass through the center of A, Sec. 170.) Turning now to the vector diagram on the right make $DE = \frac{W}{2}$ and $EL = \frac{w}{2}$; then draw DG horizontally to meet EG, which is parallel to F_1 in G. The length EG represents the force F_1 in the arm BC, while DG represents the tension on the weight W which is balanced by the other half of the governor.

Next draw GJ and EJ parallel respectively to R and P, whence these forces are found. If the slope of P has been properly assumed, the point J will be on the horizontal line through L, and if J does not lie on this line a second trial slope of P must be made and the process continued until J does fall on the horizontal through L.

The length LJ then represents $\frac{C}{2} = \frac{w}{2g} r\omega^2$ from which ω is readily computed, and from it the speed n in revolutions per minute.

The dotted lines show the case where the mechanism passes through the same position but with the balls moving inward and from the length LJ' the value of ω' and of n' may be found. If only the relation between n and n' is required, then $\frac{n}{n'} = \sqrt{\frac{LJ}{LJ'}}$. The meaning of this is that if the balls were moving outward due to a decreased load on the prime mover to which the governor was connected then they would pass through the position shown when the spindle turned at n revolutions per minute, but if the load were again increased causing the balls to move inward the speed of the spindle would have to fall to n' before the balls would pass through the position shown. Evidently the best governor is one in which n and n' most nearly agree, and the device would be of little value where they differed much.

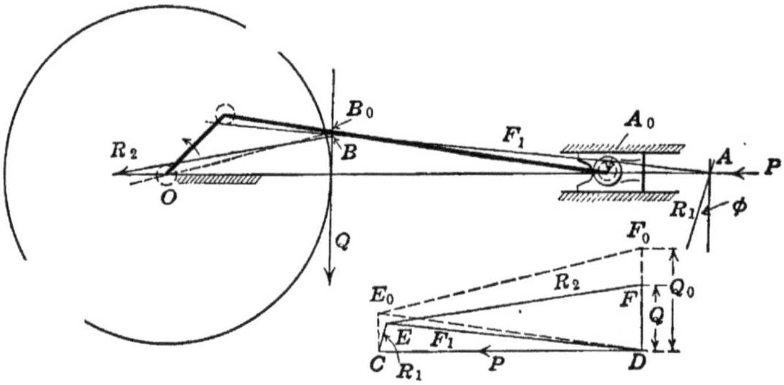

FIG. 121.

In reading this problem reference should also be made to the chapter on governors.

174. Machine with Turning and Sliding Pairs.—This chapter may be very well concluded by giving an example where both turning and sliding pairs are used, although there should be no difficulty in combining the principles already laid down in any machine. The machine considered is the steam engine, the barest outlines of which are shown in Fig. 121. The piston is assumed combined with the crosshead and only the latter is shown, and in the problem it has been assumed that the engine is lifting a weight from a pit by means of a vertical rope on a drum, the resistance of the weight being Q lb. Friction of the rope is not considered. The indicator diagram gives the information necessary for finding the pressure P acting through the piston on the crosshead, and the problem is to find Q and the efficiency.

From the principles already laid down, the direction of R_1 the pressure on the crosshead is known, also the line of action of F_1 and of R_2. For equilibrium the forces F_1, R, and P must intersect at one point which is evidently A, as P, the force due to the steam pressure, is taken to act along the center of the piston rod. On the crankshaft there is the force F_1 from the connecting rod, and the force Q due to the weight lifted, and if there were no friction, their resultant would pass through their point of intersection B and also through O the center of the crankshaft. To allow for friction, however, R_2 must be tangent to the friction circle at the crankshaft and must touch the top of the latter, hence the position of R_2 is fixed. Thus the locations of the five forces, P, F_1, R_1, R_2 and Q are known.

Now draw the vector diagram, laying off $CD = P$ and drawing CE and DE parallel respectively to R_1 and F_1, which gives these two forces, next draw EF parallel to R_2 and DF parallel to Q which thus determines the magnitude of Q.

If there were no friction, F_1 would be along the axis of the rod, and R_1 normal to the guides, both forces passing through A_0 the center of the wristpin. Further, R_2 would pass through B_0 the intersection of F_1 and Q, it would also pass through O as shown dotted, so that the lines of action of all of the forces are known and the vector diagram CE_0F_0D may be drawn obtaining the resistance $Q_0 = DF_0$, which could be overcome by the pressure P on the piston if there were no friction. The efficiency of the machine in this position is then $\eta = \frac{Q}{Q_0}$, and may be found in a similar way for other positions.

If desired, the value of the efficiency for a number of positions of the machine may be found and a curve plotted similar to a velocity diagram, Chapter *III*, from which the efficiency per cycle is obtained.

In all illustrations the factor f is much exaggerated to make the constructions clear and in many actual cases the efficiency will be much higher than the cuts show. Where the efficiency is very close to unity, the method is not as reliable as for low efficiencies, but many of the machines have such high efficiency that such a construction as described herein is not necessary, nor is any substitute for it needed in such cases.

QUESTIONS ON CHAPTER XI

1. In the engine crosshead, Fig. 110, if the friction factor is 0.05, what size is the friction angle? If the piston pressure is 5,000 pds., and the connecting rod is at 12° to the horizontal, what is the pressure in the rod and the efficiency of the crosshead, neglecting friction at the wristpin?

2. Of two 12-in. journals one has a friction factor 0.002 and the other 0.003. What are the sizes of the friction circles?

3. What would be the efficiency of the crank in Fig. 118 if the scale of the drawing is one-quarter and the pins are 1½ in. diameter?

4. Determine the direction of the force in the side rod of a locomotive in various positions.

5. A thrust bearing like Fig. 1(*b*) has five collars, the mean bearing diameter of which is 10 in. If the shaft runs at 120 revolutions per minute and has a bearing pressure of 50 lb. per square inch of area, find the power lost if the friction factor is 0.05.

6. In the engine of Fig. 121, taking the scale of the drawing as one-sixteenth and the friction factor as 0.06, find the value of Q when $P = 2{,}500$ pds. the diameters of the crank and wristpins being 3½ and 3 in. respectively.

7. In a Scotch yoke, Fig. 6, the crank is 6 in. long and the pin 2 in. diameter, the slot being 3 in. wide. With a piston pressure of 500 pds., find the efficiency for each 45° crank angle, taking $f = 0.1$.

PART II

MECHANICS OF MACHINERY

CHAPTER XII

GOVERNORS

175. Methods of Governing.—In all prime movers, which will be briefly called engines, there must be a continual balance between the energy supplied to the engine by the working fluid and the energy delivered by the machine to some other which it is driving, *e.g.*, a dynamo, lathe, etc., allowance being made for the friction of the prime mover. Thus, if the energy delivered by the working fluid (steam, water or gas) in a given time exceeds the sum of the energies delivered to the dynamo and the friction of the engine, then there will be some energy left to accelerate the latter, and it will go on increasing in speed, the friction also increasing till a balance is reached or the machine is destroyed. The opposite result happens if the energy coming in is insufficient, the result being that the machine will decrease in speed and may eventually stop.

In all cases in actual practice, the output of an engine is continually varying, because if a dynamo is being driven by it for lighting purposes the number of lights in use varies from time to time; the same is true if the engine drives a lathe or drill, the demands of these continually changing.

The output thus varying very frequently, the energy put in by the working fluid must be varied in the same way if the desired balance is to be maintained, and hence if the prime mover is to run at constant speed some means of controlling the energy admitted to it during a given time must be provided.

Various methods are employed, such as adjusting the weight of fluid admitted, adjusting the energy admitted per pound of fluid, or doing both of these at one time, and this adjustment may be made by hand as in the locomotive or automobile, or it may be automatic as in the case of the stationary engine or the water turbine where the adjustment is made by a contrivance called a **governor.**

A governor may thus be defined as a device used in connection with prime movers for so adjusting the energy admitted with the

working fluid that the speed of the prime mover will be constant under all conditions. The complete governor contains essentially two parts, the first part consisting of certain masses which rotate at a speed proportional to that of the prime mover, and the second part is a valve or similar device controlled by the part already described and operating directly on the working fluid.

It is not the intention in the present chapter to discuss the valve or its mechanism, because the form of this is so varied as to demand a complete work on it alone, and furthcr because its design depends to some extent on the principles of thermodynamics and hydraulics with which this book does not deal. This valve always works in such a way as to control the amount of energy entering the engine in a given time and this is usually done in one of the following ways:

(*a*) By shutting off a part of the working fluid so as to admit a smaller weight of it per second. This method is used in many water wheels and gas engines and is the method adopted in the steam engine where the length of cutoff is varied as in high-speed engines.

(*b*) By not only altering the weight of fluid admitted, but by changing at the same time the amount of energy contained in each pound. This method is used in throttling engines of various kinds.

(*c*) By employing combinations of the above methods in various ways, sometimes making the method (*a*) the most important, sometimes the method (*b*) The combined methods are frequently used in gas engines and water turbines.

The other part of the governor, that is the one containing the revolving masses driven at a speed proportional to that of the prime mover, will be dealt with in detail because of the nature of the problems it involves, and it will in future be briefly referred to as the governor.

176. Types of Governors.—Governors are of two general classes depending on the method of attaching them to the prime mover and also upon the disposition of the revolving masses, and the speed at which these masses revolve. The first type of governors, which is also the original type used by Watt on his engines, has been named the rotating-pendulum governor because the revolving masses are secured to the end of arms pivoted to the rotating axis somewhat similar to the method of construction of a clock pendulum, except that the clock pendulum swings in one plane, while the governor masses revolve. In this type there are three subdivisions: (*a*) gravity weighted, in which the centrifugal force due to the revolving masses or balls is largely balanced by gravity; (*b*) spring weighted, in which the same force is largely balanced by springs; and (*c*) combination governors in which both methods are used. Governors of this general class are usually mounted on a separate frame and driven by belt or gears from the engine, but they are, at times, made on a part of the main shaft.

The second type is the **inertia governor** which is usually made on the engine power shaft, although it is occasionally mounted separately. The name is now principally used to designate a class of governor with its revolving masses differently distributed to the former class; its equilibrium depends on centrifugal force but during the changes in position the inertia of the masses plays a prominent part in producing rapid adjustment. The name **shaft-governor** is also much used for this type.

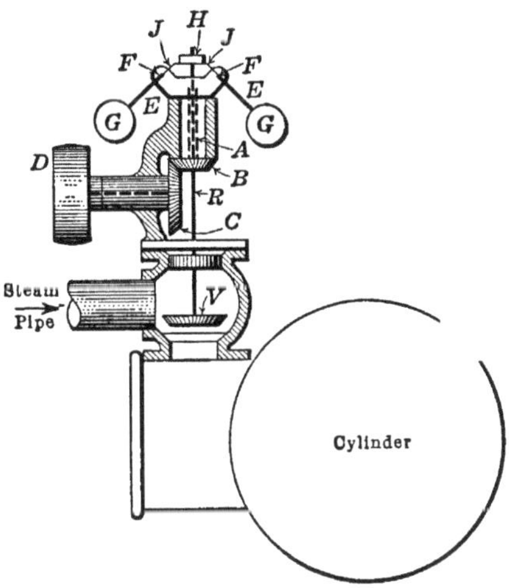

FIG. 122.—Simple governor.

177. Revolving-pendulum Governor.—Beginning with the revolving-pendulum type, an illustration of which is shown at Fig. 122 connected up to a steam engine, it is seen that it consists essentially of a spindle A, caused to revolve by means of two bevel gears B and C, the latter being driven in turn through a pulley D which is connected by a belt to the crankshaft of the engine; thus the spindle A will revolve at a speed proportional to that of the crankshaft of the engine. To this spindle at F two balls G are attached through the ball arms E, and these arms are connected by links J to the sleeve H, fastened to the rod R, which rod is free to move up and down inside the spindle A as directed by the movement of the balls and links. The sleeve H with its rod R is connected in some manner with the valve V, in this illustration a very direct connection being indicated, so that a movement of the sleeve will open or close the valve V.

The method of operation is almost self-evident; as the engine increases in speed the spindle A also increases proportionately and therefore there is an increased centrifugal force acting on the balls G causing them to move outward. As the balls move outward the sleeve H falls and closes the valve V so as to prevent as much steam from getting in and thus causing the speed of the engine to decrease, upon which the reverse series of operations takes place and the valve opens again. It is, of course, the purpose of the device to find such a position for the valve V that it will just keep the engine running at uniform speed, by admitting just the right quantity of steam for this purpose.

178. Theory of Governor.—Several different forms of the governor are shown later in the present chapter and will be discussed subsequently, but it may be well to begin with the simplest form shown in Fig. 123, where the connection of the sleeve to the valve is not so direct as in Fig. 122 but must be made through suitable linkage. The left-hand figure shows a governor with the arms pivoted on the spindle, while the right-hand figure shows the pivots away from the spindle, and the same letters are used on both. Let the total weight of the two balls be w lb., each ball therefore weighing $\frac{w}{2}$ lb., and let these be rotated in a circle of radius r ft., the spindle turning at n revolutions per minute corresponding to $\omega = \frac{2\pi n}{60}$ radians per second. For the present, friction will be neglected.

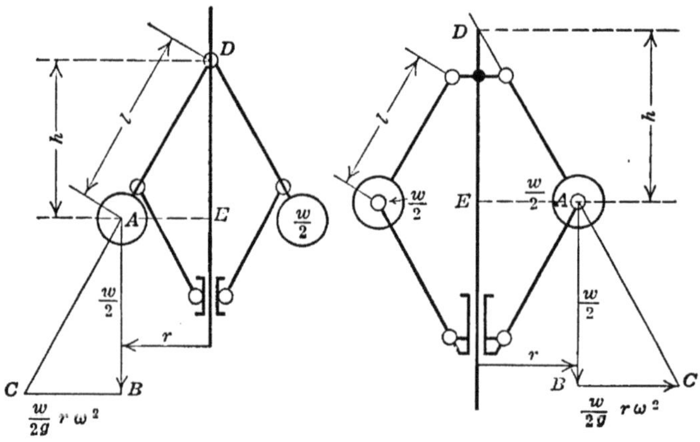

FIG. 123.

Three forces act upon **each** ball and determine its position of equilibrium. These are: (*a*) The attraction of gravity, which will act vertically downward and will therefore be parallel with the spindle in a governor where the spindle is vertical as in the illustration shown. The magnitude of this force is $\frac{w}{2}$ pds. (*b*) The second force is due to the centrifugal effect and acts radially and at right angles to the spindle, its amount being $\frac{\frac{1}{2}w}{g} \cdot r . \omega^2$ pds. (*c*) The third force is due to the pull of the ball arm, and will be in the direction of the line joining the center of gravity of the ball to the pivot on the spindle, which direction may be briefly called the direction of the ball arm.

These three forces must be in equilibrium so that the vector triangle ABC may be drawn where $AB = \frac{w}{2}$, $BC = \frac{w}{2g} r\omega^2$ and ω must be such that AC is parallel to the arm. Now let D be the point at which the ball arm intersects the spindle and draw AE perpendicular to the spindle DE; then $AE = r$, the radius of rotation of the balls and the distance $DE = h$ is called the **height** of the governor.

The triangles DAE and ACB are similar and therefore:

$$\frac{DE}{EA} = \frac{AB}{BC}$$

or

$$\frac{h}{r} = \frac{\frac{w}{2}}{\frac{w}{2g} r\omega^2}$$

which gives

$$h = \frac{g}{\omega^2}.$$

Thus, the height of the governor depends on the speed alone and not on the weight of the balls. The investigation assumes that the resistance offered at the sleeve is negligible as indeed is the case with many governors and gears, but allowance will be made for this in problems discussed later.

179. Defects of this Governor.—Such a governor possesses several serious defects. In the first place, the sleeve must move in order that the valve may be operated, and this movement of the sleeve will evidently correspond with a change in the height and hence with a change in speed ω. Thus, each position of the balls, corresponding to a given valve position, means a different speed of the governor and therefore of the engine; this is what the governor tries to prevent, for its purpose is to keep the speed of the engine constant, although the valve may have to be opened various amounts corresponding to the load which the engine carries. This defect may be briefly expressed by saying that the governor is not isochronous, the meaning of isochronism being that the speed of the governor will not vary during the entire range of travel of the sleeve, or in other words the valve may be moved into any position to suit the load, and yet the engine and therefore the governor, will always run at the same speed.

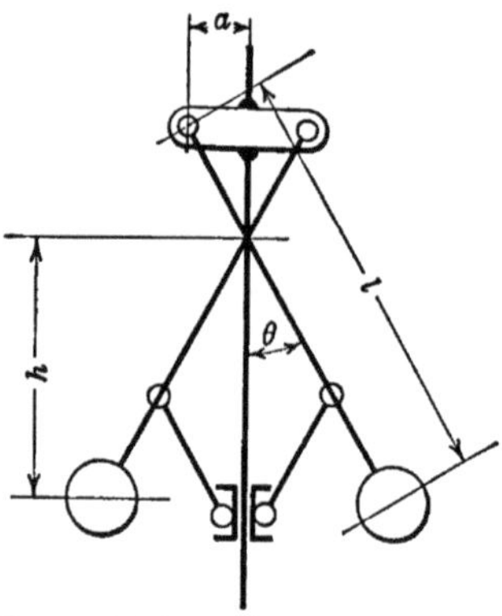

FIG. 124.—Crossed-arm governor.

The second defect is that for any reasonable speed h is extremely small. To show this let the governor run at 120 revolutions per minute so that $\omega = \frac{2\pi n}{60} = 12.57$ radians per second; then $h = \frac{g}{\omega^2} = \frac{32.16}{(12.57)^2} = 0.2036$ ft. or 2.44 in., a dimension which is so small, that if the balls were of any reasonable size, it would make the practical construction almost impossible.

180. Crossed-arm Governor.—Now it is the desire of all builders to make their governors as nearly isochronous as is consistent with other desirable characteristics, which means that the height h must be constant, and to serve this end the crossed-arm governor shown in Fig. 124, has been built somewhat extensively. The proportions which will produce isochronism may be found mathematically thus:

Inspection of the figure shows that

$$h = l \cos \theta - a \cot \theta.$$

For isochronism h is to remain constant for changes in the angle θ or

$$\frac{dh}{d\theta} = 0 = -l \sin \theta + a \operatorname{cosec}^2 \theta.$$

From which

$$a = l \sin^3 \theta$$
$$h = l \cos^3 \theta$$

and therefore $a = l \sin^3 \theta = h \tan^3\theta = \frac{g}{\omega^2} \tan^3 \theta$; which formulas give the relations between a, l and θ, and it will be noticed that the weight w does not enter into the calculation any more than it does into the time of swing of the pendulum.

As an example let the speed be $\omega = 10$ radians per second (corresponding to 97 revolutions per minute) and let $\theta = 30°$. Then the formulas give $a = 0.0618$ ft. or 0.74 in., $l = 0.495$ ft. or 5.94 in. and the value of h corresponding to $\theta = 30°$ is 0.322 ft. With these proportions the value of h when θ becomes 35° will be 0.317 ft., a decrease of 1.56 per cent., corresponding to a change of speed of about 0.8 per cent.

With a governor as shown at Fig. 123 and $\omega = 10$ as before, a change from 30° to 35° produces a change in speed of about 3 per cent.

It is possible to design a governor of this type which will maintain absolutely constant speed for all positions of the balls, and the reader may prove that for this it is only necessary to do away with the ball arms, and place the balls on a curved track of parabolic form, so that they will always remain on the surface of a paraboloid of revolution of which the spindle is the axis. In such a case, h and therefore ω will remain constant.

A perfectly isochronous governor, however, has the serious defect that it is **unstable** or has no definite position for a given speed, and thus the slightest disturbing force will cause the balls to move to one end or other of their extreme range and the governor will **hunt** for a position where it will finally come to rest. Such a condition of instability is not admissible in practice and designers always must sacrifice isochronism to some extent to the very necessary feature of **stability,** because the hunting of the balls in and out for their final position means that the valve is being opened and closed too much and hence the engine is changing its speed continually, or is **racing.** In the simple governor quoted in Sec. 178 it is evident that while it is not isochronous it is **stable,** for each position of the balls corresponds to a different but definite speed belonging to the corresponding value of the height h.

181. Weighted or Porter Governor.—In order to obviate these difficulties Charles T. Porter conceived the idea of placing on the sleeve a heavy central weight, free to move up and down on the spindle and having its center of gravity on the axis of rotation. This modified governor is shown in Fig. 125, with the arms pivoted on the spindle, although sometimes the arms are crossed and when not crossed they are frequently sus-

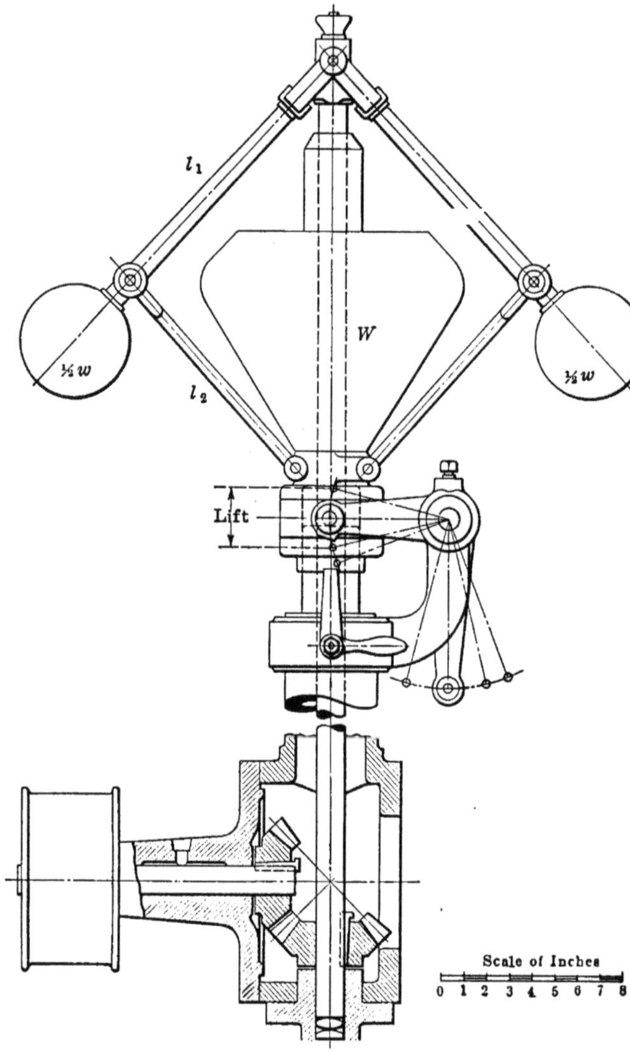

FIG. 125.—Porter governor.

pended by pins not on the spindle. In Fig. 126 a similar governor is shown diagrammatically, with the pivots O to one side.

To study the conditions of equilibrium of such a governor find the image Q' of the point Q where the link l_2 is attached to the central weight W. Then by the propositions of Chapter IX the half of the weight which acts at Q may be transferred to Q', and let it be assumed that l_1 and l_2 are of equal length; then by taking moments of the weights and centrifugal force about O the equation is

$$\frac{W}{2} \cdot 2l_1 \sin \theta + \frac{w}{2} l_1 \sin \theta - \frac{w}{2g} r\omega^2 l_1 \cos \theta = 0.$$

From which it follows that

$$h = \frac{2W + w}{w} \cdot \frac{g}{\omega^2}$$

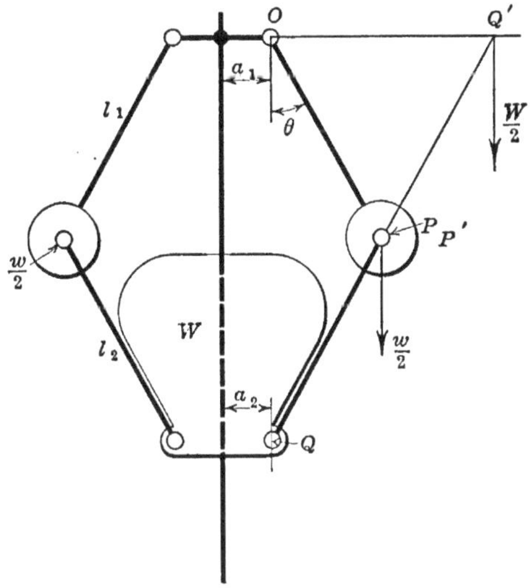

FIG. 126.—Porter or weighted governor.

For example, let $l_1 = l_2 = 9$ in. or 0.75 ft., speed 194 revolutions per minute for which $\omega = 20$ radians per second, and let each ball weigh 4 lb., *i.e.*, $w = 8$ lb., $\theta = 45°$ and $a_1 = a_2 = O$. Then by measuring from a drawing, or by computation, h is found to be 0.53 ft., and

$$\frac{2W + w}{w} = h \cdot \frac{\omega^2}{g}$$

gives

$$2W + w = wh\frac{\omega^2}{g} = 8 \times 0.53 \times \frac{20^2}{32.16} = 52.8 \text{ lb.}$$

or

$$W = 22.4 \text{ lb.}$$

182. Advantages of Weighted Governors.—The first advantage of such a governor is that the height h may be varied within wide limits at any given speed by a change in the central weight W, and thus the designer is left much freedom in proportioning the parts. In the numerical example above quoted the height would be 6.6 times as great as for an unweighted governor running at the same speed, since $\frac{2W + w}{w} = 6.6$.

Again the variation in height h corresponding to a given change in speed is much increased by the use of the central load, with the result that the sleeve will move through a certain height with smaller change in speed. Now the travel of the sleeve, or the **lift** as it is often called, is fixed by the valve and its mechanism, and the above statement means that for a given lift the variation in speed will be decreased, or the governor will become more **sensitive.** By **sensitiveness** is meant the proportional change of speed that occurs while the sleeve goes through its complete travel, the governor being most sensitive which has the least variation.

To prove this property let h', h, ω' and ω represent the heights and speeds corresponding to the highest and lowest positions of the sleeve.

Then

$$h' = \frac{2W + w}{w} \cdot \frac{g}{\omega'^2} \quad \text{and} \quad h = \frac{2W + w}{w} \cdot \frac{g}{\omega^2}$$

or

$$\frac{h'}{h} = \left(\frac{\omega}{\omega'}\right)^2 \quad \text{or} \quad \frac{\omega}{\omega'} = \sqrt{\frac{h'}{h}}.$$

But since h and h' are much greater in the weighted than in the unweighted governor, therefore $\frac{\omega}{\omega'}$, is more nearly unity in the former case.

Again,

$$\frac{\omega'^2}{\omega^2} = \frac{h}{h'} \quad \text{or} \quad \frac{\omega'^2 - \omega^2}{\omega^2} = -\frac{h' - h}{h'}.$$

Therefore,

$$\frac{\omega' - \omega}{\omega} \cdot \frac{\omega' + \omega}{\omega} = -\frac{h' - h}{h'}.$$

Now usually ω' and ω do not differ very much, so that $\omega' + \omega = 2\omega$ nearly, and therefore,

$$2\frac{\omega' - \omega}{\omega} = -\frac{h' - h}{h'}.$$

The relation $\frac{\delta\omega}{\omega}$ is evidently the sensitiveness of the governor[1] and the smaller the ratio the more sensitive is the governor. For an isochronous governor $\delta\omega = 0$.

To compare the weighted and unweighted governors in regard to sensitiveness take the angular velocity $\omega = 10$ radians per second and let $W = 60$ lb. and $W = 8$ lb. Let the change in height necessary to move the sleeve through its entire lift be ½ in.

(*a*) **Unweighted Governor.**—For the data given $h = 3.86$ in., and, therefore,[2]

$$\frac{\delta h}{h} = \frac{0.5}{3.86} = 0.129.$$

Hence, $2\frac{\delta\omega}{\omega} = 0.129$ or $\frac{\delta\omega}{\omega} = 0.064$ or 6.4 per cent., so that the variation in speed will be 6.4 per cent.

(*b*) **Weighted Governor.**—For this governor

$$h = \frac{2W + w}{w} \times 3.86 = \frac{2 \times 60 + 8}{8} \times 3.86 = 61.76 \text{ in.}$$

and

$$\frac{\delta h}{h} = \frac{0.5}{61.76} = 0.008$$

or

$$2\frac{\delta\omega}{\omega} = 0.008 \text{ giving } \frac{\delta\omega}{\omega} = 0.004 \text{ or } 0.4 \text{ per cent.}$$

the variation in speed being only 0.4 per cent. Such a governor would therefore be very nearly isochronous.

A third property of this weighted governor is that it is **power-**

[1] This may be simply shown by the calculus thus:

$$h = \frac{2W + w}{w} \cdot \frac{g}{\omega^2}$$

and differentiating,

$$\delta h = -\frac{2W + w}{w} g \cdot \frac{2\delta\omega}{\omega^3}; \qquad \therefore \frac{\delta h}{h} = -2\frac{\delta\omega}{\omega}$$

or

$$2\frac{\delta\omega}{\omega} = -\frac{\delta h}{h}$$

where $\delta\omega$ and δh represent the small changes taking place in ω and h.

[2] The negative sign appearing before δh on the preceding formula merely means that an increase in speed corresponds to a decrease in height.

ful, that is to say, that as the central weight is very heavy the equilibrium of the device is very little affected by any slight disturbing force, such as that required to operate the valve gear or to overcome friction. **Powerfulness** is a very desirable feature, for it is well known in practice that the force required to operate the valve gear is not constant and therefore produces a variable effect on the governor mechanism, which, unless the governor is powerful, is sufficient to move the weights, causing hunting.

The Porter governor thus enables the designer to make a very sensitive governor, of practical proportions and one which may be made as powerful as desired, so that it will not easily be disturbed by outside forces.

THE CHARACTERISTIC CURVE

183. A number of the results and properties of governors may be graphically represented by means of characteristic curves, and it will be convenient at this stage to explain these curves in connection with the Porter governor. Let Fig. 127(a) represent the right-hand part of a Porter governor, the letters having the same significance as before. Choose a pair of axes, OC in the direction of the spindle and OA at right angles to the spindle, and let the centrifugal force on the ball be plotted vertically along OC, as against radii of rotation of the balls, which are plotted along OA, r_1 and r_2 representing respectively the inner and outer limiting radii, the resulting figure will usually be a curved line somewhat similar to C_1CC_2 in Fig. 127(a).

Let the angular velocities corresponding to the radii r_1 and r_2 be ω_1 and ω_2 radians per second respectively, and let $\omega = \frac{1}{2}(\omega_1 + \omega_2)$ represent the **mean angular velocity** to which the corresponding radius of rotation is r ft.

Then

$$C_1 = \frac{w}{g} r_1 \omega_1^2$$

$$C_2 = \frac{w}{g} r_2 \omega_2^2$$

and

$$C = \frac{w}{g} r \omega^2$$

where the forces C, C_1 and C_2 are the total centrifugal forces acting on the **two** balls. The properties of this curve, which may be briefly called the C **curve,** may now be discussed.

1. Condition for Isochronism.—If the governor is to be isochronous then the angular velocity for all positions of the balls must be the same, that is $\omega = \omega_1 = \omega_2$ and hence the centrifugal force depends only on the radius of rotation (see formulas above) or

$$\frac{C_1}{r_1} = \frac{C_2}{r_2} = \frac{C}{r} = \text{a constant},$$

a condition which is fulfilled by a C curve forming part of a straight line passing through O. Thus any part of OC would satisfy this condition and the part ED corresponds to the radii r_1 and r_2 in the governor selected.

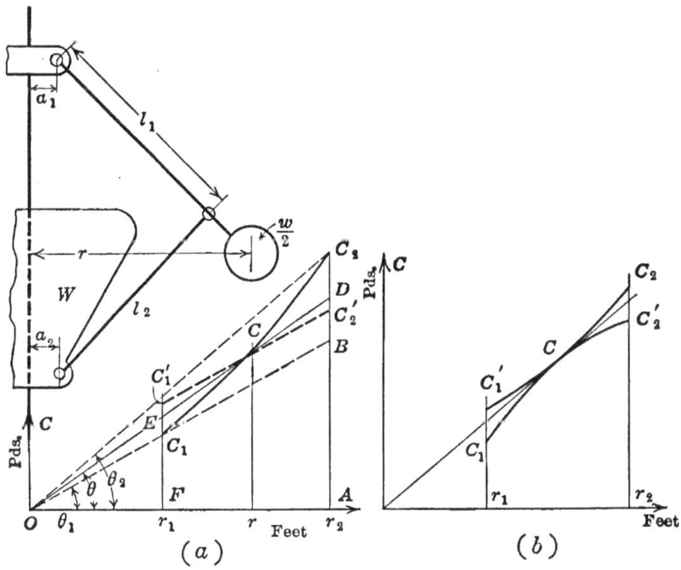

FIG. 127.—Characteristic curve.

2. Condition for Stability.—Although the curve ED will give an isochronous governor, it produces instability. The curve C_1CC_2 indicates that the speeds are not the same for the various positions of the balls, and a little consideration will show that C_1 corresponds to a lower speed and C_2 to a higher speed than C. This is evident on examining the conditions at radius r_1, for the point E corresponds to the same speed as C, but since E and C_1 are both taken at the same radius, and since the centrifugal force FE is greater than FC_1 it is evident that the angular velocity ω_1 corresponding to C_1 is less than the angular velocity ω corresponding to E. Thus a curve such as C_1CC_2, which is steeper than the isochronous curve where they cross, indicates that the speed of the governor will increase when the balls move out, and it may similarly be shown that such a curve as $C_1'CC_2'$, which is flatter than the isochronous curve, shows that the speed of the governor decreases as the balls move out.

Now an examination of these curves shows that the one C_1CC_2 belongs to a governor that is stable, for the reason that when the ball is at radius r_1 it has a definite speed and in order to make it move further out the centrifugal force must increase. But on account of the nature of the curve the centrifugal force must increase faster than the radius or the speed must increase as the ball moves out, and thus to each radius there is a corresponding speed. On the other hand, the curve $C_1'CC_2'$ shows an entirely different state of affairs, for at the radius r_1 the centrifugal force is greater than FE or the ball has a higher speed than ω and thus as the ball moves out the speed will decrease. Any force that would disturb the governor would cause the ball to fly outward under the action of a resultant force $C_1'E$, and if it were at radius r_2 any disturbance would cause the ball to move inward.

Another way of treating this is that for the curve C_1CC_2 the energy of the ball due to the centrifugal force is increasing due both to the increase in r and in the speed, and as the weights W and w are being lifted, the forces balance one another and there is equilibrium; whereas with the curve $C_1'CC_2'$ there is a decrease in speed and also in the energy of the balls while the weights are being lifted and the forces are therefore unbalanced and the governor is unstable.

Thus, for stability the C curve must be steeper than the line joining any point on it to the origin O. Sometimes governors have curves such as those shown at Fig. 127(b) and curve C_1CC_2 indicates a stable governor, $C_1'CC_2'$ an unstable governor, C_1CC_2' partly stable and partly unstable and finally $C_1'CC_2$ partly unstable and partly stable.

3. Sensitiveness.—The shape of the curve is a measure of the sensitiveness of the governor. If S indicates the sensitiveness, then by definition $S = \frac{\omega_2 - \omega_1}{\omega}$.

Now

$$\frac{\omega_2 - \omega_1}{\omega} = \frac{(\omega_2 - \omega_1)(\omega_2 + \omega_1)}{\omega(\omega_2 + \omega_1)} = \frac{\omega_2^2 - \omega_1^2}{2\omega^2}$$

since $\omega_2 + \omega_1 = 2\omega$ nearly.
Therefore

$$S = \frac{1}{2} \cdot \frac{\omega_2^2 - \omega_1^2}{\omega^2}$$

But

$$C_2 = \frac{w}{g} r_2 \omega_2^2$$

and

$$C_1 = \frac{w}{g} r_1 \omega_1^2$$

and

$$C = \frac{w}{g} r\omega^2.$$

Hence, by substituting in the formula for S, the result is

$$S = \frac{1}{2}\left[\frac{\dfrac{C_2}{\frac{w}{g}r_2} - \dfrac{C_1}{\frac{w}{g}r_1}}{\dfrac{C}{\frac{w}{g}r}}\right] = \frac{1}{2}\left[\frac{\dfrac{C_2}{r_2} - \dfrac{C_1}{r_1}}{\dfrac{C}{r}}\right].$$

Referring now to Fig. 127 it is seen that

$$\frac{C_2}{r_2} = \tan \theta_2 = \frac{C_2A}{OA}; \quad \frac{C_1}{r_1} = \tan \theta_1 = \frac{BA}{OA}$$

and

$$\frac{C}{r} = \tan \theta = \frac{DA}{OA}$$

or

$$S = \frac{1}{2}\left[\frac{\tan \theta_2 - \tan \theta_1}{\tan \theta}\right] = \frac{1}{2}\left[\frac{C_2A - BA}{DA}\right] = \frac{1}{2}\frac{C_2B}{DA}.$$

Thus the C curve is also valuable in showing the sensitiveness of the governor. For an isochronous governor C_2, B and D coincide and $S = O$. Evidently the more stable the governor is the less sensitive it is, and in a general way an unstable governor is more sensitive than a stable one. At C_1, Fig. 127(a), the stable governor is most nearly isochronous, and evidently a fair degree of stability and sensitiveness could both be obtained in a governor having a reverse curve with point of inflexion near C, the part CC_2 being concave to OA, the part C_1C convex to OA.

4. Powerfulness.—The C curve also shows the powerfulness of the governor, since in this curve vertical distances represent the centrifugal forces acting on the balls, while horizontal distances represent the number of feet the balls move horizontally in the direction of the forces. Thus, an elementary area represents the product $C.\delta r$ ft.-pds. and the whole area between the C curve and the axis OA gives the work done by the balls in moving over their entire range, and is therefore the work available to move the valve gear and raise the weights. The higher the curve is above OA the greater is the available work, and this clearly corresponds to increased speed in a given governor.

5. Friction.—The effect of friction has been discussed in the previous chapter and need not be considered here. Some writers treat friction as the equivalent of an alteration to the central weight, and if this is done the effect is very well shown in Fig. 128 where the C curve for the frictionless governor is shown at C_1CC_2. As the weight W is lifted the effect of friction when treated in this way is to increase W by the friction f with the result that the C curve is raised to 3-4, whereas when the weight W is falling the friction has the effect of decreasing the weight W and to lower the C curve to 5-6. The effects of these changes are evident without discussion.

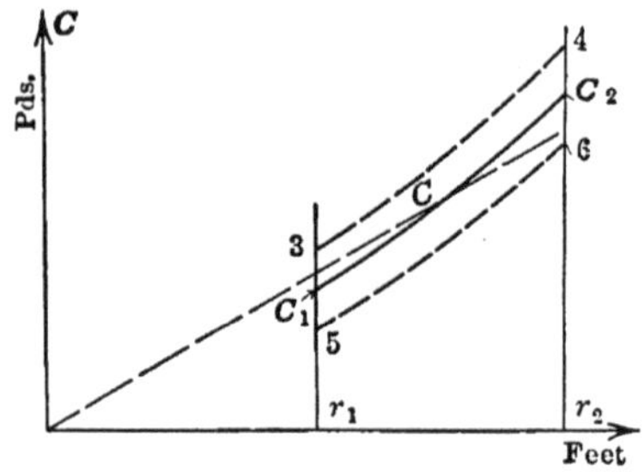

FIG. 128.

184. Relative Effects of the Weights of the Balls and the Central Weight.—For the purpose of further understanding the governor and also for the purpose of design, it is necessary to analyze the effects of the weights separately. Referring to Fig. 129 and finding the phorograph by the principles of Chapter IV the image of D is at D' and taking moments about A, remembering that $\frac{W}{2}$ may be transferred from D to its image D',

$$\tfrac{1}{2}Wb + \tfrac{1}{2}we - \tfrac{1}{2}Ch = 0.$$

(Sec. 151, Chapter IX). Now let C_W be the part of C necessary to support W and C_w the corresponding quantity for w, so that.

$$C_W + C_w = C \text{ where } C = \frac{w}{g} r\omega^2.$$

But

$$W\frac{b}{h} + w\frac{e}{h} = C.$$

That is $$C_W = W\frac{b}{h} \text{ and } C_w = w\frac{e}{h}.$$

The graphical construction is shown in Fig. 129. Draw JH and LG horizontally at distances below A to represent w and W respectively, then join AE, the line $D'E$ being a vertical through D'. Then it may be easily shown that

$$C_W = AF \quad \text{and } C_w = AK.$$

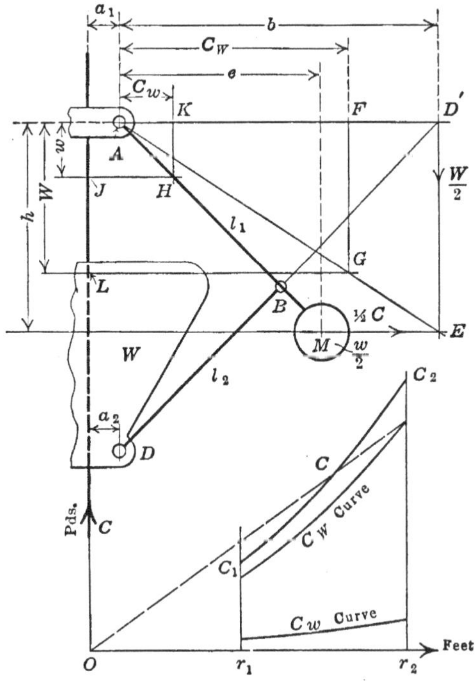

FIG. 129.—Governor analysis.

Making this construction for various positions and plotting for the complete travel of the balls the two curves are as drawn in Fig. 129.

185. Example.—The following dimensions are taken from an actual governor and refer to Fig. 129. $a_1 = 0$, $a_2 = 1\frac{1}{2}$ in., $AB = 12\frac{1}{2}$ in., $AM = 16$ in. and $BD = 10\frac{1}{2}$ in., while the travel of the sleeve is $2\frac{1}{2}$ in. and the point D is $15\frac{5}{8}$ in. below A when the sleeve is at the top of its travel. Each ball weighs 15 lb. so that $w = 30$ lb., also $W = 124$ lb.

Then drawing the governor mechanism in the upper, the mean

and the lowest positions of the sleeve, the following table of results is obtained, since for the ball $\frac{w}{g} = \frac{30}{32.16} = 0.933$.

RESULTS ON PORTER GOVERNOR

Sleeve position	b feet	$r = e$ feet	h feet	$W\frac{b}{h} = C_W$ pounds	$w\frac{e}{h} = C_w$ pounds	$C_W + C_w = C$ pounds	$\omega^2 = \frac{C}{\frac{w}{g}r}$	n rev. per min.
1. Upper..	1.51	0.98	0.90	208.2	32.8	241.0	263.7	155.2
2. Mean...	1.40	0.92	0.97	178.3	27.5	206.8	241.3	148.5
3. Lower...	1.25	0.83	1.04	149.1	24.0	173.1	222.8	142.3

The corresponding C curve and the two components C_W and C_w are plotted at Fig. 130 from which it is clear that the governor

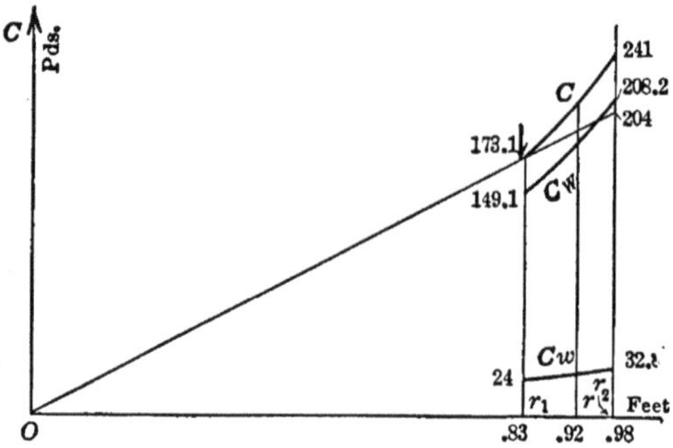

FIG. 130.

is stable. From this curve it appears that the sensitiveness is $\frac{1}{2} \times \frac{37}{222} = 0.0856$ or 8.56 per cent., which checks very well with the speeds as shown in the last column, and which indicates a sensitiveness of 8.65 per cent.

If it is desired to find the position of the balls for a speed of 150 revolutions per minute, then $\omega = 15.7$ radians per second and the force $C = \frac{w}{g} r\omega^2 = 0.933 \times r \times 246.5 = 230\, r$. Then draw the line for which the tangent is $\frac{C}{r} = 230$ and where it cuts the C curve is the radius of the balls corresponding to this speed.

Assuming the mean height of the C curve to be 207 pds. the work done in the entire travel of the balls is 207 $(0.98 - 0.83) =$ 31 ft.-pds.

186. Design of a Porter Governor.—These curves may be conveniently used in the design of a Porter governor to satisfy given conditions. Let it be required to design a governor of this type to run at a mean speed of 200 revolutions per minute with an overall variation of less than 10 per cent. for the extreme range. The sleeve is to have a travel of 2 in. and the governor is to have a powerfulness represented by 20 ft.-pds.

From general experience select the dimensions a_1, a_2, l_1, l_2 and BM in Fig. 129. Thus take $l_1 = l_2 = 10$ in., and $BM = 3$ in.; also make $a_1 = a_2 = 1$ in. Draw the governor in the central position of the sleeve with the arms at 90°, as this angle gives greater uniformity than other angles, and measure the extreme radii and also that for the central position of the sleeve. The C curve may now be constructed and at Fig. 131 the three radii are marked, which are $r_1 = 9.5$ in., $r = 10.22$ in. and $r_2 = 10.82$ in. Now the power of the governor is 20 ft.-pds., and dividing this by $r_2 - r_1 = 0.11$ ft. gives the mean height of the C curve as 182 pds. Plot this at radius r making $HG = 182$ pds. and join to O; it cuts r_2 at D.

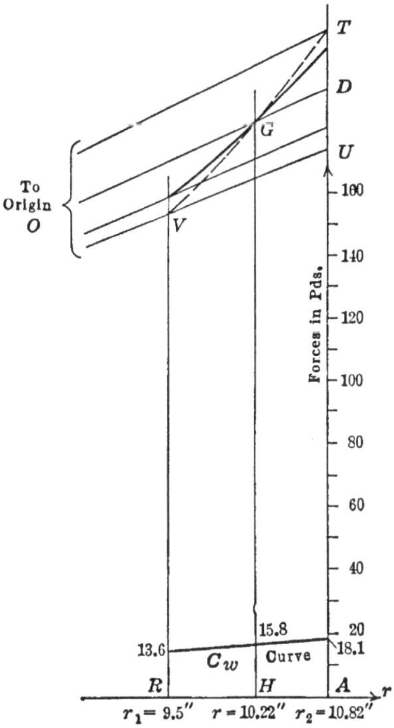

FIG. 131.—Governor design.

Now the sensitiveness is to be 10 per cent., so that T and U are found such that $DT = DU = 0.10 \times AD = 19.30$ pds. Join T and U to O, thus locating V and the resulting C curve will be VGT shown dotted.

Next, since the centrifugal force $GH = 182$ pds. corresponds to a radius $r = 10.22$ in. and a speed of 200 revolutions, the weight w may be found from the formula $C = \frac{w}{g} r\omega^2$ and gives $w = 15.75$ lb. By the use of such a diagram as Fig. 129 the three values of C_w are measured for the three radii and the C_w curve is drawn in Fig. 131, and then the values of C_W are found. Thus,

RV scales off as 153 pds. and hence $C_{W_1} = 153 - 13.6 = 139.4$ pds., and similarly the other values of C_W are found, and from them $W_1 = 104.6$ pds., $W = 107.6$ pds. and $W_2 = 110.6$ pds. are obtained, as shown at Fig. 129.

As a trial assume the mean of these values $W = 107.6$ pds., as the value of the central weight and proceeding as in Sec. 185 find the three new values of C_W and also of $C = C_W + C_w$ and lay these off at the various radii giving the plain curve in Fig. 131. This will be found to correspond to a range in speeds from 192

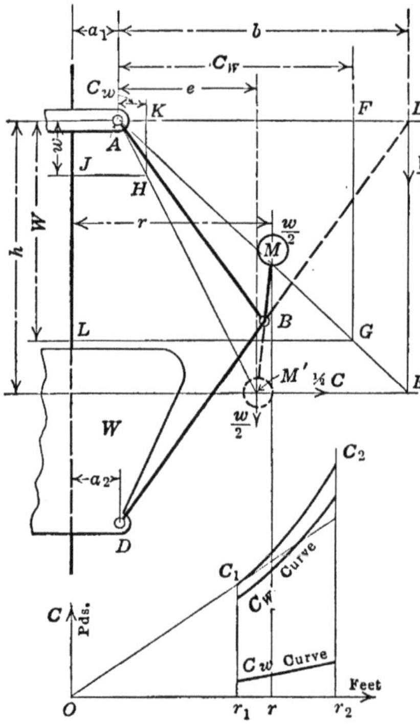

FIG. 132.—Proell governor.

to 207.5 revolutions per minute, and as this gives less than a 5 per cent. variation either way from the mean speed of 200 revolutions it would usually be satisfactory. If it is desired to have the exact value of 5 per cent., then it will be better to start with a little larger variation of say 6 per cent. and proceed as above.

187. Proell Governor.—The method already described may be applied to more complicated forms of governor with the same ease as is used in the Porter governor, the phorograph making these cases quite simple. As an illustration, the Proell governor is shown in Fig. 132 and is similarly lettered to Fig. 129, the difference between these governors being that in the Proell the ball is fastened to an extension of the lower arm *DB* instead of the upper arm *AB* as in the Porter governor.

As before, *AB* is chosen as the link of reference and the images found on it of the points *D* and *M* by the phorograph, Chapter IV. The force $\frac{1}{2}W$ is then transferred to D' and $\frac{1}{2}C$ and $\frac{1}{2}w$ to M' from Chapter IX, but in computing *C* the radius is to be measured from the spindle to *M* and not to M', since the former is the radius of rotation of the ball. The meanings of the letters will appear from the figure and by taking moments about *A* the same relation is found as in Sec. 184. The results for the complete travel of the balls is shown on the lower part of Fig. 132.

SPRING GOVERNORS

188. Spring governors have been made in order to eliminate the central weight and to make possible the use of a nearly isochronous and yet sensitive and powerful governor. These governors always run at high speed and are sometimes mounted on the main engine shaft, but more frequently on a separate spindle.

189. Analysis of Hartnell Governor.—One form of this governor, frequently ascribed to Hartnell of England, is shown in Fig. 133 and the action of the governor may now be analyzed. Let the total weight of the two balls be w lb., as before, and let W denote the force on the ball arms at *BB*, due to the weight of the central spring and any additional weight of valve gear, etc. In this case W will remain constant as in the loaded governor. Now let F be the pressure produced at the points *B* by the spring, F clearly increasing as the spring is compressed due to the outward motion of the balls.

In dealing with governors of this class it is best to use the moments of the forces about the pin *A* in preference to the forces themselves, and hence in place of a *C* curve for this governor a moment or *M* curve will be plotted in its place, the radius of rotation of the balls being used as the horizontal axis. The symbols M, M_F, M_w and M_W indicate, respectively, the moments about *A* of the centrifugal force *C*, the spring force F, the weight of the balls w and the dead weight W along the spindle. Then $M = M_F + M_W + M_w$

or $$Ca \cos \theta = Fb \cos \theta + Wb \cos \theta - wa \sin \theta.$$

The moment curves may be drawn and take the general shapes shown in Fig. 133 and similar statements may be made about these curves as about those for the Porter governor. If it is desired, the corresponding C curves may readily be drawn from the formula

$$Ca \cos \theta = Fb \cos \theta + Wb \cos \theta - wa \sin \theta$$

or

$$C = F\frac{b}{a} + W\frac{b}{a} - w \tan \theta$$

$$= C_F + C_W + C_w$$

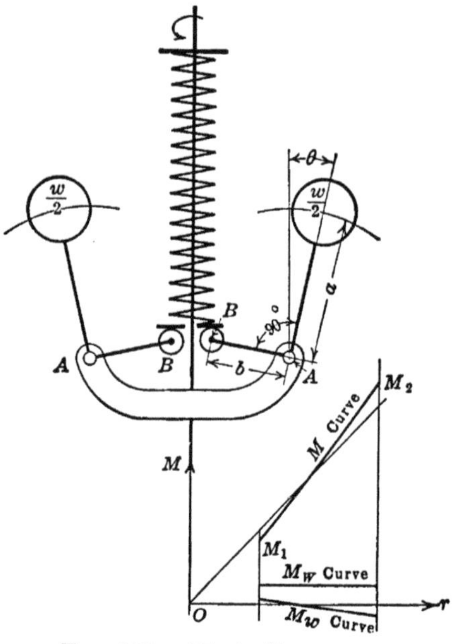

Fig. 133.—Hartnell governor.

and a graphical method for finding these values is easily devised. The curve for W is evidently a horizontal line since W, b and a are all constant, while that for w is a sloping line cutting the axis of r under the pin A and the C_F curve may be found by differences.

190. Design of Spring.—The data for the design of the spring may be worked out from the C_F curve found as above. Evidently $C_F = F\frac{b}{a}$ or $F = C_F \times \frac{a}{b} = C_F \times$ a constant, and thus from the curve for C_F it is possible to read forces F to a suitable scale.

These forces F may now be plotted as at Fig. 134 which gives the values of F for the different radii of rotation. As the line *EGL* thus found is slightly curved, no spring could exactly fulfil the requirements, but by joining *E* and *L* and producing to *H*, a solution may be found which will fit two points, *E* and *L*, and will nearly satisfy other points. Draw *EK* horizontally; then *LK* represents the increase in pressure due to the spring while the balls move out *EK* in., and hence the spring must be such that $\frac{LK}{EK} \times \frac{a}{b}$ pds. will compress it 1 in., and further the force produced by the spring when the balls are in is *EJ* pds., that is the spring must be compressed through *HJ* in., for the inner position of the balls.

In ordinary problems it is safe to assume for preliminary calculations that the effect of the weights *W* and *w* can be neglected and the spring may be designed to balance the centrifugal force alone. In completing the final computations the results may be modified to allow for these. In the diagrams here shown their effects have been very much exaggerated for clearness in the cuts.

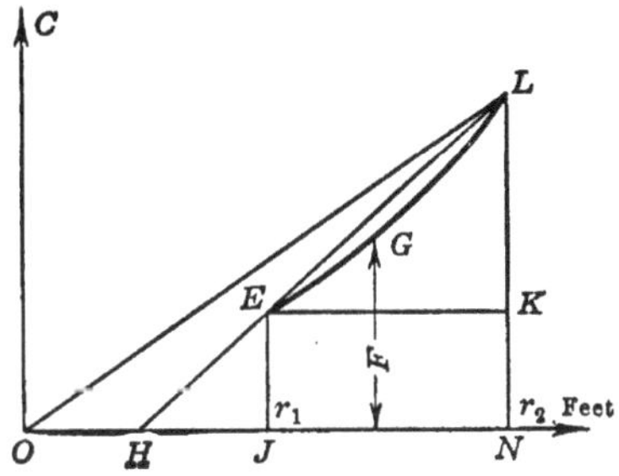

Fig. 134.

191. Governors with Horizontal Spindle.—Spring governors are powerful, as the complete computations in the next case will show, and are therefore well adapted to cases where the movement of the valve gear is difficult and unsteady.

When such a governor is placed with horizontal spindle such as Fig. 135 the effects of the weights are balanced and the spring alone balances the centrifugal force.

192. Belliss and Morcom Governor.—One other governor of this general type may be discussed in concluding this section. It is a form of governor now much in use and the one shown in the illustration, Figs. 136 and 137, is used by Belliss and Morcom of Birmingham, England, in connection with their high-speed engine. The governor is attached to the crankshaft, and therefore the weights revolve in a vertical plane, so that their gravity effect is zero. There are two revolving weights *W* with their centers of gravity at *G* and these are pivoted to the spindle by pins *A*.

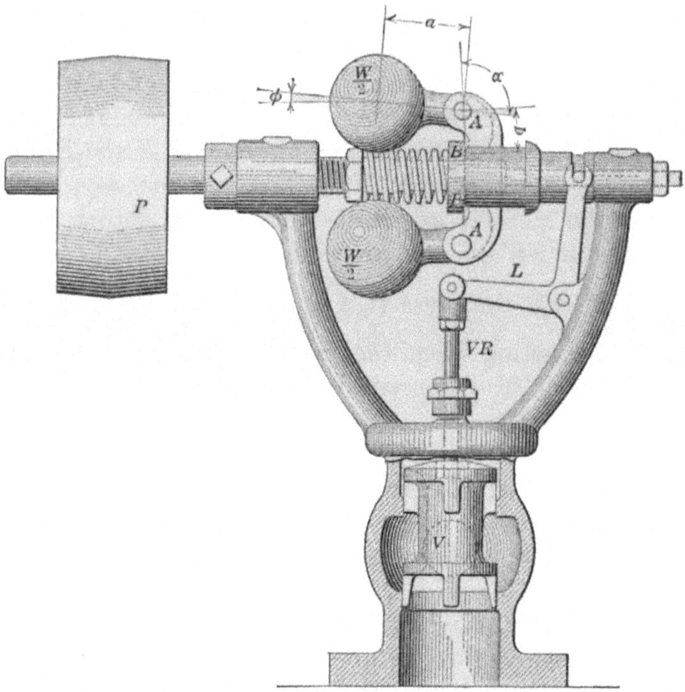

FIG. 135.—Governor with gravity effect neutralized.

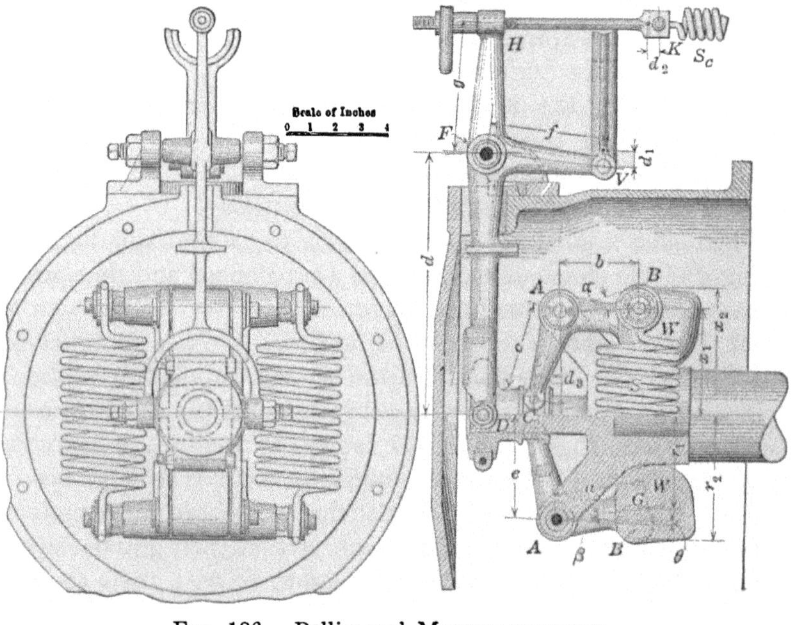

FIG. 136.—Belliss and Morcom governor.

Between the weights there are two springs *S* fastened to the former by means of pins at *B*. The balls operate the collar *C*, which slides along the spindle, thus operating the bell-crank lever *DFV*, which is pivoted to the engine frame at F and connected at *V*, by means of a vertical rod, to the throttle valve of the engine. There is an additional compensating spring S_c with its right-hand end attached to the frame and its left-hand end connected to the bell-crank lever *DFV* at *H*, there being a hand wheel at this connection so that the tension in the spring may be changed within certain limits and thus the engine speed may be varied to some extent. This spring will easily allow the operator to run the engine at nearly 5 per cent. above or below normal.

The diagrammatic sketch of the governor, shown in Fig. 137, enables the different parts to be distinctly seen as well as the colations of the various points. It will be noticed that this governor differs from all the others already described in that part of the centrifugal force is directly taken up by the springs *S*, while the forces acting on the sleeve are due to the dead weight of the valve *V* and its rod, and the slightly unbalanced steam pressure (for the valve is nearly balanced against steam pressure) on the valve, and in addition to these forces there is the pressure due to the spring S_c. The governor is very efficiently oiled and it is found by actual experiment that the frictional effect may be practically neglected.

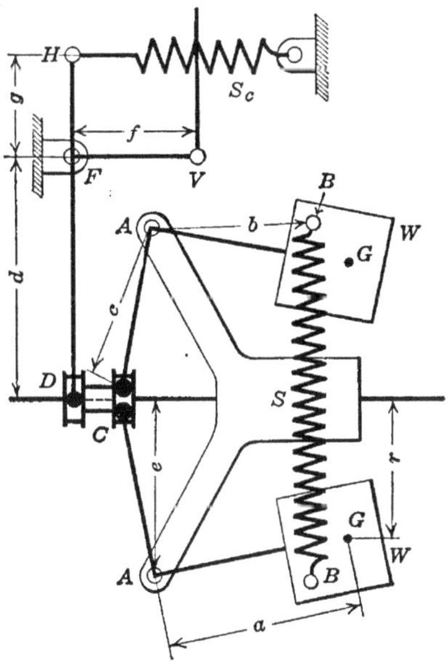

FIG. 137.—Belliss and Morcom governor.

In this case it will be advisable to draw the moment curve for the governor as well as the *C* curve and from the latter the usual information may be obtained. As this moment curve presents no difficulties it seems unnecessary to put the investigation in a mathematical form as the formulas become lengthy on account of the disposition of the parts. An actual case has been worked out and the results are given herewith and show the effects of the various parts of the governor.

193. Numerical Example.—The governor here selected, is attached to the crankshaft of an engine which has a normal mean speed of 525 revolutions per minute although the actual speed depends upon the load and the adjustment of the spring S_c. The governor spindle also rotates at the same speed as the engine. The two springs S together require a total force of 112 pds. for each inch of extension, while the spring S_c requires 220 pds. per inch of extension, the springs having been found on calibration to be extremely uniform. Each of the revolving masses has an effective weight of 10.516 lb. and the radius of rotation of the center of gravity varies from 4.20 in. to 4.83 in. The other dimensions are: e = 4 in. radius, b = 3.2 in., c = 3.5 in., a = 3.56 in., d = 10.31 in., f = 4.67 in., g = 5.31 in.

The weight of the valve spindle, valve and parts together with the unbalanced steam pressure under full-load normal conditions is 20 lb.

The following table gives the results for the governor for four different radii of the center of gravity G, all the moments being expressed in inch-pound units, when reduced to the equivalent moment about the pivots A of the balls.

Belliss and Morcom Governor

Speed, revolutions per minute	Radius of rotation of G, inches	Centrifugal force, pounds	Moment about A, inch-pounds	(1) Moment of main springs about A, inch-pounds	(2) Moment of spring S_c about A, inch-pounds	(3) Moment due to valve weight about A, inch-pounds	Sum of (1), (2) and (3), inch-pounds
508	4.47	345.5	1,220	1,083	115	16	1,214
526	4.73	392.0	1,364	1,196	143	16	1,355
529	4.78	400.0	1,388	1,223	145	16	1,384
532	4.83	410.0	1,418	1,250	148	16	1,414

In examining the table it will be observed that the sum of columns (1), (2) and (3) is always a little less than the moment due to the centrifugal force. As the results are all computed from measurements made on the engine during operation, there is possibly a slight error in the dimensions, and further the effect of centrifugal force on the springs S will make some difference.

The results agree very well, however, and show that the calculations agree with actual conditions.

The results are plotted in Fig. 138, the left-hand part of the curves being dotted. The reason of this is that the observations below 508 revolutions per minute were taken when the engine was being controlled partly by the throttle valve, and do not therefore show the action of the governor fairly; the points are, however, useful in showing the tendency of the curves and represent actual positions of equilibrium of the governor.

The effect of the weight of the valve and unbalanced steam pressure are almost negligible, so that the power of the governor

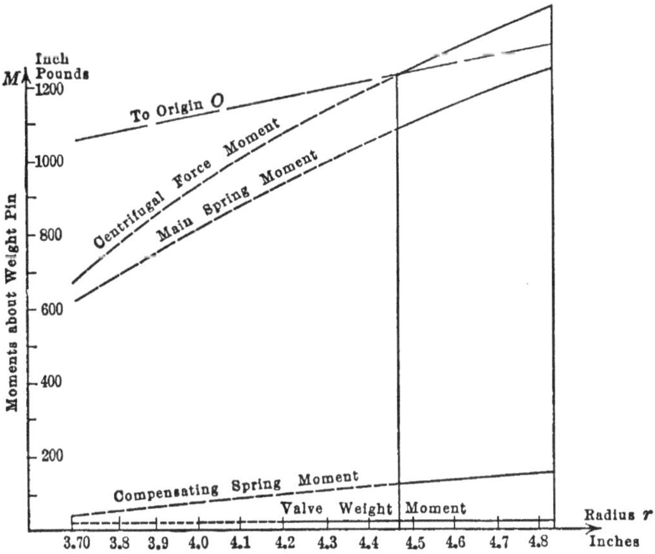

Fig. 138.

does not need to be large, but the spring S_c produces an appreciable effect amounting to about 11 per cent. of the total at the highest speed. If the compensating spring S_c were removed, the governor would run at a lower speed.

Joining any point on the moment curve to the origin O, as has been done on the figure, shows that the governor is stable.

The sensitiveness and powerfulness may be found from the C curve shown at Fig. 139. At the radius 4.47 in. the centrifugal force is 345.5 pds., and if the origin be joined to this point and the line produced it will cut the radius 4.83 in. at 373.5 pds., whereas the actual C is 410 pds. The sensitiveness then is

$\frac{\frac{1}{2}(410 - 373.5)}{\frac{1}{2}(410 + 373.5)} = 0.0465$ or 4.65 per cent. From the speeds the corresponding result would be $\frac{(532 - 508)}{\frac{1}{2}(532 + 508)} = 0.0442$ or 4.42 per cent., which agrees quite closely with the former value.

The moment curves cannot be used directly for the determination of the power of the governor because areas on the diagram do not represent work done. If the power is required, then the base must be altered either so as to represent equal angles passed through by the ball arm, or more simply by use of the C curve plotted on Fig. 139. It will be seen that the C curve differs

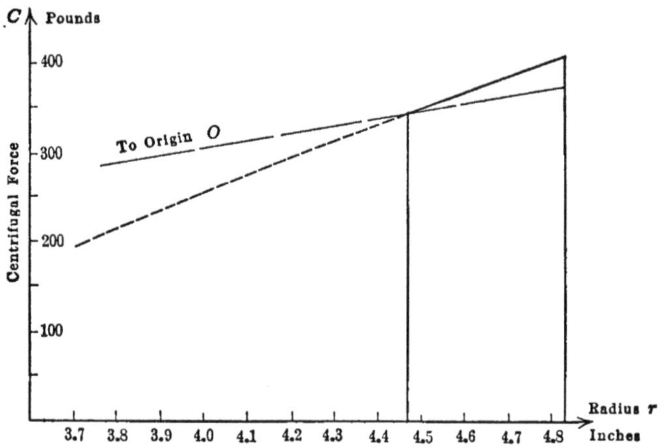

Fig. 139.

very little in character from the moment curve. The power of the governor is only about 11.6 ft.-pds.

The computations on this one governor will give a good general idea of the relative effects of the different parts in this style of governor, and also show that spring governors of this class possess some advantages.

THE INERTIA GOVERNOR, FREQUENTLY CALLED THE SHAFT GOVERNOR

194. Reasons for Using this Type.—The shaft governor was probably originally so named because it is usually secured to the crankshaft of an engine and runs therefore at the engine speed. In recent practice, however, certain spring governors, such as the Belliss and Morcom governor, are attached to the crankshaft and yet these scarcely come under the name of shaft governors. The term is more usually restricted to a governor in which the controlling forces differ to some extent from those already discussed. This type of governor is not nearly so old as the others and was introduced into America mainly as an adjunct to the high-speed engine.

On this continent builders of high (rotative)-speed engines have almost entirely governed them by the method first mentioned at the beginning of this chapter, that is by varying the point of cut-off of the steam, and in order to do this they have usually changed the angle of advance and also the throw of the eccentric by means of a governor which caused the center of the eccentric to vary in position relative to the crank according to the load, the result will be a change in all the events of the stroke. The eccentric's position is usually directly controlled by the governor, and hence it is necessary to have a powerful governor or else the force required to move the valve may cause very serious disturbances of the governor and render it useless. Again as the governor works directly on the eccentric, it is convenient to have it on the crankshaft.

Governors of this class also possess another peculiarity In those already described the pins about which the balls swung were in all cases perpendicular to the axis of rotation, so that the balls moved out and in a plane passing through this axis. In the shaft governor, on the other hand, the axis of the pins is parallel with the axis of rotation and the weights move out and in in the plane in which they rotate. While this may at first appear to be a small matter, it is really the point which makes this class of governor distinct from the others and which brings into play inertia forces during adjustment that are absent in the other types. Such governors may be made to adjust themselves to their new positions very rapidly and are thus very valuable on machinery subjected to sudden and frequent changes of load.

195. Description.—One make of shaft governor is shown at Fig. 140, being made by the Robb Engineering Co., Amherst, Nova Scotia, and is similar to the Sweet governor. In this make there is only one rotating weight W, the centrifugal effect of which is partly counteracted by the flat leaf spring S, to which the ball is directly attached. The eccentric E is pivoted by the pin P to the flywheel, and an extension of the eccentric is attached by the link b to the ball W. The wheel rotating in the sense shown, causes the ball to try to move out on a radial line, which movement is resisted by the spring S. As the ball moves out, due to increased speed, the eccentric sheave swings about P, and thus the center of the eccentric will take up a position depending upon the speed. Two stops are provided to limit the extreme movement of the eccentric and ball.

Other forms of governor are shown later at Fig. 143 and at Fig. 147, these having somewhat different dispositions of the parts.

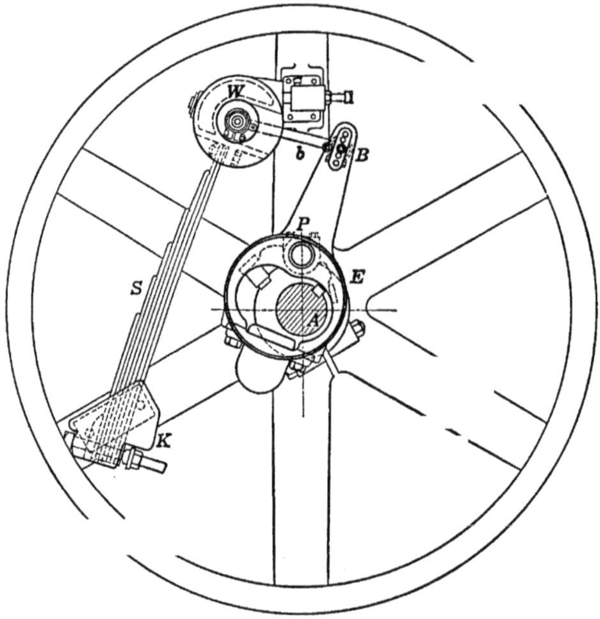

FIG. 140.—Robb inertia governor.

Powerfulness in such governors is obtained by the use of heavy weights moving at high speed, for example in one governor the revolving weight is 80 lb. and it revolves in a circle of over 29 in. radius at 200 revolutions per minute, dimensions which should be compared with those in the governors already discussed.

196. Conditions to be Fulfilled.—The conditions to be fulfilled are quite similar to those in other spring governors so that only a brief discussion will be necessary, which may be illustrated in the following example.

Let A, Fig. 141, represent a disc rotating about a center O at n revolutions per minute, and let this disc have a weight w

mounted on it so that it may move in and out along a radial line as indicated, and further let the motion radially be resisted by a spring S which is pivoted to the disc at E. Let the spring pull per foot of extension be S pds., and let the weight be in equilibrium at distance r ft. from O, the extension of the spring at this instant being a ft. The centrifugal force on the ball is $C = \frac{w}{g} r\omega^2$ pds. where ω is the angular velocity of the disc in radians per second, and since $\omega = \frac{2\pi n}{60}$, therefore

$$C = \frac{w}{g} r \left(\frac{2\pi n}{60}\right)^2 = 0.000341\ wrn^2$$

where r is in feet. For the same position the spring pull will be Sa pds., so that for equilibrium $Sa = C$ or

$$Sa = 0.000341\ wrn^2,$$

that is,

$$S = 0.000341\ w \frac{r}{a} n^2.$$

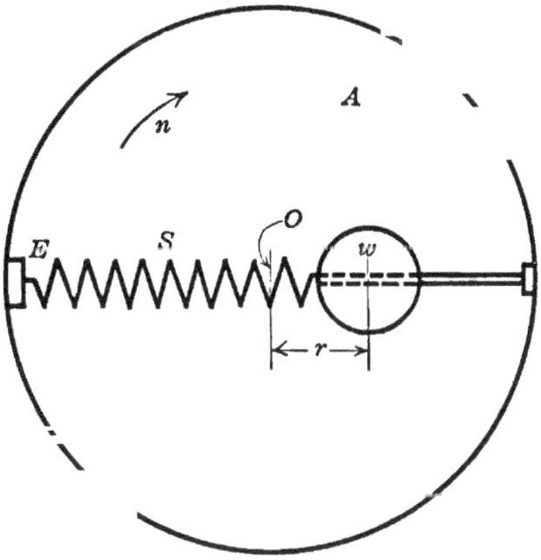

FIG. 141.

To make the meaning of this clear it will be well to take a numerical example, and let it be assumed that the weight $w = 25$ lb. and the speed is 200 revolutions per minute. Three cases may be considered, according to whether r is equal to, greater than or less than a, and these will result as follows:

1. $r = a = 1$ ft., $S = 0.000341 \times 25 \times \frac{1}{1} \times 200^2 = 341$ pds.

2. $r = 1$ ft., $a = 0.57$ ft., $S = 0.000341 \times 25 \times \frac{1}{0.57} \times 200^2 = 600$ pds.

3. $r = 1$ ft., $a = 1.19$ ft., $S = 0.000341 \times 25 \times \frac{1}{1.19} \times 200^2 = 288$ pds.

So that, as the formula shows, the spring strength depends upon the relation of r to a.

The resulting conditions when the ball is 10 in., 12 in. and 14 in. respectively from the center of rotation with the three springs, are set down in the following table and in Fig. 142.

Radius r, Inches	Centrifugal force at 200 rev.	Spring pull		
		$S = 288$	$S = 341$	$S = 600$
10	284	293	284	241
12	341	341	341	341
14	398	389	398	441

For the spring $S = 288$ pds. per foot of extension it is seen that at the smaller radius the spring pull is higher than the centrifugal force or the disc must run at a higher speed than 200 revolutions for equilibrium, while at the outer radius the spring pull is too low and the speed must be below 200 revolutions for equilibrium,

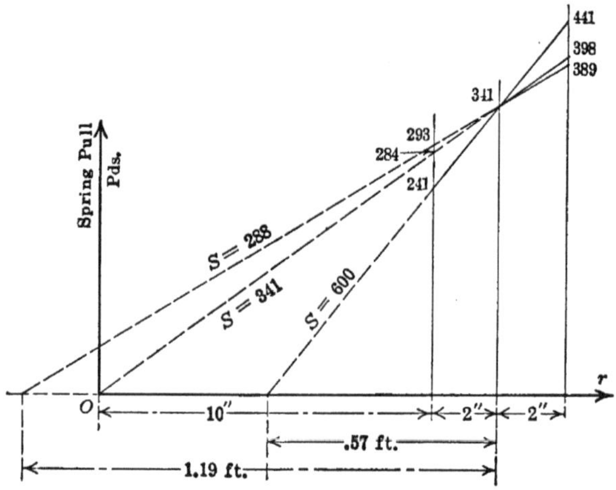

Fig. 142.

that is the speed should decrease as the ball moves out. With spring $S = 600$ exactly the reverse is true, or the speed must increase as the ball moves out, while for spring $S = 341$ the speed will be constant for all positions of the ball.

The spring $S = 341$ is properly designed and set for isochronism, but evidently there is no force holding the ball anywhere and the slightest push would send it oscillating along the scale, that is, it lacks stability. The spring $S = 288$ also gives an unstable arrangement for the reason that the centrifugal force increases and decreases faster than the spring pull, and thus if the ball happened to be 12 in. from the center and was disturbed it would instantly fly to the inner or outer extreme stop. However, spring $S = 600$ gives a stable arrangement, because whenever the ball is at say 12 in. from the center and any force pushes it away it immediately tries to return to this position, and will do so on account of the preponderating effect of the spring force acting upon it, unless there should be a change of speed forcing it to the new position, but to each speed there is a definitely fixed position of the ball. It is to be noticed that the curve for $S = 600$, is always steeper than any line from it to the origin O, which has been already given as a condition of stability (Sec. 183(2)).

The gain in stability is, however, made at a sacrifice in sensitiveness. For the spring $S = 600$ the speed changes from 184 to 211 revolutions per minute or the sensitiveness is $\frac{211 - 184}{\frac{1}{2}(211 + 184)} = 0.136$ or 13.6 per cent., while with spring

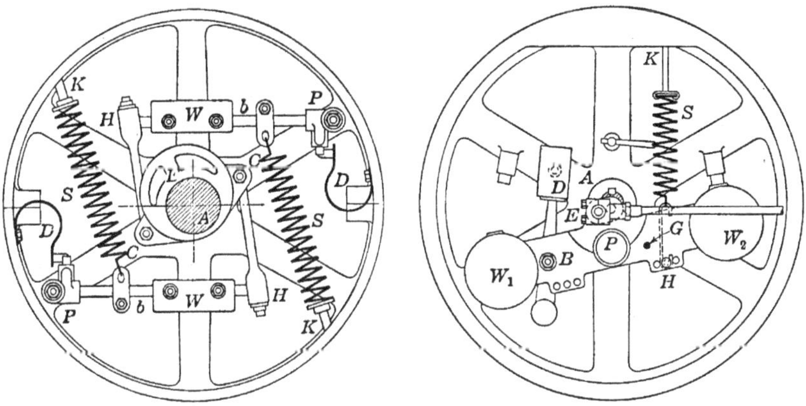

FIG. 143.—Buckeye and McEwen governors.

$S = 288$ the range of speeds is from 198 to 203 revolutions per minute or the sensitiveness is 2.4 per cent.

197. Analysis of the Governor.—Having now discussed the conditions of stability and isochronism and the effect the design of the spring has on them, a complete analysis of the governor may be made.

Two forms of governor are shown in Fig. 143 and these show a somewhat different disposition of the revolving weights. The one on the left is used by the Buckeye Engine Co. and has two revolving weights W connected by arms b to the pivots P. The centrifugal force is resisted by springs S attached to b and to the flywheel rim at K. The ends of the links b are connected at H to links attached to the eccentric E at C and the operation of the weights revolves E and changes the steam distribution. Auxiliary springs D oppose springs S at inner positions of weights W.

The right-hand figure shows the McEwen governor having two unequal weights W_1 and W_2 cast on a single bar, the combined center of gravity being at G, and the pivot connection to the wheel is P. There is a single spring S attached to the weights at H and to the wheel at K. There is a dashpot at D attached to the wheel and to the weight at B; this consists of a cylinder and piston, the latter being prevented from moving rapidly in the cylinder. The purpose of the dashpot is to prevent oscillations

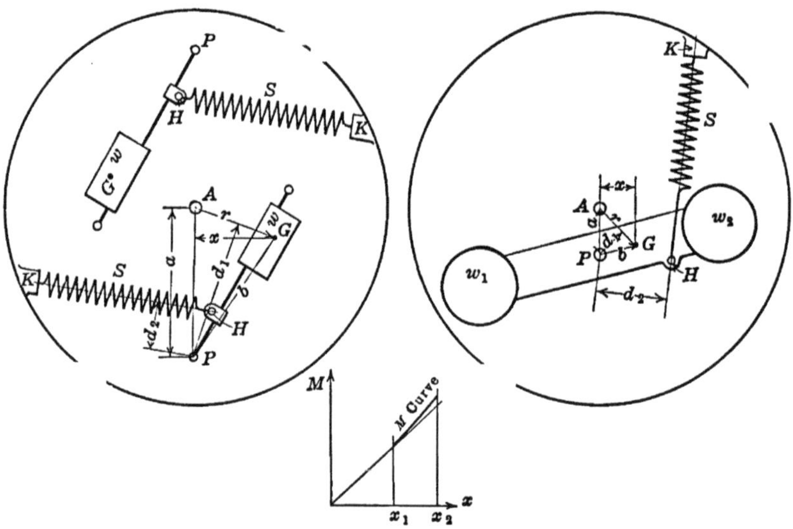

FIG. 144.

of the weight during adjustment and to keep it steady, but after adjustment has been made D has no effect on the conditions of equilibrium. In this governor a frictionless pin is provided at P by the use of a roller bearing. The valve rod is at E.

A diagrammatic drawing of these two governors, which may be looked upon as fairly representative of this class, is given at Fig. 144, similar letters being used in both cases.

Let the wheel revolve about A, Fig. 143, with angular velocity ω radians per second and let F denote the spring pull when the center of gravity G is at radius r from A; further, let d_1 and d_2 in. represent the shortest distances from the weight pivot P to the directions of r and S respectively. Then for equilibrium the moments about P due to the centrifugal force C and to the spring pull S must be equal if, for the present, the effect of gravity and of the forces required to move the valve are neglected. That is:

$$Cd_1 = Sd_2 \text{ in.-pds.}$$

or

$$\frac{w}{g}r\omega^2 d_1 = Sd_2 \text{ in.-pds.}$$

In such an arrangement as shown the effect of the forces required to move the valves is frequently quite appreciable and is generally also variable, as is also the effect of friction and gravity, although usually gravity is relatively so small that it may be neglected. If it is desired to take these into account then

$$\frac{w}{g}r\omega^2 d_1 = Sd_2 + \text{moment due friction, valve motion and gravity.}$$

Denote the distance AP by a and the shortest distance from G to AP by x; thus a is constant but x depends on the position of the balls. From similar triangles it is evident that $rd_1 = ax$ and therefore

$$\frac{w}{g}\omega^2 rd_1 = \frac{w}{g}\omega^2 ax.$$

Thus the moment due to the centrifugal force is, for a given speed, variable only with x and hence the characteristics of the governor are very well shown on a curve[1] in which the base represents values of x and vertical distances the centrifugal moments $\frac{w}{g}\omega^2 ax$.

Such a curve is shown below, Fig. 144, and the shape of the curve here represents a stable governor since it is steeper at all points than the line joining it to the origin O. From this curve information may be had as to stability and sensitiveness, but the power of the governor cannot be determined without either placing the curve on a base which represents the angular swing of the balls about P or else by obtaining a C curve on an r base as in former cases.

If ω is constant, or the governor is isochronous, M varies directly with x or the moment curve is a straight line passing through the origin O.

Having obtained the M curve in this way the moment curves

[1] For more complete discussion of this method see TOLLE, "Die Regelung der Kraftmaschinen."

about P corresponding to gravity, friction of the valves and parts and also those necessary to operate the valves are next found, these three curves also being plotted on the x base, and the difference between the sum of these three moments and the total centrifugal moment will give the moment which must be provided by the spring which is Sd_2 in.-pds. From the curve giving Sd_2 the force S may be computed by dividing by d_2 and these values of S are most conveniently plotted on a base of spring lengths, from which all information for the design of the spring may readily be obtained (see Sec. 190).

In order that the relative values of the different quantities may be understood, Fig. 145 shows these curves for a Buckeye

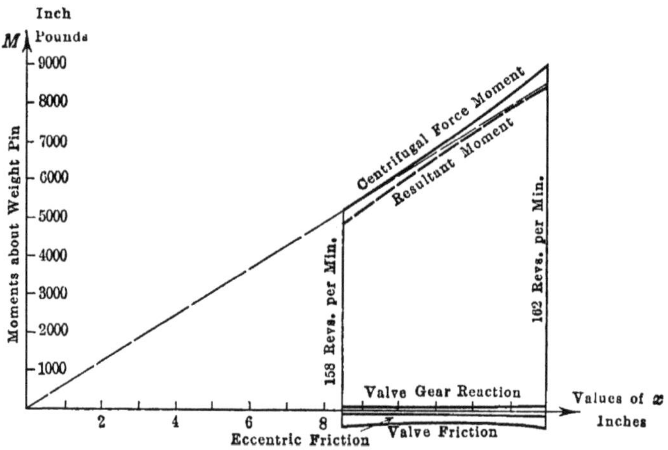

FIG. 145.

governor, in which the gravity effect is balanced by using two revolving weights symmetrically located. Friction of the valve and eccentric and the moment required to move the valve are all shown and the curves show how closely the spring-moment and centrifugal-moment curves lie together. The curves are drawn from the table given by Trinks and Housom, in whose[1] treatise all the details of computing the results is shown so as to be clearly understood. The governor has a powerfulness of nearly 600 ft.-pds.

RAPIDITY OF ADJUSTMENT

198. The inertia or shaft governor is particularly well adapted to rapid adjustment to new conditions and it is often made so

[1] TRINKS and HOUSOM, "Shaft Governors."

that it will move through its entire range in one revolution, which often means only a small fraction of a second. The rapidity of this adjustment depends almost entirely upon the distribution of the revolving weight and not nearly so much upon its magnitude. For a given position of the parts the only force acting is centrifugal force already discussed but during change of position the parts are being accelerated and forces due to this also come into play. Fig. 146 represents seven different arrangements of the weights; in five of these the weight is concentrated into a ball with center of gravity at G and hence with very small moment of inertia about G, so that the torque required to revolve such a weight at any moderate acceleration will not be great; the opposite is true of the two remaining cases, however, the weight

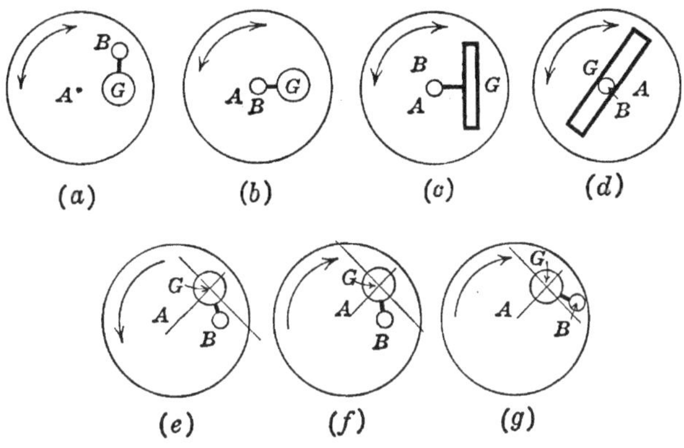

FIG. 146.

being much elongated and having a large moment of inertia about G.

Assuming a sudden increase of speed in all cases, then at (*a*) this only increases the pressure on the pin B because BC is normal to the radius AG, at (*b*) an increase in speed will produce a relatively large turning moment about the pin which is shown at A. Comparing (*c*) and (*d*) with (*a*) and (*b*) it is seen that the torque in the former cases is increased at (*c*) because in addition to the acceleration of G there is also an angular acceleration about G, whereas at (*d*) G is stationary and yet there is a decided torque due to its angular acceleration. At (*e*), (*f*) and (*g*) the sense of rotation is important and if an increase in speed occur in the first and last cases the accelerating forces assist in moving the weights out rapidly to their new positions, whereas at (*f*) the accelerating forces oppose the movement.

Space prevents further discussion of this matter here, but it will appear that the accelerating forces may be adjusted in any desired way to produce rapid changes of position, the weights being first determined from principles already stated and the distribution of these depending on the inertia effects desired. Chapter XV will assist the reader in understanding these forces more definitely.

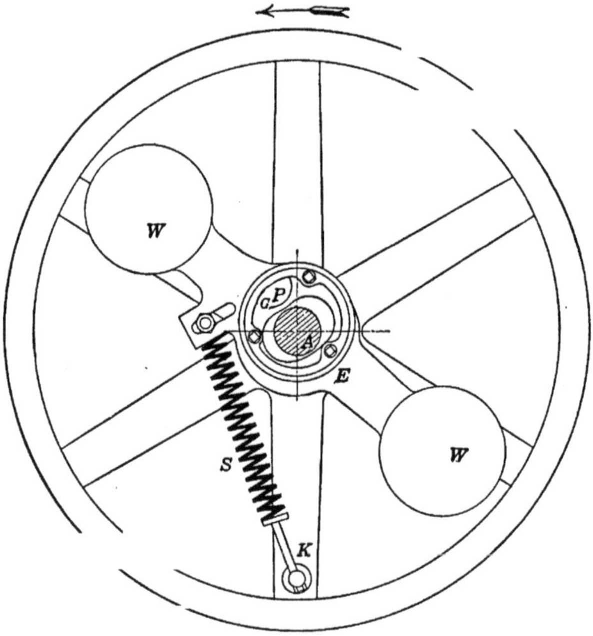

FIG. 147.—Rites governor.

A form of governor made by Rites, in which the inertia forces play a prominent part during adjustment is shown at Fig. 147. The revolving weights are heavy and are set far apart, but their center of gravity G is fairly close to A so that the centrifugal moment is relatively small. In a governor for a 10 by 10-in. engine the weight W was over 120 lb. and the two weight centers were 32 in. apart.

QUESTIONS ON CHAPTER XII

1. Define a governor. What is the difference between the functions of a governor and a flywheel?

2. What is the height of a simple governor running at 95 revolutions per minute?

3. What is meant by an isochronous governor? Is such a governor desirable or not? Why?

4. Explain fully the terms stability and powerfulness.

5. Prove that in a governor where the balls move in a paraboloid of revolution, h is constant and the governor is isochronous.

6. What are the advantages of the Porter governor?

7. Using the data, $n_1 = 100$, $n_2 = 110$, prove that $\frac{\delta h}{h} = -2\frac{\delta\omega}{\omega}$.

8. Compare the sensitiveness of a simple and a Porter governor at 115 revolutions per minute and with a sleeve travel of ¾ in., taking $W = 120$ lb. and $w = 15$ lb.

9. Analyze the following governor for sensitiveness and power (see Fig. 129):

$n = 130$, $W = 110$, $w = 12$, $l_1 = 12\frac{1}{2}$, $BM = 3\frac{1}{2}$, $l_2 = 10\frac{1}{2}$, $a_1 = 0$, $a_2 = 2\frac{1}{2}$, sleeve travel 2½ in.

10. Design a Porter governor for a speed of 170 revolutions per minute with a speed variation of 5 per cent. each way, travel 2½ in., power 35 ft.-pds.

11. In a governor of the type of Fig. 133, $a = 2.1$ in.; $b = 0.75$ in., distance between pivots 2¾ in. inner radius of ball 1.6 in., weight per ball 1¾ lb., travel ¼ in. and speed 250. Design the spring for 5 per cent. variation.

12. What are the advantages of the shaft governor? Show how the distribution of the weight affects the rapidity of adjustment.

CHAPTER XIII

SPEED FLUCTUATIONS IN MACHINERY

199. Nature of the Problem.—The preceding chapter deals with governors which are used to prevent undue variations in speed of various classes of machinery, the governor usually controlling the supply of energy to the machine in a way to suit the work to be done and so as to keep the **mean** speed of the machine constant. The present chapter does not deal with this kind of a problem at all, but in the discussion herein, it is assumed that the mean speed of the machine is constant and that it is so controlled by a governor or other device as to remain so.

In addition to the variations in the mean speed there are variations taking place during the cycle of the machine and which may cause just as much trouble as the other. Everyone is familiar with the small direct-acting pump, and knows that although such a pump may make 80 strokes per minute, for example, and keep this up with considerable regularity, yet the piston moves very much faster at certain times than others, and in fact this variation is so great that larger pumps are not constructed in such a simple way. With the larger pumps, on which a crank and flywheel are used, an observer frequently notices that, although the mean speed is perfectly constant, yet the flywheel speed during the revolution is very variable. Where a steam engine drives an air compressor, these variations are usually visible, at certain parts of the revolution the crankshaft almost coming to rest at times. These illustrations need not be multiplied, but those quoted will suffice. The speed variations which occur in this way during the cycle are dealt with in this chapter.

200. Cause of Speed Fluctuations.—The flywheel of an engine or punch or other similar machine is used to store energy and to restore it to the machine according to the demands. Consider, for example, the steam engine; there the energy supplied by the steam at different parts of the stroke is not constant, but varies from time to time; at the dead centers the piston is stationary and hence no energy is delivered by the working fluid, whereas when the piston has covered about one-third of its stroke, energy is being delivered by the steam to the piston at about its maximum rate, since the piston is moving at nearly its maximum speed and the steam pressure is also high, as cutoff has not usually taken place. Toward the end of the stroke the rate of delivery of the energy by the steam is small because the steam pressure is low on account of expansion and the piston is moving at slow speed. During the return stroke the piston must supply energy to the steam in order to drive the latter out of the cylinder.

Now the engine above referred to may be used to drive a pump or an air compressor or a generator or any other desired machine, but in order to illustrate the present matter it will be assumed to be connected to a turbine pump, since, in such a case, the pump offers a constant resisting torque on the crankshaft of the engine. The rate of delivering energy by the working fluid is variable, as has already been explained; at the beginning of the revolution it

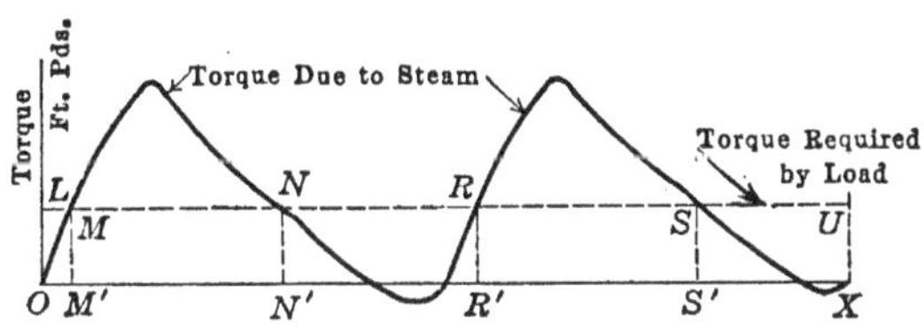

FIG. 148.

is much less than that required to drive the pump, a little further on it is much greater than that required, while further on again the steam has a deficiency of energy, and so on.

At this point it will be well to refer again to Fig. 101 which has been reproduced in a modified form at Fig. 148 and shows in a very direct and clear way these important features. During the first part of the outstroke it is evident that the crank effort due to the steam pressure is less than that necessary to drive the load; this being the case until M is reached, at which point the effort due to the steam pressure is just equal to that necessary to drive the load; thus during the part OM' of the revolution the input to the engine being less than the output the energy of the links themselves must be drawn upon and must supply the work represented by OML. But the energy which may be obtained from the links will depend upon the mass and velocity of them, the energy being greater the larger the mass and the greater the velocity, the result is that if the energy of the links is decreased by drawing from them for any purpose, then since the mass of the links is fixed by construction, the only other thing which may happen is that the speed of the links must decrease.

In engines the greater part of the weight in the moving parts is in the flywheel and hence, from what has been already said if energy is drawn from the links then the velocity of the flywheel will decrease and it will continue to decrease so long as energy is drawn from it. Thus during OM' the speed of the flywheel will fall continually but at a decreasing rate as M' is approached, and at this point the wheel will have reached its minimum speed. Having passed M' the energy supplied by the steam is greater than that necessary to do the external work, and hence there is a balance left for the purpose of adding energy to the parts and speeding up the flywheel and other links, the energy available for this purpose in any position being that due to the height of the torque curve above the load line. In this way the speed of the parts will increase between M' and N' reaching a maximum for this period at N'.

From N' to R' the speed will again decrease, first rapidly then more slowly, reaching a minimum again at R' and from R to S', there is increasing speed with a maximum at S'. The flywheel and other parts will, under these conditions, be continually changing their speeds from minimum to maximum and *vice versa*, producing much unsteadiness in the motion during the revolution. The magnitude of the unsteadiness will evidently depend upon the fluctuation in the crank-effort curve, if the latter curve has large variations then the unsteadiness will be increased; it will also depend on the weights of the parts.

In the case of the punch the conditions are the reverse of the engine, for the rate of energy supplied by the belt is nearly constant but that given out is variable. While the punch runs light, no energy is given out (neglecting friction), but when a hole is being punched the energy supplied by the belt is not sufficient, and the flywheel is drawn upon, with a corresponding decrease in its speed, to supply the extra energy, and then after the hole is punched, the belt gradually speeds the wheel up to normal again, after which another hole may be punched. To store up energy for such a purpose the flywheel has a large heavy rim running at high speed.

It will thus be noticed that a flywheel, or other part serving the same purpose, is required if the supply of energy to the machine, or the delivery of energy by the machine, *i.e.*, the load, is variable; thus a flywheel is required on an engine driving a dynamo or a reciprocating pump, or a compressor, or a turbine pump; also a flywheel is necessary on a punch or on a sheet metal press. It is not, however, in general necessary to have a flywheel on a steam turbo-generator, or on a motor-driven turbine pumping set, or on a water turbine-driven generator set working with constant load, because in such cases the energy supplied is always equal to that given out.

The present investigation is for the purpose of determining the variations or fluctuations in speed that may occur in a given machine, when the methods of supplying the energy and also of loading are known. Thus, in an engine-driven compressor, hoth the steam- and air-indicator diagrams are assumed known, as well as the dimensions and weights of the moving parts.

THE KINETIC ENERGY OF MACHINES

201. Kinetic Energy of Bodies.—In order to determine the speed fluctuations in a machine it is necessary, first of all, to find the kinetic energy of the machine itself in any given position and this will now be determined.

If any body has plane motion at any instant, this motion may be divided into two parts:

(*a*) A motion of translation of the body.

(*b*) A motion of rotation of the body about its center of gravity. Let the weight of the body be w lb., then its mass will be $m = \frac{w}{g}$, where g is the attraction due to gravity and is equal to 32.16 in pound, foot and second units, and let the body be moving in a plane, the velocity of its center of gravity at the instant being v ft. per second. Further, let the body be turning at the same instant at the rate of ω radians per second, and assume that the moment of inertia of the body about its center of gravity is I, the corresponding radius of gyration being k ft., so that $I = mk^2$.

Then it is shown in books on mechanics that the kinetic energy of the body is $E = \frac{1}{2}mv^2 + \frac{1}{2} I\omega^2 = \frac{1}{2}mv^2 + \frac{1}{2}mk^2\omega^2$ ft.-pds., and, hence, in order to find the kinetic energy of the body it is necessary to know its weight and the distribution of the latter because of its effect on k, and in addition the velocity of the center of gravity of the weight and also the angular velocity of the body.

202. Application to Machines.—Let Fig. 149 represent a mechanism with four links connected by four turning pairs, the links being a, b, c and d, of which the latter is fixed, and let I_a, I_b and I_c represent the moments of inertia of a, b and c respectively about their centers of gravity, the masses of the links being m_a, m_b and m_c. Assuming that in this position the angular

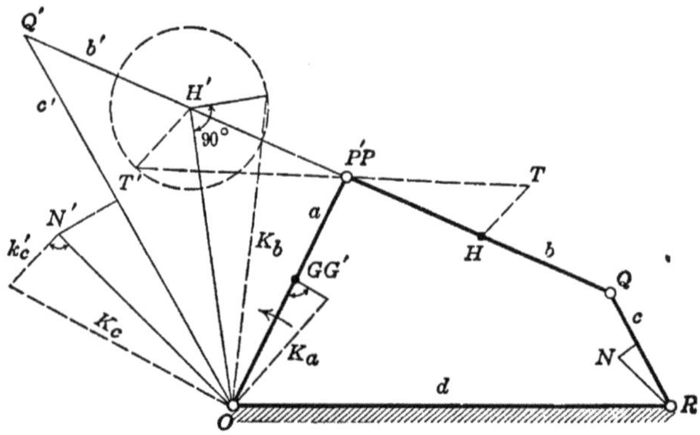

FIG. 149.

velocity ω of the link a is known, it is required to find the corresponding kinetic energy of the machine.

Find the images of P, Q, a, b, c, d and of G, H and N, the centers of gravity of a, b and c respectively, by means of the phorograph discussed in Chapter IV. Now if v_G, v_H and v_N be used to represent the linear velocities of G, H and N and also if ω_b and ω_c be used to denote the angular velocities of the links b and c, it is at once known, from the phorograph (Secs. 66 and 68), that: $v_G = OG'.\omega$; $v_H = OH'.\omega$ and $v_N = ON'.\omega$ ft. per second, and $\omega_b = \frac{b^1}{b}\cdot\omega$ and $\omega_c = \frac{c'}{c}\cdot\omega$ radians per second, so that all the necessary linear and angular velocities are found from the drawing.

203. Reduced Inertia of the Machine.—The investigation will be confined to the determination of the kinetic energy of the link b, which will be designated by E_b, and having found this quantity the energy of the other links may be found by a similar process. Since for any body the kinetic energy at any instant is given by the formula:

$$E = \tfrac{1}{2} mv^2 + \tfrac{1}{2} I\omega^2 \text{ ft.-pds.}$$

Therefore, $$E_b = \tfrac{1}{2} m_b.v^2_H + \tfrac{1}{2} I_b\omega_b^2 \text{ ft.-pds.}$$

Now, $$I_b\omega_b^2 = m_b k_b^2 \omega_b^2 = m_b k_b^2 \left(\frac{b'}{b}\omega\right)^2 = m_b \left(\frac{b'}{b}k_b\right)^2 \omega^2.$$

Following the notation already adopted, it will be convenient to write k'_b for $\frac{b'}{b}k_b$, since the length $\frac{b'}{b}k_b$ is the length of the image of k_b on the phorograph. The magnitude of k'_b is found by drawing a line HT in any direction from H to represent k_b and finding T' by drawing $H'T'$ parallel to HT to meet TP produced in T' as indicated in Fig. 149; then $H'T'$ is the corresponding value of k'_b.

Hence $$E_b = \tfrac{1}{2} m_b.v_H^2 + \tfrac{1}{2} I_b\omega_b^2$$

$$= \tfrac{1}{2} m_b \left[\overline{OH'}^2 + \overline{k_b'}^2\right] \omega^2 \text{ ft.-pds.}$$

Let the quantity in the square bracket be denoted by K_b^2; then evidently K_b may be considered as the radius of gyration of a body, which if secured to the link a and having a mass m_b would have the same kinetic energy as the link b has at this instant. It is evidently a very simple matter to find K_b graphically since it is the hypotenuse of the right angled triangle of which one side is OH' and the other k_b'; this construction is shown in Fig. 149.

Thus $$E_b = \tfrac{1}{2} m_b K_b^2 \omega^2 \text{ ft.-pds.}$$

Similarly $$E_a = \tfrac{1}{2} m_a K_a^2 \omega^2 \text{ ft.-pds.}$$

and $$E_c = \tfrac{1}{2} m_c K_c^2 \omega^2 \text{ ft.-pds.};$$

constructions for K_a and K_c are shown in the figure.

For the whole machine the kinetic energy is

$$\begin{aligned} E &= E_a + E_b + E_c \\ &= \tfrac{1}{2} m_a K^2_a \omega^2 + \tfrac{1}{2} m_b K^2_b \omega^2 + \tfrac{1}{2} m_c K^2_c \omega^2 \\ &= \tfrac{1}{2}[m_a K^2_a + m_b K^2_b + m_c K^2_c]\omega^2 \\ &= \tfrac{1}{2}[I'_a + I'_b + I'_c]\,\omega^2 \\ &= \tfrac{1}{2} J\omega^2 \text{ ft.-pds.} \end{aligned}$$

A study of these formulas and a comparison with the work just covered, shows that I'_a is the moment of inertia of the mass with center of gravity at O and rotating with angular velocity ω which will have the same kinetic energy as the link a actually has; in other words, I'_a may be looked upon as the reduced moment of inertia of the link a, while similar meanings may be attached to I'_b and I'_c. Note that I'_a and I_a differ because the former is the inertia of the corresponding mass with center of gravity at O, whereas I_a is the moment of inertia about the center of gravity G of the actual link. The quantity J is, on the same basis, the reduced inertia of the entire machine, by which is meant that the kinetic energy of the machine is the same as if it were replaced by a single mass with center of gravity at O, and having a moment of inertia J about O, this mass rotating at the angular speed of the primary link. It will be readily understood that J differs for each position of the machine and is also a function of the form and weight of the links.

The foregoing method of reduction is of the greatest importance in solving the problems under consideration, because it makes it possible to reduce any machine, no matter how complex, down

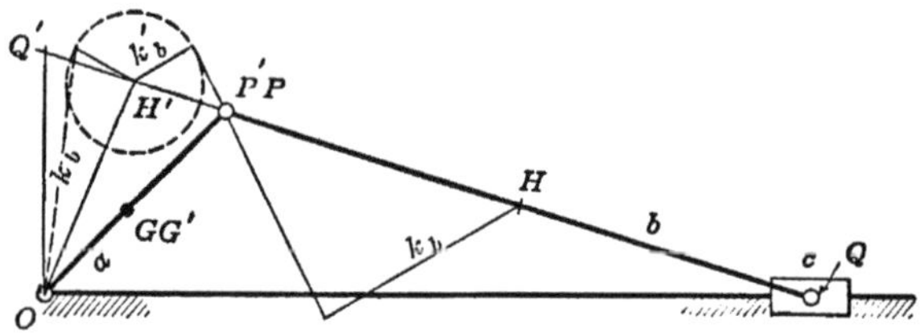

FIG. 150.

to a single mass, rotating with known speed, about a fixed center, so that the kinetic energy of the machine is readily found from the drawing.

204. Application to Reciprocating Engine.—The method may be further illustrated in the common case of the reciprocating engine, which in addition to the turning pairs contains also a sliding pair. The mechanism is shown in Fig. 150 and the same notation is employed as was used in the previous case, and the only peculiarity about the mechanism is the treatment of the link c.

The link c has a motion of translation only and therefore $\omega_c = 0$ and $I_c\omega_c^2 = 0$ so that the kinetic energy of the link is

$E_c = \frac{1}{2} m_c \cdot v_c^2 = \frac{1}{2} m_c\, OQ'^2 \omega^2$ or $I_c' = m_c OQ'^2$ since the point Q has the same linear velocity as all points in the link. The remainder of the machine is treated as before.

Lack of space prevents further multiplication of these illustrations, but it will be found that the method is easily applied to any machine and that the time required to work out the values of J for a complete cycle is not very great.

SPEED FLUCTUATIONS

205. Conditions Affecting Speed Variations.—One of the most useful applications of the foregoing theory is to the determination of the proper weight of flywheel to suit given running conditions and to prevent undue fluctuations in speed of the main shaft of a prime mover. Usually the allowable speed variations are set by the machine which the engine or turbine or other motor is driving and these variations must be kept within very narrow limits in order to make the engine of value. When a dynamo is being driven, for example, fluctuations in speed affect the lights, causing them to flicker and to become so annoying in certain cases that they are useless. The writer has seen a particularly bad case of this kind in a gas engine driven generator. If alternators are to be run in parallel the speed fluctuation must be very small to make the arrangement practicable.

In many rolling mills motors are being used to drive the rolls and in such cases the rolls run light until a bar of metal is put in, and then the maximum work has to be done in rolling the bar. Thus, in such a case the load rises suddenly from zero to a maximum and then falls off again suddenly to zero. Without some storage of energy this would probably cause damage to the motor and hence it is usual to attach a heavy flywheel somewhere between the motor and the rolls, this flywheel storing up energy as it is being accelerated after a bar has passed through the rolls, and again giving out part of its stored-up energy as the bar enters and passes through the rolls. The electrical conditions determine the allowable variation in speed, but when this is known, and also the work required to roll the bar and the torque which the motor is capable of exerting under given conditions, then it is necessary to be able to determine the proper weight of flywheel to keep the speed variation within the set limits.

In the case of a punch already mentioned, the machine runs

light for some time until a plate is pushed in suddenly and the full load is thrown on the punch. If power is being supplied by a belt a flywheel is also placed on the machine, usually on the shaft holding the belt pulley, this flywheel storing up energy while the machine is light and assisting the belt to drive the punch through the plate when a hole is being punched. The allowable percentage of slip of the belt is usually known and the wheel must be heavy enough to prevent this amount of slip being exceeded.

The present discussion is devoted to the determination of the speed fluctuations with a given machine, and the investigation will enable the designer to devise a machine that will keep these fluctuations within any desired limits, although the next chapter deals more particularly with this phase.

206. Determination of Speed Variation in Given Machine.—Let E_1 and E_2 be the kinetic energies, determined as already explained, of any machine at the beginning and end of a certain interval of time corresponding with a definite change of position of the parts. Then the gain in energy, $E_2 - E_1$, during the interval under consideration represents the difference between the energy supplied with the working fluid and the sum of the friction of the machine and of the work done at the main shaft on some other machine or object during the same interval, because the kinetic energy of the machine can only change from instant to instant if the work done by the machine differs from the work done on it by the working fluid. In order to simplify the problem friction will be neglected, or assumed included in the output.

Consideration will show that $E_2 - E_1$ will be alternately positive and negative, that is, during the cycle of the machine its kinetic energy will increase to a maximum and then fall again to a minimum and so on. As long as the kinetic energy is increasing the speed of the machine must also increase in general, so that the speed will be a maximum just where the kinetic energy begins to decrease, and conversely the speed will be a minimum just where the kinetic energy begins to increase again. But the kinetic energy of the machine will increase just so long as the energy put into the machine is greater than the work done by it in the same time; hence the maximum speed occurs at the end of any period in which the input to the machine exceeds its output and *vice versa*.

The method of computing this speed fluctuation will now be considered.

It has already been shown that the kinetic energy of the machine is given by

$$E = \tfrac{1}{2} J\omega^2 \text{ ft.-pds.}$$

from which there is obtained by differentiation[1]

$$\delta E = \tfrac{1}{2} \{2J.\ \omega.\delta\omega + \omega^2.\delta J\} \text{ ft.-pds.}$$

or

$$\delta\omega = \frac{\delta E - \tfrac{1}{2}\,\omega^2.\delta J}{J\omega}$$

where $\delta\omega$ is the change in speed in radians per second in the interval of time in which the gain of energy of the machine is δE and that in J is δJ. Of course, any of these changes may be positive or negative and they are not usually all of the same sign. The values of J and ω used in the formula may, without sensible error, be taken as those at the beginning or end of the interval or as the average throughout the interval, the latter being preferable.

207. Approximate Value of Speed Variation.—The calculation is frequently simplified by making an approximation on the assumption that the variation in J may be neglected, *i.e.*, that $\delta J = 0$. The writer has not found that there is enough saving in time in the work involved to make this approximation worth while, but since it is often assumed, it is placed here for consideration and a slightly different method of deducing the resulting formula is given. Let E_1, E_2, ω_1 and ω_2 have the meanings already assigned, at the beginning and end of the interval of time and let the reduced moment J be considered constant.

[1] To those not familiar with the calculus the following method may be of value.

Let E, J and ω be the values of the quantities at the beginning of the interval of time and $E + \delta E$, $J + \delta J$ and $\omega + \delta\omega$, the corresponding values at the end of the same interval.

Then

$$E = \tfrac{1}{2} J\omega^2$$

$$E + \delta E = \tfrac{1}{2}(J + \delta J)\ (\omega + \delta\omega)^2$$

$$= \tfrac{1}{2}(J\omega^2 + 2J.\omega\delta\omega + \omega^2\delta J)$$

where in the multiplication such terms as $(\delta\omega)^2$, and $\delta J.\delta\omega$ are neglected as being of the second order of small quantities.

By subtraction, then,

$$E + \delta E - E = \delta E = \tfrac{1}{2} \{2J\omega\delta\omega + \omega^2\delta J\} \text{ as above.}$$

Then

$$E_2 - E_1 = \tfrac{1}{2}J(\omega_2^2 - \omega_1^2)$$

or

$$\frac{E_2 - E_1}{J} = \tfrac{1}{2}(\omega_2^2 - \omega_1^2)$$
$$= \tfrac{1}{2}(\omega_2 + \omega_1)(\omega_2 - \omega_1)$$
$$= \omega(\omega_2 - \omega_1)$$

where ω has been written for $\frac{\omega_2 + \omega_1}{2}$, a substitution which causes little inaccuracy in practice.

Therefore $$\omega_2 - \omega_1 = \frac{E_2 - E_1}{J\omega}$$

or $$\delta\omega = \frac{\delta E}{J\omega}.$$

This is the same result as would have been obtained from the former formula by making $\delta J = 0$.

208. Practical Application to the Engine.—The meanings of the different quantities can best be explained by an example

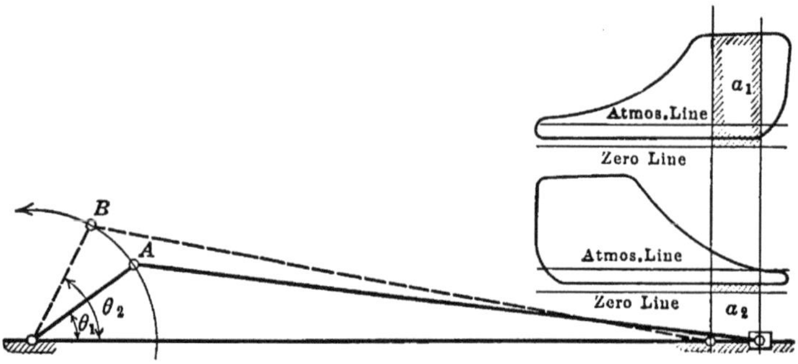

FIG. 151.

which will now be worked out. The steam engine has been selected, because all the principles are involved and the method of selecting the data in this case may be rather more readily understood. The computations have all been made by the exact formula, which takes account of variations in J.

Consider the double-acting engine, which is shown with the indicator diagrams in Fig. 151; it is required to find the change of speed of the crank while passing from A to B. Friction will be neglected.

For simplicity, it will be assumed that the engine is driving a turbine pump which offers a uniform resisting turning moment and hence the work done by the engine during any interval is proportional to the crank angle passed through in the given interval. If the work done per revolution as computed from the diagram is W ft.-pds., then the work done by the engine during the interval from A to B will be $\frac{\theta_2 - \theta_1}{360}W$ ft.-pds. To make the case as definite as possible suppose that $\theta_2 - \theta_1 = 18°$; then the work done by the engine will be $\frac{1}{20}W$ ft.-pds.

209. Output and Input Work.—Again, let A_1 and A_2 represent the areas in square inches of the head end and the crank end of the cylinder respectively, l_1 and l_2 being the lengths of the corresponding indicator diagrams in inches. The stroke of the piston is taken as L feet and the indicator diagrams are assumed drawn to scale s pds. per square inch = 1 in. in height. With these symbols the work represented by each square inch on the diagram is $sA_1\frac{L}{l_1}$ ft.-pds. for the head-end and $sA_2\frac{L}{l_2}$ ft.-pds. for the crank-end diagram.

Now suppose that during the crank's motion from A to B the area of the head-end diagram reckoned above the zero line is a_1 sq. in., see Fig. 151, and the corresponding area for the crank-end diagram a_2 sq. in. Then the energy delivered to the engine by the steam during the interval is

$$a_1 s A_1\frac{L}{l_1} - a_2 s A_2\frac{L}{l_2} \text{ ft.-pds.}$$

while the work done by the engine is

$$\frac{W}{20} \text{ ft.-pds.}$$

Note that the work W is the total area of the two diagrams in square inches multiplied by their corresponding constants to bring the quantities to foot-pounds.

Then the input work exceeds the output by

$$a_1 s A_1\frac{L}{l_1} - a_2 s A_2\frac{L}{l_2} - \frac{W}{20} \text{ ft.-pds.}$$

which amount of energy must be stored up in the moving parts during the interval. That is, the gain in energy during the period is

$$E_2 - E_1 = a_1 s A_1\frac{L}{l_1} - a_2 s A_2\frac{L}{l_2} - \frac{W}{20} \text{ ft.-pds.}$$

so that the gain in energy is thus known.

Again, the method described earlier in the chapter enables the values of J_1 and J_2 to be found and hence the value of $J_2 - J_1$. Substituting these in the formula,

$$\delta\omega = \frac{\delta E - \frac{1}{2}\omega^2 \delta J}{J\omega}$$

the gain in angular velocity is readily found. The values are $E_2 - E_1 = \delta E$, $J_2 - J_1 = \delta J$, $\frac{1}{2}(J_2 + J_1) = J$ and for ω no error will result in practice by using the mean speed of rotation of the crank.

During a complete revolution the values of $\delta\omega$ will sometimes be positive and sometimes negative, and in order that the engine may maintain a constant mean speed the algebraic sum of these must be zero. Should the algebraic sum for a revolution be positive, the conclusion would be that there is a gain in the mean speed during the revolution, that is the engine would be steadily gaining in speed, whereas it has been assumed that the governor prevents this.

210. Numerical Example on Single-cylinder Engine.—A numerical example taken from an actual engine will now be given.

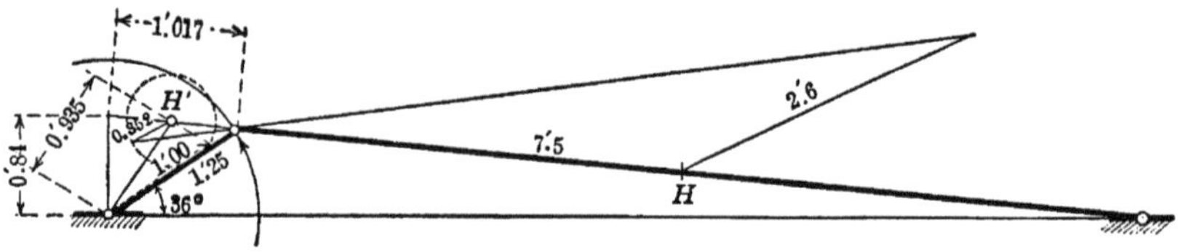

FIG. 152.

The engine used in this computation had a cylinder $12\frac{1}{16}$ in. diameter with a piston rod $1\frac{7}{8}$ in. diameter and a stroke of 30 in. The connecting rod was 90 in. long, center to center, weighed 175 lb. and had a radius of gyration about its center of gravity of 31.2 in. The piston, crosshead and other reciprocating parts weighed 250 lb., while the flywheel weighed 5,820 lb. and had a moment of inertia about the shaft of 2,400, using pound and foot units. The mean speed of rotation was 86 revolutions per minute.

Using the notation employed in the earlier discussion, the data may be set down as follows:

$$a = 1.25 \text{ ft.},\ b = 7.5 \text{ ft.},\ k_b = 2.60 \text{ ft.},\ I_a = 2{,}400 \text{ pd.(ft.)}^2,$$
$$m_a = 181,\ m_b = 5.44,\ m_c = 7.78.$$

The speed $\omega = \frac{2\,\pi n}{60} = \frac{2 \times \pi \times 87}{60} = 9$ radians per second.

Using the above data the following quantities were measured directly from the drawing, Fig. 152:

θ degrees	b' feet	OH' feet	k'_b feet	K_b feet	OQ' feet
36	1.017	0.935	0.352	1.00	0.84
54	0.741	1.123	0.257	1.15	1.11

From these the following quantities are obtained by computation:

θ degrees	$I'_b = m_b.K_b^2$	$I'_c = m_c.OQ'^2$	I_a	$J = I_a + I'_b + I'_c$	δJ
36	5.442	5.488	2,400	2,410.9	+5.9
54	7.200	9.578	2,400	2,416.8	

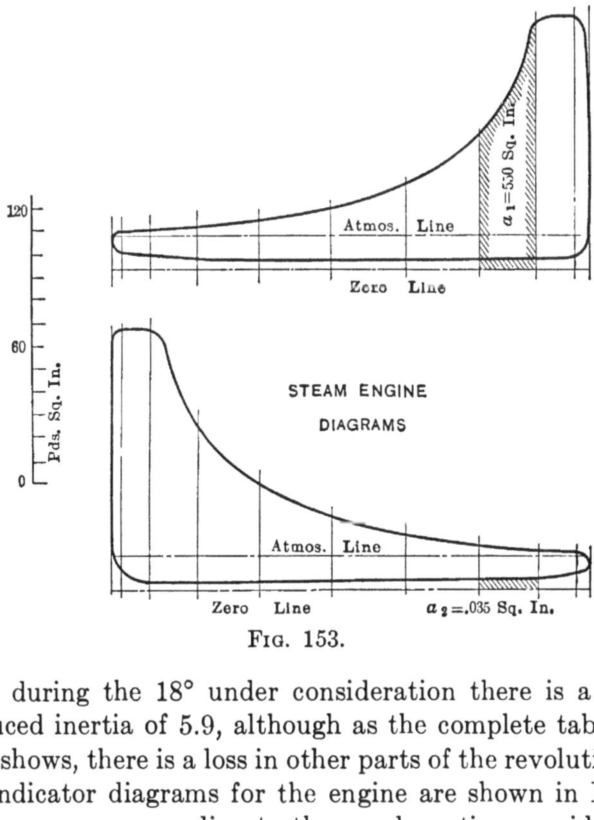

FIG. 153.

Thus, during the 18° under consideration there is a gain in the reduced inertia of 5.9, although as the complete table given later on shows, there is a loss in other parts of the revolution.

The indicator diagrams for the engine are shown in Fig. 153 and the areas corresponding to the crank motion considered are shown hatched and marked a_1 and a_2. These areas were measured on the original diagrams which were drawn to 60-pd. scale, although these have been somewhat reduced in reproduction.

Data for computations from the indicator diagrams are as follows:

Cylinder areas: Head end, $A_1 = 114.28$ sq. in. Crank end, $A_2 = 111.52$ sq. in.

Diagram lengths: Head-end diagram, $l_1 = 3.55$ in. Crank-end diagram, $l_2 = 3.58$ in.

Stroke of piston, $L = 2.5$ ft.

Hence, each square inch on the diagrams represents

$$sA_1\frac{L}{l_1} = 60 \times 114.28 \times \frac{2.5}{3.55} = 4{,}829 \text{ ft.-pds. for the head end,}$$

and

$$sA_2\frac{L}{l_2} = 60 \times 111.52 \times \frac{2.5}{3.58} = 4{,}673 \text{ ft.-pds. for the crank end.}$$

The original full-sized diagrams give $a_1 = 0.550$ sq. in. and $a_2 = 0.035$ sq. in., from which the corresponding work done will be:

$$0.550 \times 4{,}829 = 2{,}656 \text{ ft.-pds. for the head end,}$$

and

$$0.035 \times 4{,}673 = 163 \text{ ft.-pds. for the crank end.}$$

It is assumed that the engine is driving a turbine pump or electric generator which offers a constant resisting torque, so that the corresponding work output is $\frac{18}{360} = \frac{1}{20}$ of the total work represented by the two diagrams, and is 1,079 ft.-pds.

The quantities are set down in the table below.

θ degrees	Diagram areas		Work done on piston			Work done by crank, ft.-pds.	Net work producing change of kinetic energy, ft.-pds.
	Head a_1, sq. in.	Crank a_2, sq. in.	Head, ft.-pds.	Crank, ft.-pds.	Total, ft.-pds.		
36 54	0.550	0.035	2,656	163	2,493	1,079	1,414

The total combined areas of the two diagrams represent 21,584 ft.-pds., and since the speed was 86 revolutions per minute the indicated horsepower was $\frac{21{,}584}{33{,}000} \times 86 = 56.2$ hp.

The quantity in the last column is the difference between 2,493 and 1,079 and would evidently cause the machine to speed up.

Then the work available for increasing the energy is 1,414 ft.-pds. and this must represent the gain in kinetic energy of the machine, or

$$\delta E = +\ 1{,}414 \text{ ft.-pds.}$$

The gain in angular velocity may now be computed. The average value of J is

$$J = \tfrac{1}{2}\,[2{,}410.9 + 2{,}416.8] = 2{,}413.8$$

hence $\quad J.\omega = 2{,}413.8 \times 9 = 21{,}724.6$

and $\quad \tfrac{1}{2}\,\omega^2.\delta J = \tfrac{1}{2} \times 9^2 \times 5.9 = 238.9$

therefore

$$\delta\omega = \frac{\delta E - \tfrac{1}{2}\,\omega^2 \delta J}{J\omega} = \frac{1{,}414 - 238.9}{21{,}724.6} = 0.0541 \text{ radians per second}$$

which is the gain in velocity during the period considered. Similarly the results may be obtained for other periods, and thus for the whole revolution. These results are set down in the table given on page 257.

211. Speed-variation Diagram.—The values of $\delta\omega$ thus obtained are then plotted on a straight-line base, Fig. 154, which has been divided into 20 equal parts to represent each 18° of crank angle. If it is assumed that the speed variation is small, as it always must be in engines, then no serious error will be made by assuming that these crank angles are passed through in equal times, and hence that the base of the diagram on which the values of $\delta\omega$ are plotted is also a time base, equal distances along which represent equal intervals of time.

If desired, the equal angle base may be corrected for the variations in the velocity, using the values of $\delta\omega$ already found, so as to make the base exactly represent time intervals, but the author does not think it worth the labor and has made no correction of this kind on the diagram shown.

Attention should here be drawn to the fact that the height of the original base used for plotting the speed-variation curve has to be chosen at random, but after the curve has been plotted, it is necessary to find a line on this diagram representing the mean speed of rotation, $\omega = 9$. This may be readily done by finding the area under the curve by a planimeter, or otherwise, and then locating the line $\omega = 9$ so that the positive and negative areas

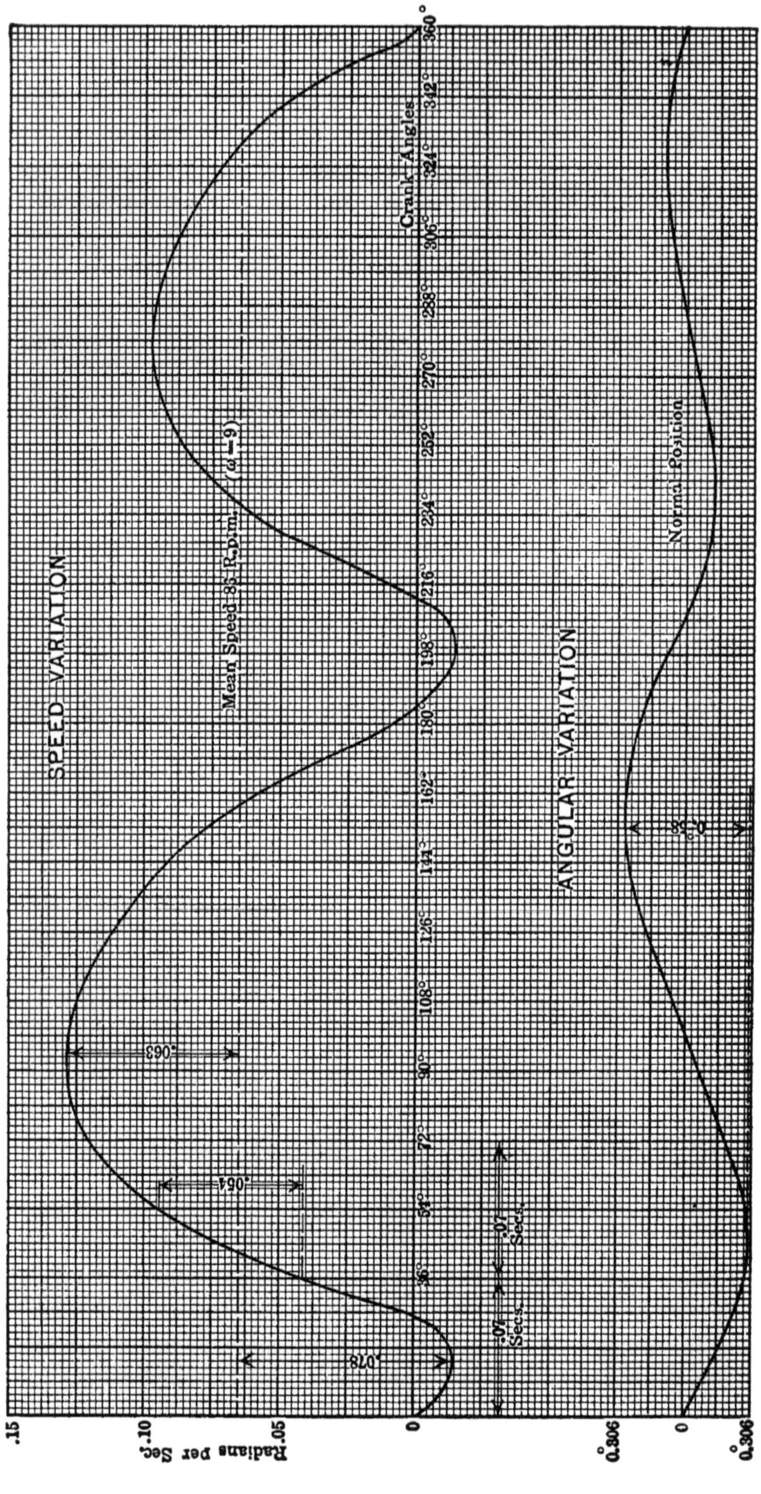
SPEED VARIATION
ANGULAR VARIATION
Mean Speed 85 R.p.m. (ω=9)
Crank Angles
Normal Position
Radians per Sec.
.15
.10
.05
0
.063
.064
.078
.07 Secs.
.07 Secs.

FIG. 154.

TABLE OF RESULTS FOR 12 1/16 BY 30-IN. ENGINE FOR DETERMINING SPEED FLUCTUATIONS

θ, deg.	K_b, ft.	$I'_b = m_b k^2_b$	OQ', ft.	$I'_c = m_c \times OQ^2$	I_a	$J = I_a + I'_b + I'_c$	δJ	Work done by steam on piston			Work output to generator, ft.-pds.	Net work producing change of kinetic energy, δE ft.-pds.	$\frac{1}{2}\omega^2 \delta J$	$\delta E - \frac{1}{2}\omega^2 \delta J$	Angular velocity variation $\delta\omega$ radians per second	
								Positive work, ft.-pds.	Negative work, ft.-pds.	Net work, ft.-pds.					+	−
0	0.76	3.15	0.00	0.00	2,400	2,403.2	+2.2	893	47	846	In each interval $= \frac{21,584}{20} = 1,079$	− 233	+ 93.1	− 326.1	......	0.0151
18	0.84	3.84	0.45	1.58	2,400	2,405.4	+5.5	2,583	117	2,466		+1,387	+222.7	+1,164.3	0.0536	
36	1.00	5.44	0.84	5.49	2,400	2,410.9	+5.9	2,656	163	2,493		+1,414	+238.9	+1,175.1	0.0541	
54	1.15	7.20	1.11	9.58	2,400	2,416.8	+3.7	1,956	178	1,778		+ 699	+149.8	+ 549.2	0.0252	
72	1.24	8.37	1.25	12.15	2,400	2,420.5	+0.2	1,424	159	1,265		+ 186	+ 8.1	+ 177.9	0.0082	
90	1.25	8.51	1.25	12.15	2,400	2,420.7	−2.2	1,038	150	888		− 191	− 89.1	− 101.9	......	0.0047
108	1.18	7.58	1.13	9.93	2,400	2,418.5	−5.8	749	131	618		− 461	−234.9	− 226.1	......	0.0104
126	1.06	6.12	0.92	6.58	2,400	2,412.7	−4.8	531	98	433		− 646	−194.4	− 451.6	......	0.0208
144	0.93	4.71	0.64	3.19	2,400	2,407.9	−2.5	314	93	221		− 858	−101.2	− 756.8	......	0.0350
162	0.81	3.57	0.33	0.85	2,400	2,404.4	−1.2	72	70	2		−1,077	− 48.6	−1,028.4	......	0.0475
180	0.76	3.15	0.00	0.00	2,400	2,403.2	+1.2	607	48	559		− 520	+ 48.6	− 568.6	......	0.0263
198	0.81	3.57	0.33	0.85	2,400	2,404.4	+2.5	1,878	121	1,757		+ 678	+101.2	+ 576.8	0.0266	
216	0.93	4.71	0.64	3.19	2,400	2,407.9	+4.8	2,584	145	2,439		+1,360	+194.4	+1,165.6	0.0537	
234	1.06	6.12	0.92	6.58	2,400	2,412.7	+5.8	1,986	169	1,817		+ 738	+234.9	+ 503.1	0.0231	
252	1.18	7.58	1.13	9.93	2,400	2,418.5	+2.2	1,542	169	1,373		+ 294	+ 89.1	+ 204.9	0.0094	
270	1.25	8.51	1.25	12.15	2,400	2,420.7	−0.2	1,215	169	1,046		− 33	− 8.1	− 24.9	......	0.0011
288	1.24	8.37	1.25	12.15	2,400	2,420.5	−3.7	916	169	747		− 332	−149.8	− 182.2	......	0.0084
306	1.15	7.20	1.11	9.58	2,400	2,416.8	−5.9	654	121	533		− 546	−238.9	− 307.1	......	0.0141
324	1.00	5.44	0.84	5.49	2,400	2,410.9	−5.5	378	96	282		− 797	−222.7	− 574.3	......	0.0265
342	0.84	3.84	0.45	1.58	2,400	2,405.4	−2.2	93	72	21		−1,058	− 93.1	− 964.9	......	0.0446
360	0.76	3.15	0.00	0.00	2,400	2,403.2										
							Totals	....	....	21,584	....	0.00	0.00	....	0.2539	0.2545

between this new line and the velocity-variation curve are equal. It is to be remembered that the computation gives the gain in velocity in each interval, and the result is plotted from the end of the curve, and not from the base line in each case.

212. Angular-space Variation.—Now since the space traversed is the product of the corresponding velocity and time, the angular-space variation, $\delta\theta$ in radians, is found by multiplying the value of $\delta\omega$ by the time t in seconds required to turn the crank through the corresponding 18°, that is

$$\delta\theta = t.\delta\omega \text{ radians.}$$

But $t.\delta\omega$ is evidently an area on the curve of angular-velocity variation, so that the angular space variation in radians up to any given crank angle, say 54°, is simply the area under the angular-velocity variation curve up to this point, the area being taken with the mean angular velocity as a base and not with the original base line. In this case the area between the mean speed line, $\omega = 9$, and the speed-variation curve, from 0 to 54°, when reduced to proper units, represents 0.275 radian as plotted in the lower curve of Fig. 154.

The upper curve shows that the minimum angular velocity was 8.922 radians per second; while the maximum was 9.063 radians per second, a variation of 0.141 radian per second, or 1.57 per cent.

The lower curve shows the angular swing of the flywheel about its mean position, and shows that the total swing between the two extremes was 0.58°, although the swing from the mean position would be only about one-half of this.

The complete computations which have been given here in full for an engine, will, it is hoped, clearly illustrate the method of procedure to be followed in any case. The method is not as lengthy as would appear at first, and the results for an engine may be quickly obtained by the use of a slide rule and drafting board.

In the case of engines, all moving parts have relatively high velocity, and it is generally advisable to take account of the variations in the reduced inertia, J. In other machines, such for example as a belt-driven punch, all parts are very slow-moving with the exception of the shaft carrying the belt pulleys and flywheel, and in such a case it is only necessary to take account of the inertia of the high-speed parts. Wherever the parts are of large size or weight or run at high speed, account must be taken of their effect on the machine.

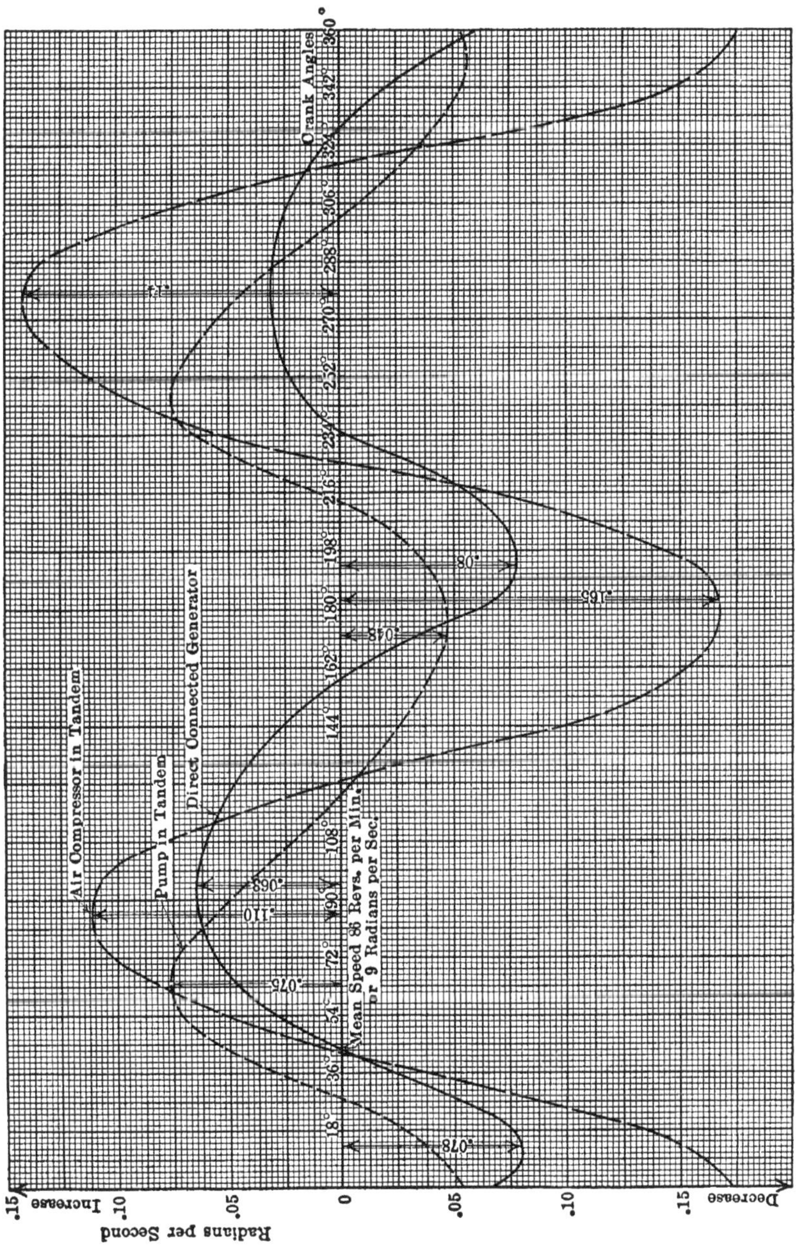

Frequently only the angular-velocity variation is required, but usually the space variation is also necessary, as in the case of alternators which are to work in parallel.

213. Factors Affecting the Speed Fluctuations.—A general discussion has been given earlier in this chapter of the factors that affect the magnitude of the speed fluctuations in machinery and as an illustration here Fig. 155 has been drawn. This figure shows three speed-fluctuation curves for the engine just referred to, and for the same indicator diagrams as are shown in Fig. 153, but in each case the engine is used for a different purpose. The curve in the plain line is an exact copy of the upper curve in Fig. 154 and represents the fluctuations which occur when the engine is direct-coupled to an electric generator, the total fluctuation being 0.14 radian per second or about 1.57 per cent.

The dotted curve corresponds to a water pump connected in tandem with the engine, a common enough arrangement, although the piston speed is rather too high for this class of work. The speed fluctuation here would be less than before, amounting to 0.123 radian or about 1.37 per cent., this being due to the fact that the unbalanced work is not so great in this class of resistance as in the generator.

The broken line corresponds to an air-compressor cylinder in tandem with the steam cylinder and the resulting variation is 0.305 radian per second or 3.38 per cent., which is over twice as much as the first case.

QUESTIONS ON CHAPTER XIII

1. A 12-in. round cast-iron disk 2 in. thick has a linear velocity of 88 ft. per second; find its kinetic energy. What would be its kinetic energy if it also revolved at 100 revolutions per minute?

2. A straight steel rod 2 ft. long, 1½ in. diameter, rotates about an axis normal to its center line, and 6 in. from its end, at 50 revolutions per minute. What is its kinetic energy?

3. Find the kinetic energy of a wheel 12 in. diameter, density 2.14, at 500 revolutions per minute.

4. What is the kinetic energy of a cast-iron wheel 3 ft. diameter, 1½ in. thick, rolling on the ground at 8 miles per hour?

5. If the side rod of a locomotive is 5 ft. long and of uniform section 2½ by 5 in., with drivers 60 in. diameter, and a stroke of 24 in., find the kinetic energy of the rod in the upper and lower positions.

6. Show how to find the kinetic energy of the tool sliding block of the Whitworth quick-return motion.

7. Suppose the wheel in question 3 is a grinder used to sharpen a tool and that its speed is decreased in the process to 450 revolutions in 1 sec.; what is the change in kinetic energy?

8. Plot the speed and angular velocity-variation curves for two engines like that discussed in the text with cranks at 90°, only one flywheel being used.

9. Repeat the above with cranks at 180°.

CHAPTER XIV

THE PROPER WEIGHT OF FLYWHEELS

214. Purpose of Flywheels.—In the preceding chapter a complete discussion has been given as to the causes of speed fluctuations in machinery and the method of determining the amount of such fluctuation. In many cases a certain machine is on hand and it is the province of the designer to find out whether it will satisfy certain conditions which are laid down. This being the case the problem is to be solved in the manner already discussed, that is, the speed fluctuation corresponding to the machine and its methods of loading are to be determined.

Frequently, however, the converse problem is given, that is, it is required to design a machine which will conform to certain definite conditions; thus a steam engine may be required for driving a certain machine at a given mean speed but it is also stipulated that the variation in speed during a revolution must not exceed a certain amount. Or a motor may be required for driving the rolls in a rolling mill, the load in such a case varying so enormously, that, if not compensated for would cause great fluctuations in speed in the motor, which fluctuations might be so bad as to prevent the use of the motor for the purpose. In a punch or shear undue fluctuation in speed causes rapid destruction of the belt. In all the above and similar cases these variations must be kept within certain limits depending upon the machine.

In all machines certain dimensions are fixed by the work to be done and the conditions of loading, and are very little affected by the speed variations. Thus, the diameter of the piston of an engine depends upon the power, pressure, mean speed, etc., and having determined the diameter, the thickness and therefore the weight is fixed by the consideration of strength almost exclusively; the same thing is largely true regarding the crosshead, connecting rod and other parts, the dimensions, weights and shapes being independent of the speed fluctuations. Similar statements may be made about the motor, its bearings, armature, etc., being fixed by the loading, and in a punch the size of gear teeth and other parts are also independent of the speed fluctuation.

Each of these machines contains also a flywheel, the dimensions of which depend on the speed variations alone and not upon the power or pressures as do the other parts. The function of the flywheel is to limit these variations; thus on a given size and make of engine the weight of flywheel will vary greatly with the conditions of working; in some cases the wheel would be very heavy, while in other cases there might be none at all on the same engine.

Ordinarily the flywheel is made heavy and run with as high a rim speed as is deemed safe; in slow-revolving engines the diameter is generally large, while in higher-speed engines the diameter is smaller, as in automobile engines, etc. The present chapter is devoted to the method of determining the dimensions of flywheel necessary to keep the speed fluctuations in a given case within definitely fixed limits.

Referring to Chapter XIII, Sec. 203, the kinetic energy of a machine is given by the equation $E = \frac{1}{2}J\omega^2$, where J is the reduced inertia found as described therein. The method of obtaining E has also been fully explained; it depends upon the input and output of the machine, such, for example, as the indicator and load curves for a steam engine. E and J are thus assumed known and the above equation may then be solved for ω, thus, $\frac{1}{2}\omega^2 = \frac{E}{J}$.

215. General Discussion of the Method Used.—In order that the matter may be most clearly presented it will be simplest to apply it to one particular machine and the one selected is the reciprocating engine, because it contains both turning and sliding elements and gives a fairly general treatment. In almost all machines there are certain parts which turn at uniform speed about a fixed center and which have a constant moment of inertia, such as the crank and flywheel in an engine, while other parts, such as the connecting rod, piston, etc., have a variable motion about moving centers and a correspondingly variable reduced moment of inertia; the table in the preceding chapter illustrates this. It will be convenient to use the symbol J_a to represent the moment of inertia of the former parts, while J_b represents that of the latter, and thus J_a is constant for all positions of the machine, and J_b is variable. The total reduced inertia of the machine is $J = J_a + J_b$. Both of these quantities

J_a and J_b are independent of the speed of rotation and depend only upon the mass and shape of the links, that is upon the relative distribution of the masses about their centers of gravity.

Suppose now that for any machine the values of J are plotted on a diagram along the x-axis, the ordinates of which diagram represent the corresponding value of the energy E; this will give a diagram of the general shape shown at Fig. 156. where the curve

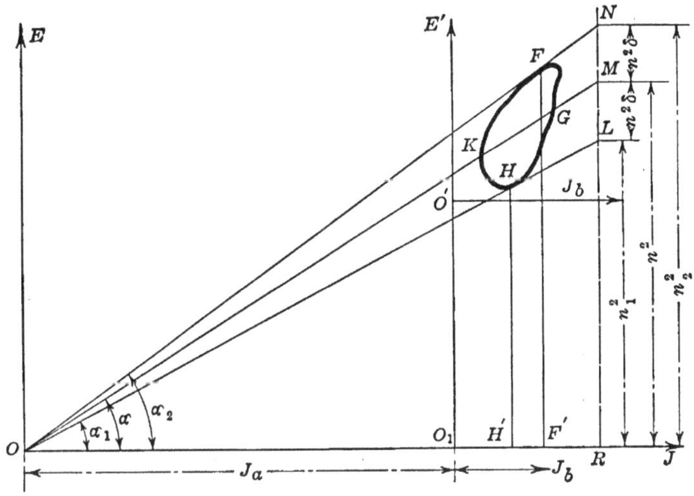

FIG. 156.

represents J for the corresponding value of E shown on the vertical line.[1]

Looking now at the figure $KFGHK$, it is evident from construction that its width depends on the values of J at the instant and is thus independent of the speed. Also, the height of this figure depends on the difference between the work put into the machine and the work delivered by the machine during given intervals, that is, it will depend on such matters as the shapes of the indicator and load curves. The shape of the input work diagrams within certain limits depends on whether the machine is run by gas or steam, and on whether it is simple or compound, etc., but for a given engine this is also, generally speaking, independent of the speed: the load curve will, of course, depend on what is being driven, whether it is dynamo, compressor, etc. Thus the height of the figure is also independent of the speed.

[1] This form of diagram appears to be due to WITTENBAUER; see "Zeitschrift des Vereines deutscher Ingenieure" for 1905.

It will further be noted that the shape of the figure does not depend on J_a, which is constant for a given machine, but only on the values of the variable J_b; hence the shape of this figure will be independent of the weight of the flywheel and speed, in so far as the input and load curves are independent of the speed, depending solely on the reciprocating masses, the connecting rod, the input-work diagrams and the load curves.

Now draw from O the two tangents, OF and OH, to $KFGH$, touching it at F and H respectively, then for OH the energy $E_1 = HH'$, and $J_1 = OH'$ and (Sec. 203), $\frac{1}{2}\omega_1^2 = \frac{E_1}{J_1} = \tan \alpha_1$, and since α_1 is the least value such an angle can have it is evident that ω_1 is the minimum speed of the engine. Similarly, $E_2 = FF'$ and $J_2 = OF'$, and $\frac{1}{2}\omega_2^2 = \frac{E_2}{J_2} = \tan \alpha_2$ and hence, ω_2 would be the maximum speed of the engine, since α_2 is the maximum value of α.

216. Dimensions of the Flywheel.—Suppose now that it is required to find the dimensions of a flywheel necessary for a given engine which is to be used on a certain class of service, the mean speed of rotation being known. The class of service will fix the variations allowable and the mean speed; in engines driving alternators for parallel operation the variation must be small, while in the driving of air compressors and plunger pumps very much larger variations are allowable. Thus, the class of service fixes the speed variation $\omega_2 - \omega_1$ radians per second, and the mean speed $\omega = \frac{\omega_2 + \omega_1}{2}$ is fixed by the requirements of the output. Experience enables the indicator diagrams to be assumed with considerable accuracy and the load curve will again depend on what class of work is being done.

The only part of the machine to be designed here is the flywheel, and as the other parts are known, and the indicator and load curves are assumed, the values of E and J_b are found as explained in Chapter *XIII* and the $E - J_b$ curve is drawn in. In plotting this curve the actual value of E is not of importance, but any point may arbitrarily be selected as a starting point and then the values of δE, or the change in E, and J_b will alone give the desired curve. Thus, in Fig. 156 the diagram $KFGH$ has been so drawn and it is to be observed that the exact position of this figure with regard to the origin O is unknown until J_a is known, but it is J_a that is sought. A little consideration will show, however, that an axis $E'O_1$ may be selected and used as the axis for plotting J_b, values of which may be laid off to the right.

Further, any horizontal axis $O' - J_b$ may be selected, and for any value of J_b a point may be arbitrarily selected to represent the corresponding value of E and the meaning of this point may be later determined. Having selected the first point, the remaining points are definitely fixed, since the change in E corresponding to each change in J_b is known. Thus, the curve may be found in any case without knowing J_a or the speed, but the origin O has its position entirely dependent upon both, and cannot be determined without knowing them. Thus the correct position of the axes of E and J are as yet unknown, although their directions are fixed.

Having settled on ω_1 and ω_2, two lines may be drawn tangent to the figure at H and F and making the angles α_1 and α_2 respectively, with the direction $O' - J_b$, where $\tan \alpha_1 = \frac{1}{2}\omega_1^2$ and $\tan \alpha_2 = \frac{1}{2}\omega_2^2$. The intersection of these two lines gives O and hence the axis of E, so that the required moment of inertia of the wheel may be scaled from the figure, since $J_a = OO_1$. It should, however, be pointed out that if the position of the axis of E is known, and also the mean speed ω, it is not possible to choose ω_1 and ω_2 at will, for the selection of either E or the speeds will determine the position of O. In making a design it is usual to select ω and $\frac{\omega_2 - \omega_1}{\omega}$, which give ω_1 and ω_2, and from the chosen values to determine the position of O and hence the axes of E and J. The mean speed ω corresponds with the angle α.

Draw a line $NMLR$ perpendicular to OJ, close to the $E - J$ diagram but in any convenient position. Then $\frac{LR}{OR} = \tan \alpha_1$, $\frac{NR}{OR} = \tan \alpha_2$ and $\frac{MR}{OR} = \tan \alpha$, so that on some scale which may be found, LR represents ω_1^2, or the square of the speed n_1 in revolutions per minute, NR represents n_2^2 and MR represents the square of the mean speed n all on the same scale. As in engines the difference between n_1 and n_2 is never large it is fairly safe to assume $2n^2 = n_2^2 + n_1^2$ or that M is midway between N and L.

217. Coefficient of Speed Fluctuation.—Using now δ to denote the **coefficient of speed fluctuation,** then δ is defined by the relation

$$\delta = \frac{n_2 - n_1}{n}.$$

Now

$$\delta = \frac{n_2 - n_1}{n} = \frac{n_2 - n_1}{\frac{1}{2}(n_2 + n_1)} = 2\frac{n_2^2 - n_1^2}{(n_2 + n_1)^2}.$$

Therefore

$$\delta = 2\frac{n_2^2 + n_1^2}{(2n)^2} = \frac{n_2^2 - n_1^2}{2n^2},$$

or

$$2\delta = \frac{n_2^2 - n_1^2}{n^2}.$$

But it has already been shown that

$$\tfrac{1}{2}\omega_1^2 = \frac{E_1}{J_1} = \tan \alpha_1 \text{ and since } \omega_1 = \frac{2\pi n_1}{60},$$

therefore

$$\frac{4\pi^2 n_1^2}{2 \times 60^2} = \tan \alpha_1,$$

or

$$n_1^2 = \frac{2 \times 60^2}{4\pi^2} \tan \alpha_1 = 182.3 \tan \alpha_1.$$

Similarly, $n_2^2 = 182.3 \tan \alpha_2$; thus the speeds depend on α only.

Since in Fig. 156 the base OR is common to the three triangles with vertices at N, M and L, it follows that

$$RL = OR \tan \alpha_1 = OR \times \frac{n_1^2}{182.3} = Cn_1^2$$

and $RN = Cn_2^2$ where $C = \dfrac{OR}{182.3}$ in both cases. Further generally, $RM = Cn^2$.

Then, referring to the formula for 2δ, which is $2\delta = \dfrac{n_2^2 - n_1^2}{n^2}$, this may be put into the following form:[1]

$$2\delta = \frac{n_2^2 - n_1^2}{n^2} = \frac{\dfrac{RN}{C} - \dfrac{RL}{C}}{\dfrac{RM}{C}} = \frac{RN - RL}{RM} = \frac{NL}{RM}.$$

Thus $NL = 2\delta \times RM$. These are marked in Fig. 156.

[1] It is instructive to compare this investigation with the corresponding one for governors given in Sec. 183 and Fig. 127*a*.

In general, $\alpha_2 - \alpha_1$ is a small angle in practice, in which case M may be assumed midway between N and L without serious error, and on this assumption

$$NM = ML = n^2 \times \delta.$$

The foregoing investigation shows that the shape of the $E - J$ diagram has a very important effect on the best speed for a given flywheel and the best weight of flywheel at the given speed. Thus, Fig. 158 shows one form of this curve for an engine to be discussed later, while Fig. 160 shows two other forms of such curves for the same engine but different conditions of loading. With such a curve as that on the right of Fig. 160, the best speed condition will be obtained where the origin O is located along the line through the long axis of the figure. In order to make this more clear, this figure is reproduced again on a reduced scale at Fig. 157 and several positions of the origin O are drawn in. This matter will now be discussed.

218. Effects of Speed and Flywheel Weight.—Two variables enter into the problem, namely the best speed and the most economical weight of flywheel. Now, the formula connecting the speed with the angle α is $\frac{1}{2}\omega^2 = \tan \alpha$, Sec. 215, so that the speed depends upon the angle α alone, and for any origin along such a line as OF there is the same mean speed since α is constant for this line. To get the maximum and minimum speeds corresponding to this mean speed, tangents are drawn from O to the figure giving the angles α_1 and α_2 and hence ω_1 and ω_2. A glance at Fig. 157 shows that the best speed corresponds to the line OF and that for any other origin such as O_1, which represents a lower mean speed, since for it α and hence $\tan \alpha$ is smaller, there will be a greater difference between ω_1 and ω_2 in relation to ω than there is for the origin at O. A few cases have been drawn in, and it is seen that even for the case O_4 which represents a higher mean speed than O the value of δ will be increased; thus the best speed corresponds to the line OF and its value is found from $\frac{1}{2}\omega^2 = \tan \alpha$.

But the speed variations also depend on the weight of the flywheel and hence upon the value of J or the horizontal distance of the origin from the axis $O'E'$. If the origin was at O_4, there would be no flywheel at all but the speed variation taken from a scaled drawing, would be prohibitive as it is excessively large. For the position O the inertia of the flywheel is represented by

$O - O_4$ and the speed variations would be comparatively small, but if the origin is moved up along OF to O_3, the speed being the same as at O, the variations will be increased very slightly, but the flywheel weight also shows a greater corresponding decrease. Similarly, O_1 corresponding to the heaviest wheel, shows a variation in excess of O and nearly equal to O_3, and O_2 with the same

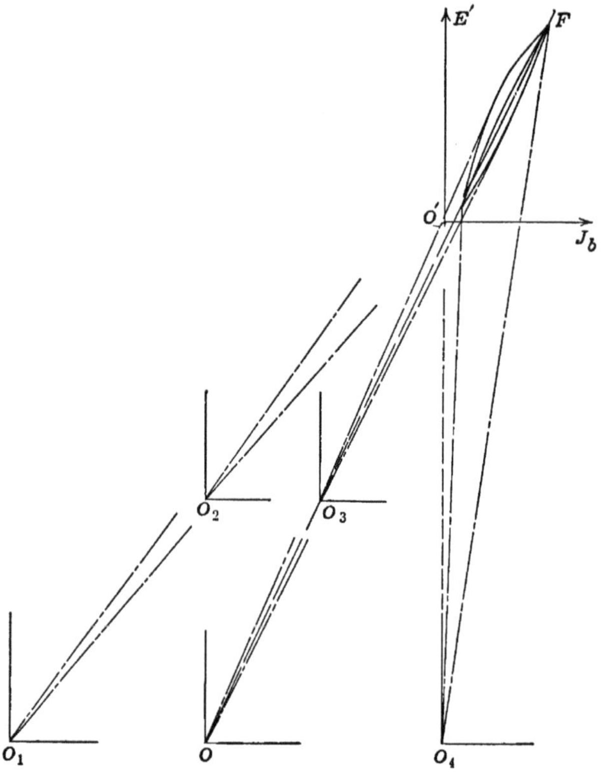

FIG. 157.—Effect of speed and weight of flywheel.

weight of wheel as at O shows nearly double the variation that O does.

Thus, increasing the weight of the wheel may increase the speed variations if the speed is not the best one, and increasing the speed may produce the same result, but at the speed represented by OF, the heavier the wheel the smaller will be the variation, although the gain in steadiness is not nearly balanced by the extra weight of the wheel beyond a certain point. Frequently the operating conditions prevent the best speed being selected, and if this is so it is clear that the weight of the wheel must be neither too large nor too small.

These results may be stated as follows: **For a given machine and method of loading there is a certain readily obtained speed which corresponds to minimum speed variations, and for this best value the variations will decrease slowly as the weight of flywheel is increased. For a certain flywheel weight the speed variations will increase as the speed changes either way from the best speed, and an increase in the weight of the flywheel does not mean smaller fluctuation in speed unless the mean speed is suitable to this condition.**

219. Minimum Mean Speed.—The above results are not quite so evident nor so marked in a curve like Fig. 158 but the same

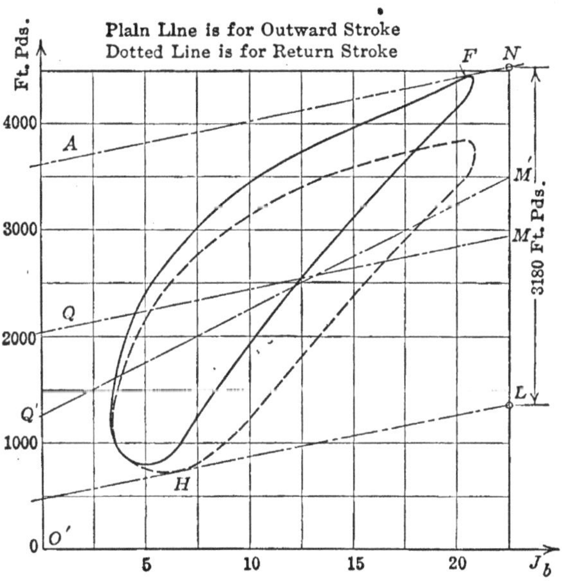

FIG. 158.—Steam engine with generator or turbine pump load.

conditions hold in this case also. The best speed is much more definitely fixed for an elongated $E-J$ curve and becomes less marked as the boundary of the curve comes most nearly to the form of a circle. The foregoing investigation further shows that no point of the $E-J$ curve can fall below the axis $O-J$, because if it should cut this axis, the machine would stop. The minimum mean speed at which the machine will run with a given flywheel will be found by making the axis $O-J$ touch the bottom of the curve, and finding the corresponding mean speed; the minimum speed will, of course, be zero, since $\alpha_2 = 0$. The minimum speed of operation may be readily computed for Figs. 158 and 160 and it is at once seen that the right-hand diagram of Fig. 160 corresponds to a larger minimum speed than any of the others, that is, when driving the air compressor the engine will stop at a higher mean speed than when driving the generator.

220. Numerical Example of a Steam Engine.—The principles already explained may be very well illustrated in the case of the steam engine used in the last chapter, which had a cylinder 12$\frac{1}{16}$ in. diameter and 30 in. stroke and a mean speed of 87 revolutions per minute for which $\omega = 9$ radians per second. The form of indicator diagrams and loading are assumed as before and the engine drives a turbine pump which is assumed to offer constant resisting torque. The weight of the flywheel is required.

Near the end of Chapter *XIII* is a table containing the values of J and δE for equal parts of the whole revolution and for convenience these results are set down in the table given herewith.

TABLE OF VALUES OF J and E for 12$\frac{1}{16}$ BY 30-IN. ENGINE

θ, degrees	J, total	$J_b = J - 2{,}400$	δE, foot-pounds
0	2,403.2	3.2	
18	2,405.4	5.4	− 233
36	2,410.9	10.9	+1,387
54	2,416.8	16.8	+1,414
72	2,420.5	20.5	+ 699
90	2,420.7	20.7	+ 186
108	2,418.5	18.5	− 191
126	2,412.7	12.7	− 461
144	2,407.9	7.9	− 646
162	2,404.4	4.4	− 858
180	2,403.2	3.2	−1,077
198	2,404.4	4.4	− 520
216	2,407.9	7.9	+ 678
234	2,412.7	12.7	+1,360
252	2,418.5	18.5	+ 738
270	2,420.7	20.7	+ 294
288	2,420.5	20.5	− 33
306	2,416.8	16.8	− 332
324	2,410.9	10.9	− 546
342	2,405.4	5.4	− 797
360	2,403.2	3.2	−1,058

Selecting the axes $O' - E'$ and $O' - J'_b$, Fig. 158, the corresponding $E - J$ curve is readily plotted as follows: The table shows that when $\theta = 0°$, $J_b = 3.2$ and when $\theta = 18°$, $J_b = 5.4$, the gain in energy which is negative, during this part of the revolution being $\delta E = -233$ ft.-pds. Starting with $J_b = 3.2$ and arbitrarily calling E at this point 1,000 ft.-pds. gives the first point on the diagram; the second point is found by remembering that when J_b has reached the value 5.4 the energy has decreased by 233 ft.-pds., so that the point is located on the line $J_b = 5.4$ and 233 ft.-pds. below the first point. The third point is at $J_b = 10.9$ and 1,387 ft.-pds. above the second point and so on.

Now draw on the diagram the line QM to represent the mean speed $\omega = 9$, its inclination to the axis of J_b being α where $\tan \alpha = \frac{1}{2}\omega^2 - \frac{1}{2} \times 9^2 = 40.5$. The actual slope on the paper is readily found by noticing that the scales are so chosen that the same length on the vertical scale stands for 1,000 as is used on the horizontal scale to represent 5, the ratio being $\frac{1{,}000}{5} = 200$; then the actual slope of QM on the paper is $\frac{40.5}{200} = 0.2025$, which enables the line to be drawn. This line may be placed quite accurately by making the perpendicular distance to it from the extreme lower point on the figure equal the perpendicular to it from F (see Figs. 156 and 158). Thus the position and direction of the mean speed line QM are known.

Now suppose the conditions of operation require that the maximum speed shall be 1.6 per cent. above the minimum speed, or that the coefficient of speed fluctuation shall be 1.6 per cent. Then, from Sec. 217, $\delta = 0.016$, that is $\delta = \frac{\omega_2 - \omega_1}{\omega} = 0.016$ and the problem also states that the mean speed shall be $\omega = 9 = \frac{\omega_2 + \omega_1}{2}$. On comparing these two results it is found that $\omega_2 = 9.072$ and $\omega_1 = 8.928$.

On substituting these two values in the equations for the angles, the results are $\tan \alpha_1 = \frac{1}{2}\omega_1^2 = \frac{1}{2} \times 79.709 = 39.854$ and $\tan \alpha_2 = \frac{1}{2}\omega_2^2 = \frac{1}{2} \times 82.301 = 41.150$ which enables the two lines HL and AF to be drawn tangent to the figure at H and F and at angles α_1 and α_2 respectively to the axis of J_b (on the paper the tangents of the slopes of these lines will be, for $AF = \frac{41.150}{200} = 0.2058$ and for $HL = \frac{39.854}{200} = 0.1993$). These lines are so nearly parallel that their distance apart vertically can be measured anywhere on the figure, and it has actually been measured along NML, the distance NL representing 3,180 ft.-pds.

Referring again to Fig. 156 it is seen that $NR = OR \tan \alpha_2$ and $LR = OR \tan \alpha_1$ and by combining these it may be shown that $OR = \frac{NL}{\tan \alpha_2 - \tan \alpha_1}$. Substituting the results for this problem give

$$OR = \frac{3,180}{41.150 - 39.854} = 2,453 = J_a + J_b = J_a + 25.$$

Hence, the moment of inertia of the flywheel should be 2,430 approximately, which gives the desired solution of the problem.

221. Method of Finding Speed Fluctuation from $E - J$ Diagram.—The converse problem, that of finding the speed variation corresponding to an assumed value of J_a, has been solved in the previous chapter but the diagram may be used for this purpose also. Thus, let $\omega = 9$, the same mean speed as before, and $J_a = 2,000$. Then, since $\frac{1}{2}\omega^2 = \frac{E}{J} = \tan \alpha$ the value of E at M is $(2,000 + 25) \times \frac{1}{2} \times 9^2 = 82,012$. The points N and L will be practically unchanged and hence at N the value of E_2 is $82,012 + \frac{1}{2}(3,180) = 83,602$ ft.-pds. and the value of $\omega_2{}^2$ may be computed from the relation $\frac{1}{2}\omega_2{}^2 = \frac{83,602}{2,025}$ and in a similar way ω_1 may be found and the corresponding speed variation $\delta = \frac{\omega_2 - \omega_1}{\omega}$.

A somewhat simpler method may be used, however, by referring to Fig. 156, from which it appears that $NL = 2n^2\delta$. Thus, $2n^2\delta$ is represented by 3,180 ft.-pds. and n^2 by 82,012 ft.-pds., from which the value of δ is found to be 0.0194 which corresponds to a speed variation of 1.94 per cent.

In order to show the effect of making various changes, let the speed of the engine be much increased to say 136 revolutions per minute for which $\omega = 14.1$, and let the speed variation be still limited to 1.6 per cent. The line QM will then take the position $Q'M'$ for which the tangent on the paper is $\frac{1}{2}$ and the distance corresponding to LN measures 2,400 ft.-pds. On completing the computations the moment of inertia of the flywheel is found to be $J_a = 740$, that is to say that if the wheel remains of the same diameter it need be less than one-third of the weight required for the speed of 87 revolutions.

The diagram Fig. 158 has been placed on the correct axis and is shown in Fig. 159 which gives an idea of the position of the origin O for the value $J_a = 2{,}400$ and $\omega = 9$.

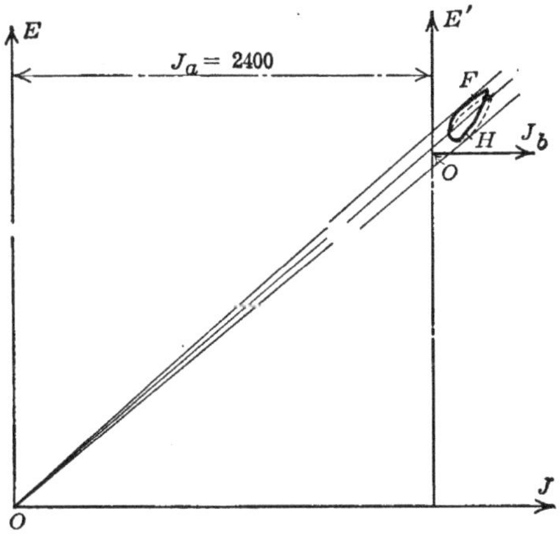

FIG. 159.

222. Effect of Form of Load Curve on Weight and Speed.—To show the effects of the form of load curve on this diagram and on the speed and weight of the flywheel, the curves shown in Fig. 160 have been drawn. The two diagrams shown there were made for the same engine and indicator diagrams as were used in Fig. 158, the sole difference is in the load applied to the engine. The left-hand diagram corresponds to a plunger pump connected in tandem with the steam cylinder, while the right-hand diagram is from an air compressor connected in tandem with the steam cylinder. The effect of the form of loading alone on the $E-J$ diagram is most marked and the air compressor especially produces a most peculiar result, the best speed here being definitely fixed and being much higher than for either of the other cases, and if the machine is run at this speed it is clear that the weight of the flywheel is not very important so long as it is not extremely small.

It is needless to say that the form of indicator diagram also produces a marked effect and both the input and output diagrams are necessary for the determination of the flywheel weight and the speed of the machine. The curves mentioned are sufficient to show that the weight of wheel and the best speed of operation depend on the kind of engine and also on the purpose for which it is used. It is frequently impossible, practically, to run an engine at the speed which gives greatest steadiness of motion and then the weight of wheel must be selected with care as outlined in Sec. 218.

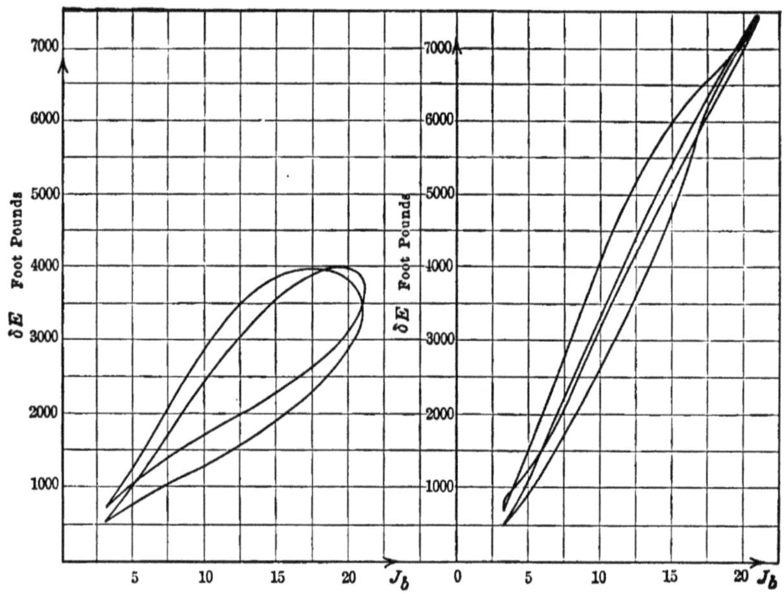

FIG. 160.—Left-hand figure is for a plunger pump in tandem with steam engine; right-hand figure is for an air compressor in tandem with steam engine.

223. Numerical Example on Four-cycle Gas Engine.—An illustration of the application to a gas engine of the four-cycle type is shown at Fig. 162, this being taken from an actual case of an engine direct-connected to an electric generator. The engine had a cylinder $14\frac{1}{2}$ in. diameter and 22 in. stroke and was single-acting; the indicator diagram for it is shown at Fig. 161. The piston and other reciprocating parts weighed 360 lb., while the weight of the connecting rod was 332 lb., and its radius of gyration about its center of gravity 1.97 ft., the latter point being 24.3 in. from the center of the crankpin, and the length of the rod between centers was 55 in.

There were two flywheels of a combined weight of 7,000 lb. and the combined moment of inertia of these and of the rotor of the generator was 1,600 (foot-pound units).

The form of the $E-J$ diagram for this case is given in Fig. 162 and differs materially in appearance from any of those yet shown, and the best speed is much more difficult to determine because of the shape of the diagram. The actual speed of the engine

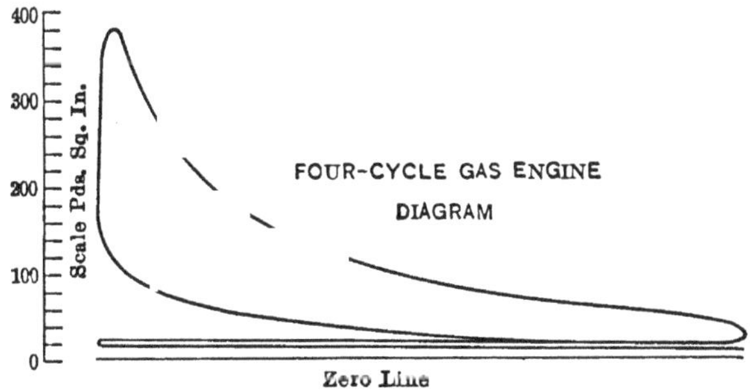

FIG. 161.

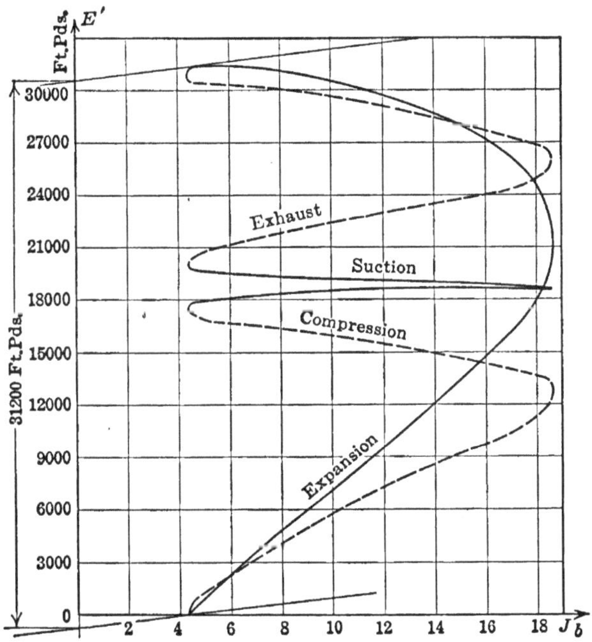

FIG. 162.—Gas engine driving dynamo.

was 172 revolutions per minute and for this value the sloping lines on the diagram have been drawn. The mean-speed line would have an inclination to the axis of J_b given by $\tan \alpha = \frac{1}{2}\omega^2 = 162$ and its slope on the paper would be $\frac{1}{1,500}$ of this, since the vertical scale is 1,500 times the horizontal; thus the tangent of the actual slope is $\frac{162}{1,500} = 0.108$ and the lines are drawn with this inclination.

The total height of the diagram is 31,200 ft.-pds. and using the value $J_a = 1,600$, the mean value of E is $162 \times 1,600 = 259,200$ ft.-pds. so that the speed variation is

$$\delta = \tfrac{1}{2} \times \frac{31,200}{259,200} = 0.0602 \text{ or } 6.02 \text{ per cent.}$$

The engine here described was installed to produce electric light and it is perfectly evident that it was entirely unsuited to its purpose as such a large speed variation is quite inadmissible. Owing to the peculiar shape of this diagram and the fact that the tangent points touch it on the left-hand side, it appears that the distance between them will not be materially changed by any reasonable change of slope of the lines, so that if the speed remains constant at 172 revolutions per minute the value of J_a or the flywheel weight is inversely proportional to the speed variation and flywheels of double the weight would reduce the fluctuation to about 3 per cent.

A change in speed would bring an improvement in conditions and the results may readily be worked out.

QUESTIONS ON CHAPTER XIV

1. Show the effect of the following: (*a*) increase in flywheel weight, constant speed; (*b*) decrease under same conditions; (*c*) increase in speed with same flywheel; (*d*) increase and decrease in both weight and speed; all with reference to a gas engine.

2. What would be the shape of the $E-J$ diagram for a geared punch, neglecting the effect of the reciprocating head?

3. If the connecting rod and piston of an engine are neglected, what would be the shape of the $E-J$ curve? What would be its dimensions in the two examples of the chapter?

4. What would be the best speed for the steam engine given in the text when driving the three different machines? At what mean speed would the engine stop in the three cases?

5. What would be the best speed for the gas engine and at what mean speed would it stop?

6. What flywheel weight would reduce the speed variation 5 per cent. for the steam engine?

7. Examine the effect on the $E-J$ diagram for the engine-driven compressor if a crank for operating the latter is set at 90° to the engine crank. What effect has this on the best speed?

CHAPTER XV

ACCELERATIONS IN MACHINERY AND DISTURBING FORCES DUE TO THE INERTIA OF THE PARTS

224. General Effects of Accelerations.—It has become a practice in modern machinery to operate it at as high a speed as possible in order to increase its output. Where the machines contain parts that are not moving at a uniform speed, such as the connecting rod of an engine or the swinging jaw of a rock crusher, the variable nature of the motion requires alternate acceleration and retardation of these parts, to produce which forces are required. These alternate accelerations and retardations cause vibrations in the machine and disturb its equilibrium; almost everyone is familiar with the vibrations in a motor boat with a single-cylinder engine, and many law-suits have resulted from the vibrations in buildings caused by machinery in shops and factories nearby.

These vibrations are very largely due to the irregular motions of the parts and to the accelerating forces due to this, and the forces increase much more rapidly than the speed, so that with high-speed machinery the determination of these forces becomes of prime importance, and they are, indeed, also to be reckoned with in slow-speed machinery, as there are not a few cases of slow-running machines where the accelerating forces have caused such disturbances as to prevent the owners operating them.

Again, in prime movers such as reciprocating engines of all classes, the effective turning moment on the crankshaft is much modified by the forces necessary to accelerate the parts; in some cases these forces are so great that the fluid pressure in the cylinder will not overcome them and the flywheel has to be drawn upon for assistance. The troubles are particularly aggravated in engines of high rotative speed and appear in a most marked way in the high-speed steam engine and in the gasoline engines used in automobiles.

The forces required to accelerate the valves of automobile engines may also be so great that the valve will not always remain in contact with its cam but will alternately leave it and return again, thus causing very noisy and unsatisfactory operation.

Specific problems involving the considerations outlined above will be dealt with later but before such problems can be solved it will be necessary to devise a means of finding the accelerations of the parts in as convenient and simple a way as possible, and this will now be discussed.

225. The Acceleration of Bodies.—The general problem of acceleration in space has not much application in machinery, so that the investigation will here be confined to a body moving in one plane, which will cover most practical cases. Let a body having weight w lb. move in a plane at any instant; its mass will be $m = \frac{w}{g}$, and by the principle of the virtual center as outlined in Chapter *III*, its motion is equivalent at any instant to one of rotation about a point in the plane of motion, which point may be near or remote according to the nature of the motion; if the point is infinitely distant the body moves in a straight line, or has a motion of translation.

226. Normal and Tangential Acceleration.—Let Fig. 163 represent a body moving in the plane of the paper and let O be its virtual center relative to the paper, O being thus the point about which the body is turning at the instant. The body is also assumed to be turning with variable speed, but at the instant when it is passing through the position shown let its angular velocity be ω radians per second. Any point P in the body will travel in a direction normal to OP, Sec. 34, in the sense indicated, as this corresponds with the sense of the angular velocity. This point P has accelerations in two directions: (*a*) Since the body is moving about O at variable angular velocity it will have an acceleration in the direction of its motion, that is, normal to OP; and (*b*) it will have an acceleration toward O even if ω is constant, since the point is being forced to move in a circle instead of a straight line. The first of these may be called the **tangential acceleration** of the point since it is the acceleration of the point along a tangent to its path, while the second is its **normal acceler-**

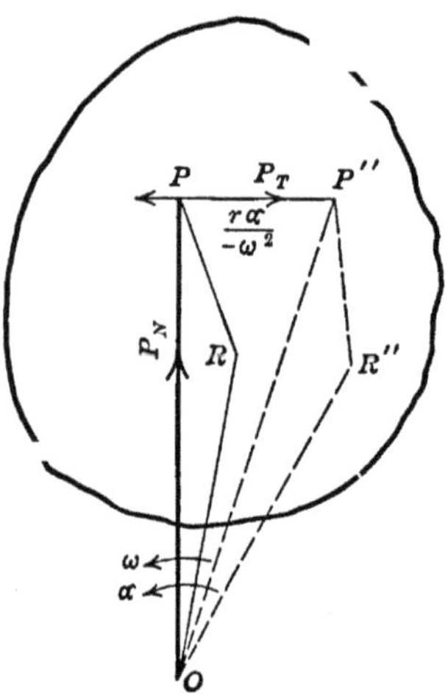

FIG. 163.

ation for similar reasons. Every point in a body rotating at a given instant has normal acceleration, no matter what kind of motion the body has, but it will only have tangential acceleration if the angular velocity of the body is variable. If the body has a motion of translation it can only have tangential acceleration.

In Fig. 164 let OP be drawn separately, its length being r ft. and at the time δt sec. later let OP be in the position OQ, the angle QOP being $\delta\theta$, so that the body has turned through the angle $\delta\theta$ radians in δt sec. The angular velocity when in the position OP is ω radians per second and in the position OQ is assumed to be $\omega + \delta\omega$ radians per second; thus the gain in angular velocity is $\delta\omega$ radians per second in the time δt, or the angular acceleration of the body is $\alpha = \frac{\delta\omega}{\delta t}$ radians per second per second. Now draw the corresponding velocity triangle as shown on the right

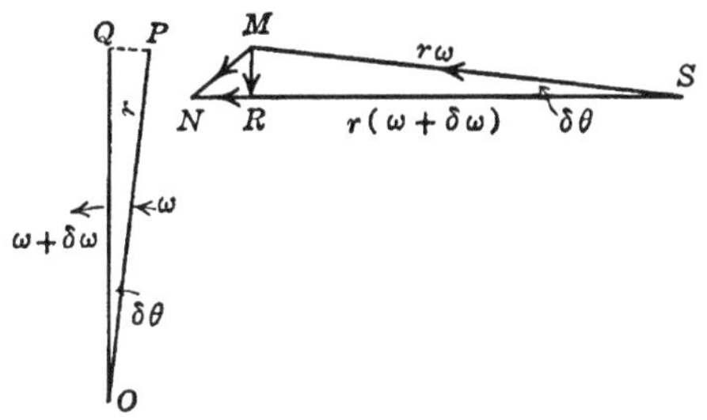

FIG. 164.

of Fig. 164, making SM normal to OP, equal to $OP \times \omega = r\omega$ ft. per second and SN normal to OQ equal to $OP(\omega + \delta\omega) = r(\omega + \delta\omega)$ ft. per second, so that the gain in linear velocity in the time δt sec. is MN ft. per second and its components in the normal and tangential directions are MR and RN ft. per second respectively.

The normal gain in velocity of the point P in the time δt is MR, so that its normal acceleration is

$$P_N = \frac{MR}{\delta t} = \frac{r\omega\delta\theta}{\delta t} = r\omega\frac{\delta\theta}{\delta t} = r\omega^2 \text{ ft. per second per second,}$$

and similarly the tangential acceleration is

$$P_T = \frac{NR}{\delta t} = \frac{SN-SR}{\delta t} = \frac{r(\omega + \delta\omega) - r\omega}{\delta t} = r\frac{\delta\omega}{\delta t} = r\alpha \text{ ft. per second}$$

per second.

The sense of P_T is determined by that of α while P_N is always radially inward toward the center O. Thus, the normal acceleration of the point is simply its instantaneous radius of rotation multiplied by the square of the angular velocity of its link while the tangential acceleration of the point is the radius of rotation of the point multiplied by the angular acceleration of its link. Where the link turns with uniform velocity $\alpha = 0$ and therefore $P_T = O$, but P_N can only be zero if ω is zero, which means that the link is at rest or has a motion of translation. In the latter case the link can only have tangential acceleration.

227. Graphical Construction.—Returning now to Fig. 163, the normal acceleration of P or P_N is $r\omega^2$ toward O, then take the length OP to represent this quantity, thus adopting the scale of $-\omega^2 : 1$; this is negative since the line OP represents the acceleration $r\omega^2$ in the direction and sense PO. Then the tangential acceleration P_T will be represented by a line normal to OP, its length will be $\frac{r\alpha}{-\omega^2}$ since the scale is $-\omega^2:1$, and its sense is to the right, since the scale is negative, hence draw $PP'' = \frac{r\alpha}{-\omega^2}$. Now if O be joined to P'' then OP'' = vector sum $OP + PP''$ or $OP'' = P_N + P_T$ which will therefore represent the total acceleration of P, that is the total acceleration of P is $P''O \times \omega^2$ in the direction and sense $P''O$. It may very easily be shown that in order to find the acceleration of any other point R on this body at the given instant it will only be necessary to locate a point R'' bearing the same relation to OP'' that R does to OP, the acceleration of R, which is represented by OR'', being $R''O.\omega^2$ and its direction and sense $R''O$. The acceleration of R with reference to P is $R''P''\omega^2$.

228. Application to Machines.—The accelerations may now be found for machines and the first case considered will be as general as possible, the machine being one of four links with four turning pairs, Fig. 165. Let the angular velocity ω and the angular acceleration α of the selected primary link a be known, it is required to find the angular accelerations of the other links as well as the linear accelerations of different points in them. From the phorograph, Chapter IV, the angular velocities of the links b and c are $\omega_b = \frac{b'}{b}\omega$ and $\omega_c = \frac{c'}{c}\omega$ respectively, and these may readily be found. Further, if α_b and α_c represent the, as yet unknown, angular accelerations in space of b and c respectively, and also if Q_N and Q_T represent respectively the normal and tangential accelerations of Q with regard to P, the point about which b is turning relative to a, and R_N and R_T have the same significance as regards R relative to Q, then the previous paragraph enables the following relations to be established:

$P_N = a\omega^2$; $P_T = a\alpha$; $Q_N = b\omega_b{}^2$; $Q_T = b\alpha_b$; $R_N = c.\omega_c{}^2$ and $R_T = c\alpha_c$.

Using the principle of vector addition the total acceleration of R with regard to O is the vector sum of the accelerations of R

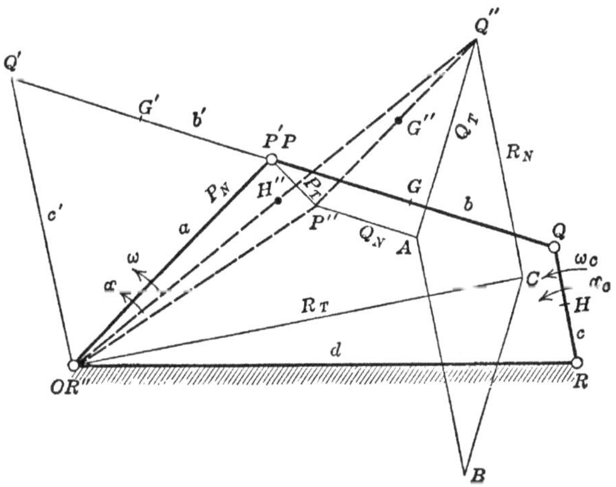

FIG. 165.

with regard to Q, of Q with regard to P and of P with regard to O. But as R and O are stationary, the total acceleration of R with regard to O is zero, hence, the sum of the above three accelerations is zero, or

$$R_T + R_N + Q_T + Q_N + P_T + P_N = 0,$$

that is, the vector polygon made up with these accelerations as its sides must close, or if the polygon be started at O it will end at O also.

Now the point P'' may be located according to the method previously given, and in order to locate Q'', giving the total acceleration of Q, proceed from P'' to O by means of the vectors $Q_N + Q_T + R_N + R_T$. The direction and sense of both Q_N and R_N are known, they are respectively QP and RQ, further, the direction, but not the sense of Q_T and of R_T is known, in each case it is normal to the link itself, or Q_T is normal to b and R_T is normal to c.

In order to represent the results graphically they may be put into the following convenient form:

$$Q_N = b\omega_b^2 = b\left(\frac{b'}{b}\omega\right)^2 = \frac{b'^2}{b} \times \omega^2$$

and

$$R_N = c\omega_c^2 = c\left(\frac{c'}{c}\omega\right)^2 = \frac{c'^2}{c} \times \omega^2$$

and remembering that the scale for the diagram is $-\ \omega^2:1$, draw $P''A = \frac{Q_N}{\omega^2} = \frac{b'^2}{b}$ and follow it with $AB = \frac{R_N}{\omega^2} = \frac{c'^2}{c}$, the negative sign having been taken into account by the sense in which these are drawn. The polygon from B to O may now be completed by adding the vectors Q_T and R_T, and as the directions of these are known, the process is evidently to draw from O the line OC in the direction R_T, that is normal to c, and from B the line BC normal to b, which is in the direction of Q_T, these lines intersecting at the point C. Then it is evident that BC represents Q_T on the scale $-\ \omega^2$ to 1, and that OC represents R_T on the same scale, so that in the diagram $OPP''ABCQ''O$ it follows that $OP = P_N$, $PP'' = P_T$, $P''A = Q_N$, $AB = R_N$, $BC = Q_T$ and $CO = R_T$, all on the scale $-\ \omega^2$ to 1. Complete the parallelogram $CBAQ''$; then $OP'' = P_N + P_T$, $P''Q'' = Q_N + Q_T$ and $Q''O = R_N + R_T$, and therefore, the vector triangle $OP''Q''R''$ gives the vector acceleration diagram of all the links on the machine.

229. Acceleration of Points.—The linear acceleration of any point such as G on b is readily shown to be represented by OG'' and to be equal to $G''O.\omega^2$, where the point G'' divides $P''Q''$ in the same way that G divides PQ, the direction and sense of the acceleration of G is $G''O\cdot$ Similarly, the acceleration of H in c is $H''O.\omega^2$ in magnitude, direction and sense where H'' divides $OQ''(R''Q'')$ in the same way as H divides RQ. In this way the linear acceleration of any point on a machine may be directly determined.

Angular Accelerations of the Links.—The angular accelerations of the links may be found as follows. Since $Q_T = AQ'' \times -\omega^2 = b\alpha_b$, then $b\alpha_b = -\ AQ''.\omega^2$ or $-\ \alpha_b = AQ'' \times \frac{\omega^2}{b}$ so that the length AQ'' represents α_b, the angular acceleration of the link

b, and similarly CO represents the angular acceleration α_c of c or $\alpha_c = -CO \times \frac{\omega^2}{c}$. The sense of these angular accelerations may be found by noticing the way one turns to them in going from the corresponding normal acceleration line; thus, in going from P_N to P_T one turns to the right, in going from $Q_N(P''A)$ to $Q_T(AQ'')$ the turn is to the left and hence α_b is in opposite sense to α, and by a similar process of reasoning α_c is in the same sense as α. Thus, in the position shown in the diagram, Fig. 165, the angular velocities are increasing for the links a and c, and that of the link b is also increasing since both α_b and ω_b are in opposite sense to α and ω.

It will be found that the method described may be applied to any machine no matter how complicated, and with comparative ease. The construction resembles the phorograph of Chapter IV, which it employs, and hence this latter chapter must be carefully read. Simple graphical methods for finding $\frac{b'^2}{b}$, etc., may be made up, one of which is shown in the applications given hereafter.

THE FORCES DUE TO ACCELERATIONS OF THE MACHINE PARTS

230. The real object of determining the accelerations of points and links in a machine is for the purpose of finding the forces which must be applied on the machine parts in order to produce these accelerations and also to learn the disturbing effects produced in the machine if the accelerations of the parts are not balanced in some way. The investigation of these disturbing effects will now be undertaken, the first matter dealt with being the forces which must be applied to the links to produce the changing velocities.

It is shown in books on dynamics, that if a body having plane motion, has a weight w lb. or mass $m = \frac{w}{g}$ and an acceleration of its center of gravity of f ft. per second per second, then the force necessary to produce this acceleration is mf pds., and this force must act through the center of gravity and in the direction of the acceleration f. In many cases the body also rotates with variable angular velocity, or with angular acceleration, in which case a torque must act on the body in any position to produce this variable rotary motion, and if the body has a moment of inertia

I about its center of gravity and angular acceleration α radians per second per second this torque must have a magnitude of $I \times \alpha$ ft.-pds. Let the mass of the link be so distributed that its radius of gyration about the center of gravity is k; then $I = mk^2$ and the torque is $mk^2\alpha$. For proof of this the reader is referred to books on dynamics.

To take a specific case let a machine with four links be selected, as illustrated in Fig. 166, and let the vector acceleration diagram

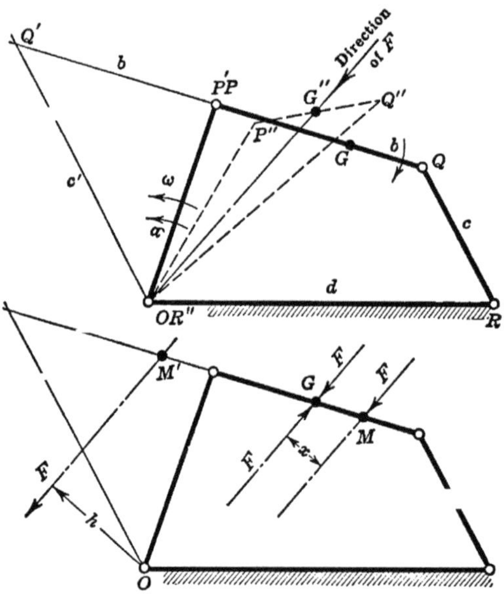

FIG. 166.—Disturbing forces due to mass of rod.

$OP''Q''O$, as well as the phorograph $OP'Q'O$ be found, as already explained; it is required to find the force which must be exerted on any link such as b to produce the motion which it has in the given position. Let G be the center of gravity of the link and let its weight be w_b lb. and its moment of inertia about G be represented by I_b in feet and pound units; then $I_b = m_b k_b^2$ where $m_b = \frac{w_b}{g}$ and k_b is the radius of gyration about the point G. From the vector diagram it is assumed that the angular acceleration α_b has been found; also the acceleration of G, which is $G''O \times \omega^2$.

To produce the acceleration of G a force must act through it of amount $F = m \times G''O \times \omega^2$ in the direction and sense $G''O$, while to produce the angular acceleration a torque T must act on the link of amount $T = I_b\alpha_b = m_b k_b^2 \alpha_b$. The torque T may be produced by a couple consisting of two parallel forces acting in opposite sense and at proper distance apart, and these forces may have any desired magnitude so long as their distance apart is adjusted to suit. For convenience let each of the forces be selected equal to F; then the distance x ft. between them will be found from the relation $T = Fx$.

Now, as this couple may act in any position on the link b let it be so placed that one of the forces passes through G and the two forces have the same direction as the acceleration of G. Further, let the force passing through G be the one which acts in opposite sense to the accelerating force F; this is shown on Fig. 166. Now the accelerating force F and one of the forces F composing the couple act through G and balance one another and thus the accelerating force and the couple producing the torque reduce to a single force F whose magnitude is $m_b.G''O.\omega^2$, whose direction and sense are the same as the acceleration of the center of gravity G of b, and which acts at a distance x from G, determined by the relation $T = Fx$, and on that side of G which makes the torque act in the same sense as the angular acceleration α_b.

The distance x of the force F from G may be found as follows:

Since $Q_T = b\alpha_b = Q''A \times \omega^2$, Fig. 165, then $\alpha_b = Q''A \times \frac{\omega^2}{b}$, because the line AQ'' represents Q_T on a scale — $\omega^2 : 1$.

Also
$$T = I_b\alpha_b = m_b k_b{}^2 \times \frac{Q''A}{b} \times \omega^2$$
and
$$F = m_b.G''O.\omega^2,$$

therefore $x = \dfrac{T}{F} = \dfrac{m_b k_b{}^2 \dfrac{Q''A}{b} \times \omega^2}{m_b.G''O.\omega^2} = \dfrac{k_b{}^2}{b} \cdot \dfrac{Q''A}{G''O}$ where $\dfrac{k_b{}^2}{b}$ is a constant, so that $x = \text{const.} \times \dfrac{Q''A}{G''O}$ which ratio can readily be found for any position of the mechanism. This gives the line of action of the single force F and, having found the position of the force, let M be its point of intersection with the axis of link b. Now find M' the image of M and move the force from M to its image M'; then the turning moment necessary on the link a to accelerate the link b is Fh, where h is the shortest distance from O to the direction of F, Fig. 166.

This completes the problem, giving the force acting on the link and also the turning moment at the link a necessary to produce this force. The same construction may be applied to each of the other links such as a and c and thus the turning moment on a necessary to accelerate the links may be found as well as the necessary force on each link itself.

DETERMINATION OF THE STRESSES IN THE PARTS DUE TO THEIR INERTIA

231. The results just obtained may be used to find the bending moment produced in any link at any instant due to its inertia.

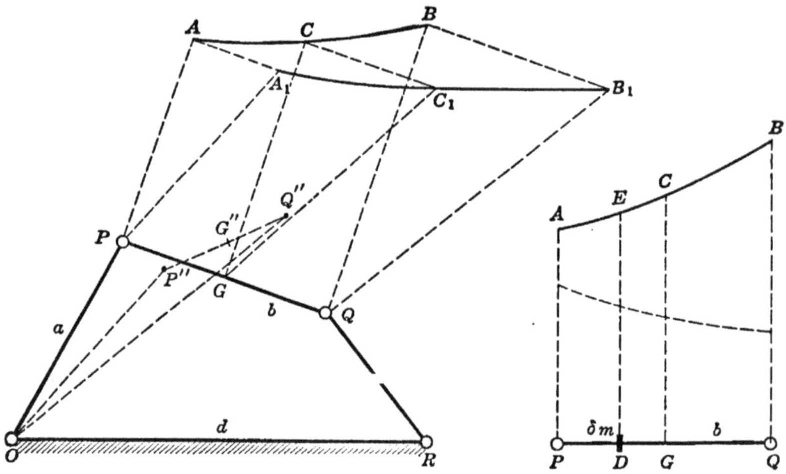

FIG. 167.—Bending forces on rod due to its inertia.

Any part such as the connecting rod of an engine is subject to stresses due to the transmission of the pressure from the piston to the crankpin, but in addition to this the rod is continually being accelerated and retarded, these changes of velocity producing bending stresses in the rod, and these latter stresses may now be determined.

To make the case as general as possible, let $OPQR$, Fig. 167, represent a machine for which the vector acceleration diagram is $OP''Q''O$, it is required to find the bending moment in the rod b due to its inertia. Lay off at each point on b the acceleration of that point; thus make PA_1, GC_1, QB_1, etc., equal and parallel respectively to OP'', OG'', OQ'', etc., obtaining in this way the curve $A_1C_1B_1$.

Now resolve the accelerations at each point in b into two parts, one normal to b and the other parallel to the link. Thus PA is the acceleration of P normal to b, and GC and QB are the corresponding accelerations for the points G and Q respectively. In this way a second curve ACB may be drawn, and the perpendicular to b drawn from any point in it to the line ACB represents the acceleration at the given point in b in the direction normal to the axis of the latter, the scale in all cases being $-\ \omega^2 : 1$. Thus the acceleration of P normal to b is $AP.\omega^2$, and so for other points.

Now let the rod be placed as shown on the right-hand side of Fig. 167 with the acceleration curve ACB above it to scale. Imagine the rod divided up into equal short lengths one of which is shown at D, having a weight δw lb. and mass $\delta m = \frac{\delta w}{g}$, and let the normal acceleration at this point be represented by DE. Should the rod be of uniform section throughout its length all the small masses like δm will be equal since all will be of the same weight δw, but if the rod is larger at the left-hand end than at the right-hand end, then the values of δm will decrease in going along from P to Q. Now the force due to the acceleration of the small mass is equal to δm multiplied by the acceleration corresponding to DE and this force may be set off along DE above D. Proceeding in this way for the entire length of the rod gives the dotted curve as shown which may be looked upon as the load curve for the rod due to its acceleration. From this load curve the bending moments and stresses in the rod may be determined by the well-known methods used in statics.

For a rod of uniform cross-section throughout the acceleration curve ACB will also be a load curve to a properly selected scale, but with the ordinary rods of varying section the work is rather longer. In carrying it out, the designer usually soon finds out by experience the position of the mechanism which corresponds to the highest position of the acceleration curve ACB, and the accelerations being the maximum for this position the rod is designed to suit them. A very few trials enable this position to be quickly found for any mechanism with which one is not familiar.

The process must, of course, be carried out on the drafting board.

232. To Find the Accelerations of the Various Parts of a Rock Crusher.—In order to get a clearer grasp of the principles involved, a few applications will be made, the first case being that of the rock crusher shown in Fig. 168. The mechanism of the crusher is shown on the left and has not been drawn closely to scale as the construction is more clear ior the proportions shown. A crank *OP* is driven at uniform speed by a belt pulley on the shaft *O* and to this crank is attached the long connecting rod *PQ*. The swinging jaw of the crusher is pivoted to the frame at *T* and connected to *PQ* by the rod *SQ*, while another rod *QR* is pivoted to the frame at *R*. As *OP* revolves *Q* swings in an arc of a circle about *R*, giving the jaw a swinging motion about *T*

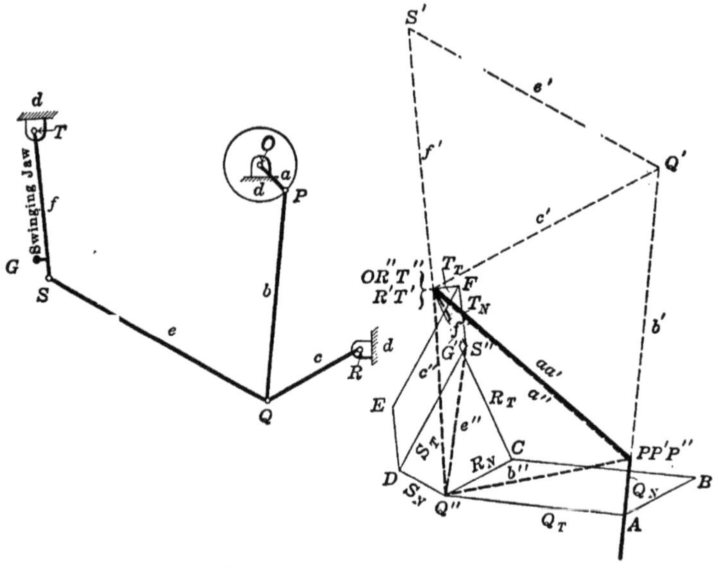

FIG. 168.—Rock crusher.

and crushing between the jaw and the frame any rocks falling into the space. In large crushers the jaw is very heavy and its variable velocity, or acceleration, sometimes sets up very serious vibrations in buildings in which it is placed.[1]

The acceleration diagram is shown on the right and there is also drawn the upper end of the rod *b* and the whole of the crank *a*. It is to be noted that the actual mechanism may be drawn to as small a scale as desired and the diagram to the right to as large a scale as is necessary, because in the phorograph and the acceleration diagram only the directions of the links are required and these may be easily obtained from the small scale drawing

[1] See article by PROF. O. P. HOOD in *American Machinist*, Nov. 26, 1908.

shown. The phorograph of the mechanism and acceleration diagram should give no difficulty because the mechanism is simply a combination of two four link mechanisms, $OPQR$ and $RQST$ exactly similar to that shown in Fig. 165 and already dealt with. The crank OP has been chosen as the primary link.

The crank OP is assumed to turn at uniform speed of ω radians per second. For the phorograph, OQ' parallel to RQ meeting b produced gives Q' and $P'Q' = b'$ and $OQ' = c'$; further OS' parallel to ST meeting $Q'S'$ parallel to QS gives S' and $Q'S' = e'$ while $S'O$ gives f'. The points R' and T' lie at O.

For the acceleration diagram P'' lies at P since a is assumed to run at uniform speed; then, following the method already described in Sec. 228, lay off $P''A = \frac{b'^2}{b}$ and AB parallel to c and of length $AB = \frac{c'^2}{c}$, and finish the diagram by making BC perpendicular to b and OC perpendicular to c, these intersecting at C. Complete the parallelogram $ABCQ''$ and join $P''Q''$ and $Q''O$; then in the acceleration diagram $OP'' = a''$, $P''Q'' = b''$ and $Q''O = c''$ which gives the vector acceleration diagram for the part $OPQR$. Then starting at Q'', which gives the acceleration of Q on the vector diagram, draw $Q''D = \frac{e'^2}{e}$ and parallel to e; this is followed by DE parallel to TS and of length $\frac{T'S'^2}{TS} = \frac{f'^2}{f}$ and the vector diagram is closed by drawing EF perpendicular to e to meet OF perpendicular to f in F. On completing the parallelogram $DEFS''$, the point S'' is found and then S'' is joined to O and to Q''. The line $S''Q''$ represents e on the acceleration diagram while $OS'' = f''$ represents f on the same figure.

The length OS'' represents the acceleration of S on a scale of $-\omega^2 : 1$ and the acceleration of any other point on f is found by locating on OS'' or $R''S''$ a point similarly situated to the desired point on ST. If the angular acceleration of the jaw is required, it may be found as described at Sec. 229 and evidently is $\alpha_f = -FO \times \frac{\omega^2}{f}$.

Calling G the center of gravity of the jaw f and locating G'' in the same way with regard to $S''T''$ that G is located with regard to ST, the acceleration of G is $G''O \times \omega^2$ and the force required to cause this acceleration and therefore shaking the machine is parallel to $G''O$ and is equal to $G''O \times \omega^2 \times \dfrac{\text{weight of jaw}}{32.16}$ pds. Or the torque required for the purpose is $I_f \times \alpha_f$ where I_f is the moment of inertia of f with regard to G.

233. Application to the Engine.—This construction and the determination of the accelerations and forces has a very useful application in the case of the reciprocating engine and this machine will now be taken up. Fig. 169 represents an engine in

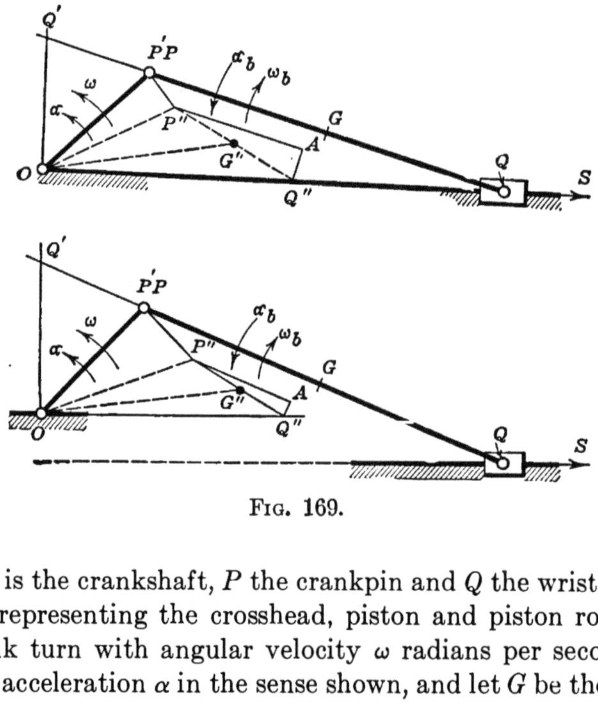

Fig. 169.

which O is the crankshaft, P the crankpin and Q the wristpin, the block c representing the crosshead, piston and piston rod. Let the crank turn with angular velocity ω radians per second and have an acceleration α in the sense shown, and let G be the center of gravity of the connecting rod b. To get the vector acceleration diagram find P'' exactly as in the former construction, OP representing the acceleration $PO.\omega^2$ and PP'' the acceleration $a\alpha$, both on the scale $-\omega^2$ to 1.

Now the motion of Q is one of sliding and thus Q has only tangential acceleration, or acceleration in the direction of sliding, in this case QS, the sense being determined later. Hence, the total acceleration of Q must be represented by a line through O in the direction, QS therefore Q'' lies on a line through the center of the crankshaft, and the diagram is reduced to a simpler form than in the more general case. Having found P'', draw $P''A$ parallel to b, of length $\frac{b'^2}{b}$, to represent Q_N, and also draw AQ'', normal to $P''A$, to meet the line $Q''O$ which is parallel to QS, in Q''. Then will AQ'' represent the value of the angular acceleration of the rod b. Since $b\alpha_b = Q''A.\omega^2$ or $\alpha_b = Q''A.\frac{\omega^2}{b}$, and since AQ'' lies on the same side of $P''A$ that PP'' does of OP, therefore α_b is in the same sense as α; thus since ω_b is opposite to ω, the angular velocity of the rod is decreasing, or the rod is being retarded.

The acceleration of the center of gravity of b is represented by OG'' and is equal to $G''O.\omega^2$, and similarly the acceleration of the end Q of the rod is represented by OQ'' and is equal to $Q''O.\omega^2$, this being also the acceleration of the piston.

It will be observed that all of these accelerations increase as the square of the number of revolutions per minute of the crank-shaft, so that while in slow-speed engines the inertia forces may not produce any very serious troubles, yet in high-speed engines they are very important and in the case of such engines as are used on automobiles, which run at as high speeds as 1,500 revolutions per minute, these accelerations are very large and the forces necessary to produce them cause considerable disturbanees. Take the piston for example, the force required to move it will depend on the product of its weight and its acceleration, so that if an engine ran normally at 750 revolutions per minute and then it was afterward decided to speed it up to 1,500 revolutions per minute, the force required to move the piston in any position in the latter case would be four times as great as in the former case.

234. Approximate Construction.—In the actual case of the engine, the calculations may be very much simplified owing to certain limitations which are imposed on all designs of engines driving other machinery, these limitations being briefly that the variations in velocity of the flywheel must be comparatively small, that is, the angular acceleration of the flywheel must not be great, and in fact, on engines the flywheels are made so heavy that α cannot be large.

To get a definite idea on this subject a case was worked out for a 10 by 10-in. steam engine, running at 310 revolutions per minute, and the maximum angular acceleration of the crank was found to be slightly less than 7 radians per second per second. For this case the normal acceleration of P is $r\omega^2 = 5/12 \times 1{,}100 =$ 458 ft. per second per second, while the tangential acceleration is $r\alpha = \frac{5}{12} \times 7 = 5.8$ ft. per second per second, which is very small compared with 458 ft. per second per second, so that on any ordinary drawing the point P'' would be very close to P. Thus without serious error $r\alpha$ may be neglected compared with $r\omega^2$ and hence P'' is at P.

With the foregoing modification for the engine, the complete acceleration diagram is shown at Fig. 170, the length PA repre-

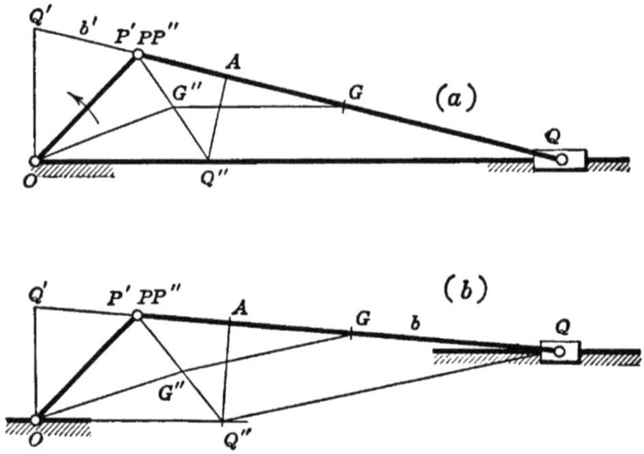

FIG. 170.—Piston acceleration.

senting $\frac{b'^2}{b}$ and AQ'' is normal to b, thus $P''Q''$ is the acceleration diagram for the connecting rod and OQ'' represents the acceleration of the piston on the scale $-\omega^2$ to 1. Two cases are shown: (a) for the ordinary construction; and (b) for the offset cylinder. The acceleration of any such point as G is found by finding G'', making the line GG'' parallel to QQ'', the acceleration then is $G''O.\omega^2$.

It should be noticed that the greater the ratio of b to a, that is the longer the connecting rod for a given crank radius, the more nearly will the point A approach to P because the distance PA represents the ratio $\frac{b'^2}{b}$ and this steadily decreases as b increases, and at the same time AQ'' becomes more nearly vertical. In the extreme case of an infinitely long rod, carried out practically as shown at Fig. 6, the point A coincides with P and AQ''

is vertical and then the acceleration of the piston which is OQ'' is simply the projection of a on the line of the piston travel or the acceleration $Q''O \times \omega^2 = a.\cos\theta \times \omega^2$ where θ is the crank angle POQ''.

235. Piston Acceleration at Certain Points.—Taking the more common form of the mechanism shown at Fig. 170(a) the numerical values of the acceleration of the piston may be found in a few special cases. When the crank is vertical, b' is zero and therefore A is at P vertically above O, so that when AQ'' is drawn, Q'' lies to the left of O showing that the piston has negative acceleration or is being retarded. For this position a circle of diameter QQ'' will pass through P and therefore $Q''O \times OQ = OP^2$ or $Q''O = \dfrac{OP^2}{QO} = \dfrac{a^2}{\sqrt{b^2 - a^2}}$ and the acceleration of the piston is $\omega^2 \times \dfrac{a^2}{\sqrt{b^2 - a^2}}$ ft. per second per second.

At both the dead centers $b' = a$ hence $P''A = \dfrac{b'^2}{b} = \dfrac{a^2}{b}$, so that for the head end, $Q''O = a + \dfrac{a^2}{b}$ and the piston has its maximum acceleration at this point, which is $\left(a + \dfrac{a^2}{b}\right)\omega^2$ toward O, while for the crank end, $Q''O = a - \dfrac{a^2}{b}$ and the acceleration is $\left(a - \dfrac{a^2}{b}\right)\omega^2$ toward O, or the piston is being retarded.

Example.—Let an engine with 7-in. stroke and a connecting rod 18 in. long run at 525 revolutions per minute. Then $a = \dfrac{3\frac{1}{2}}{12} = 0.29$ ft., $b = \dfrac{18}{12} = 1.5$ ft., and $\omega = 55$ radians per second.

At the head end the acceleration of the piston would be:

$\left(a + \dfrac{a^2}{b}\right)\omega^2 = \left(0.29 + \dfrac{0.29^2}{1.5}\right) \times 55^2 = 931$ ft. per second per second.

At the crank end the acceleration would be:

$\left(a - \dfrac{a^2}{b}\right)\omega^2 = \left(0.29 - \dfrac{0.29^2}{1.5}\right) \times 55^2 = 623$ ft. per second per second.

At the time when the crank is vertical the result is:

$\dfrac{a^2}{\sqrt{b^2 - a^2}}\omega^2 = \left[\dfrac{0.29^2}{\sqrt{1.5^2 - 0.29^2}}\right] \times 55^2 = 173$ ft. per second per second.

The angular acceleration of the connecting rod, being determined by the length AQ'', is zero at each of the dead points but when the crank is vertical it has nearly its maximum value; the formula for it is $Q''A \times \frac{\omega^2}{b}$. When the crank is vertical a diagram will show that

$$Q''A = \frac{ab}{\sqrt{b^2 - a^2}},$$

and the acceleration is

$$\alpha_b = \frac{a}{\sqrt{b^2 - a^2}}\,\omega^2.$$

For the engine already examined, when in this position,

$$\alpha_b = \left[\frac{0.29^2}{\sqrt{1.5^2 - 0.29^2}}\right]55^2 = 596 \text{ radians per second per second.}$$

APPROXIMATE GRAPHICAL SOLUTION FOR THE STEAM ENGINE

236. In the approximate method already described, in which

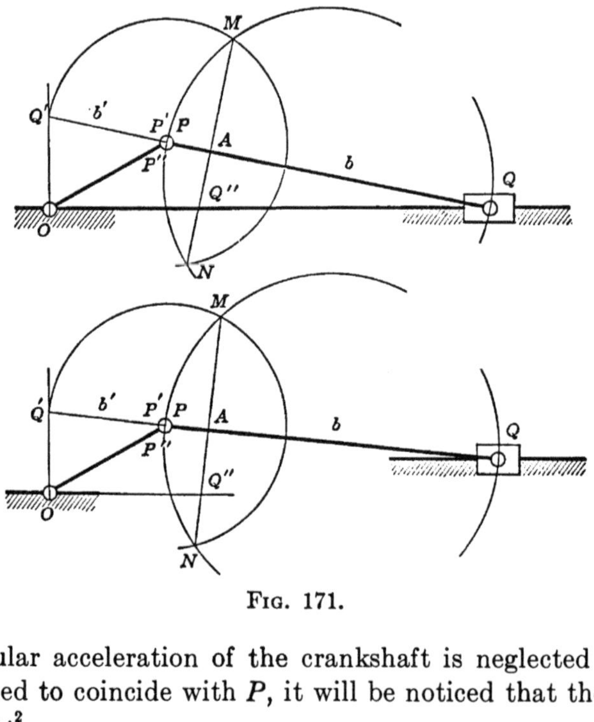

FIG. 171.

the angular acceleration of the crankshaft is neglected and P'' is assumed to coincide with P, it will be noticed that the length $P''A = \frac{b'^2}{b}$, is laid off along the connecting rod, the length $P'Q'$

representing b', and PQ the length b, and then AQ'' is drawn perpendicular to PQ. This may be carried out by a very simple graphical method as follows: With center P and radius $P'Q' = b'$ describe a circle, Fig. 171, also describe a second circle, having the connecting rod b as its diameter, cutting the first circle at M and N, and join MN. Where MN cuts b locates the point A and where it cuts the line through O in the direction of motion of Q gives Q''.

The proof is that PMQ being the angle in a semicircle is a right angle also the chord MN is normal to PQ and is bisected at A. Then in the circle $MPNQ$ there are two chords PQ and MN intersecting at A, and hence from geometry it is known that:

$$PA.AQ = MA.AN = MA^2$$

or

$$PA(PQ - PA) = MP^2 - PA^2 = b'^2 - PA^2.$$

Multiplying out the left-hand side and cancelling

$$PA.PQ = b'^2$$

that is

$$PA = \frac{b'^2}{PQ} = \frac{b'^2}{b}$$

which proves the construction.

THE EFFECTS OF THE ACCELERATIONS OF THE PARTS UPON THE FORCES ACTING AT THE CRANKSHAFT OF AN ENGINE

In order to accelerate or retard the various parts of the engine, some torque must be required or will be produced at the crankshaft, and a study of this will now be taken up in detail.

237. (*a*) **The effect produced by the piston.**

By the construction already described the acceleration of the piston is readily found and it will be seen that Q'' lies first on the cylinder side of O and then on the opposite side. When Q'' lies between O and Q, Fig. 172, then the acceleration of the piston is $Q''O \times \omega^2$, and the acceleration of the piston is in the same sense as the motion of the piston, or the piston is being accelerated. Conversely, when Q'' lies on QO produced the acceleration being in the opposite sense to the motion of the piston, the latter is being retarded. These statements apply to the motion of the piston from right to left, when the sense of motion of the piston reverses the remarks about the acceleration must also be changed. If now the accelerations for the different piston positions on the forward stroke be plotted, the diagram EJH will be obtained, Fig. 172, where the part of the diagram EJ represents positive accelerations of the piston, and the part JH negative accelerations, or retardations. The corresponding diagram for the return stroke of the piston is omitted to avoid complexity.

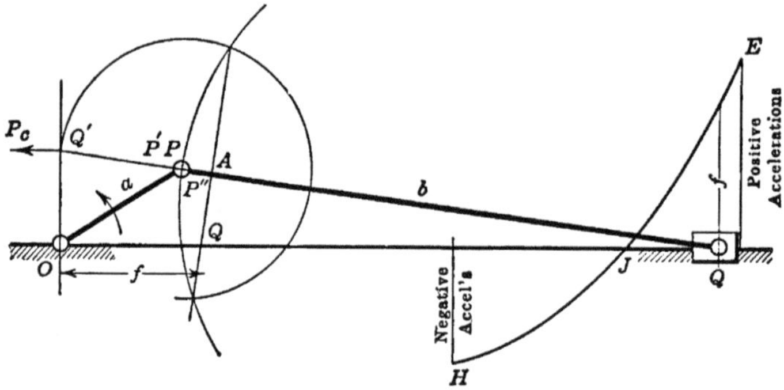

FIG. 172.—Acceleration of piston on forward stroke.

Let the combined weight of the piston, piston rod and cross-head be w_c pounds, the corresponding mass being $m_c = \frac{w_c}{g}$, and let f represent the acceleration of the piston at any instant; then the force P_c necessary to produce this acceleration is $P_c = m_c.f$.

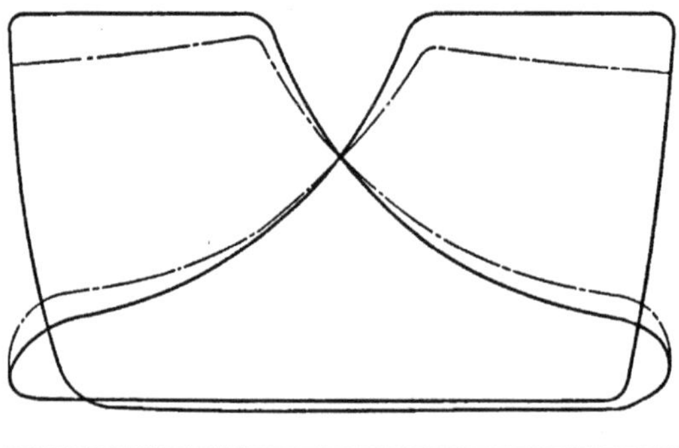

FIG. 173.—Modification of diagram due to inertia of piston.

This force will be positive if f is positive and *vice versa*, that is, if f is positive a force must be exerted on the piston in its direction of motion and if it is negative the force must be opposed to the motion. In the first case energy must be supplied by the flywheel, or steam, or gas pressure, to speed up the piston, whereas, in the latter case, energy will be given up to the flywheel due to the decreasing velocity of the piston. Since no net energy is received during the operation, therefore, the work done on the piston in accelerating it must be equal to that done by the piston while it is being retarded.

Two methods are employed for finding the turning effect of this force, P_c; the first is to reduce it to an equivalent amount per square inch of piston area by the formula $p_c = \frac{P_c}{A}$ where A is the area of the piston, and then to correct the corresponding pressures shown by the indicator diagram by this amount. In this way a reduced indicator diagram for each end is found, as shown for a steam engine in Fig. 173, where the dotted diagram is the reduced diagram found by subtracting the quantity p_c from the upper line on each diagram. The remaining area is the part effective in producing a turning moment on the crankshaft.

The second method is to find directly the turning effect necessary on the crankshaft to overcome the force P_c, and from the principles of the phorograph this torque is evidently $T_c = P_c \times OQ' = m_c \times f \times OQ'$. In the position shown in Fig. 172, P_c would act as shown, and a torque acting in the same sense as the motion of a would have to be applied.

The first method is very instructive in that it shows that the force necessary to accelerate the piston at the beginning of the stroke in very high-speed engines may be greater than that produced by the steam or gas pressure, and hence, that in such cases the connecting rod may be in tension at the beginning of the stroke, but, of course, before the stroke has very much proceeded it is in compression again. This change in the condition of stress in the rod frequently causes "pounding" due to the slight slackness allowed at the various pins.

238. (*b*) **The Effect Produced by the Connecting Rod.**—This effect is rather more difficult to deal with on account of the nature of the motion of the rod. The resultant force acting may, however, be found by the method described earlier in the chapter, Sec. 230, but in the case of the engine, the construction may be much simplified, and on account of the importance of the problem the simpler method will be described here. It consists in dividing the rod up into two equivalent concentrated masses, one at the crosshead pin the other at a point to be determined.

Referring to Fig. 174, the rod is represented on the acceleration diagram by $P''Q''$ and the acceleration of any point on it or the angular acceleration of the rod may be found by processes already explained. Let I_b be the moment of inertia of the rod about its center of gravity, k_b being the corresponding radius of gyration and m_b the mass, so that $I_b = m_b k_b^2$, and let the center of gravity G lie on PQ at distance r_1 from Q. Instead of considering the actual rod it is possible to substitute for it two concentrated masses m_1 and m_2, which, if properly placed, and if of proper weight, will have the same inertia and weight as the original rod. Let these masses be m_1 and m_2 where $m_1 = \frac{w_1}{g}$ and $m_2 = \frac{w_2}{g}$, in which w_1 and w_2 are the weights of the masses in pounds. Further, let mass m_1 be concentrated at Q, it is required

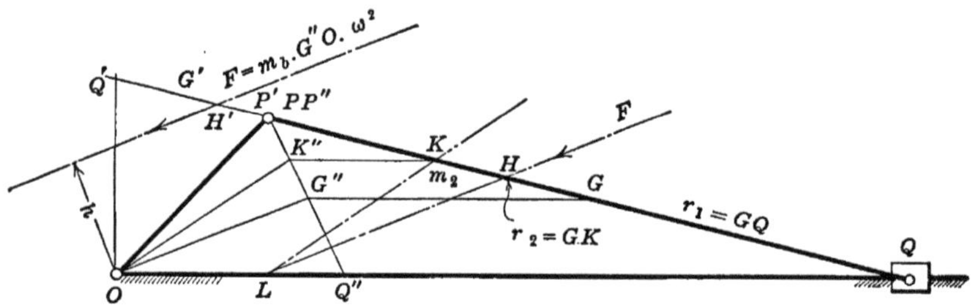

FIG. 174.

to find the weights w_1 and w_2 and the position of the weight w_2. Let r_2 be the distance from the center of gravity of the rod to mass m_2.

These masses are determined by the following three conditions:

1. The sum of the weights of the two masses must be equal to the weight of the rod, that is, $w_1 + w_2 = w_b$, or $m_1 + m_2 = m_b$.

2. The two masses m_1 and m_2, must have their combined center of gravity in the same place as before; therefore, $m_1 r_1 = m_2 r_2$.

3. The two masses must have the same moment of inertia about their combined center of gravity G as the original rod has about the same point; hence

$$m_1 r_1^2 + m_2 r_2^2 = m_b k_b^2$$

For convenience these are assembled here:

$$m_1 + m_2 = m_b \quad (1)$$
$$m_1 r_1 = m_2 r_2 \quad (2)$$
$$m_1 r_1^2 + m_2 r_2^2 = m_b k_b^2 \quad (3)$$

Solving these gives:

$$m_1 = m_b \times \frac{r_2}{r_1 + r_2} \quad \text{and} \quad m_2 = m_b \times \frac{r_1}{r_1 + r_2}$$

and $$r_1 r_2 = k_b^2 \text{ or } r_2 = \frac{k_b^2}{r_1}.$$

Thus, for the purposes of this problem the whole rod may be replaced by the two masses m_1 and m_2 placed as shown in Fig. 174. The one mass m_1 merely has the same effect as an increase in the weight of the piston and the method of finding the force required to accelerate it has already been described. Turning then to the mass m_2, which is at a fixed distance r_2 from G; the center of gravity of m_2 is K and the acceleration of K is evidently $K''O \times \omega^2$, $K''K$ being parallel to $G''G$. The direction of the force acting on m_2 is the same as that of the acceleration of its center of gravity and is therefore parallel to $K''O$, and the magnitude of this force is $m_2 \times K''O \times \omega^2$. The force acts through K, its line of action being KL parallel to $K''O$.

The whole rod may now be replaced by the two masses m_1 and m_2. The force acting on the former is $m_1 \times Q''O \times \omega^2$ through Q parallel to $Q''O$, that is, this force is in the direction of motion of Q and passes through L on $Q''O$. The force on the mass m_2 is $m_2 \times K''O \times \omega^2$, which also passes through L, so that the resultant force F acting on the rod must also pass through L. Thus the construction just described gives a convenient graphical method for locating one point L on the line of action of the resultant force F acting on the connecting rod.

Having found the point L the direction of the force F has been already shown to be parallel to $G''O$ and its magnitude is $m_b \times G''O \times \omega^2$. Let F intersect the axis of the rod at H, find the image H' of H, and transfer F to H'. The moment required to produce the acceleration of the rod is then Fh.

A number of trials on different forms and proportions of engines have shown that the point L remains in the same position for all crank angles, and hence if this is determined once for a given engine it will be only necessary to determine $G''O$ for the different crank positions; as this enables the magnitude and direction of F to be found and its position is fixed by the point L.

239. Net Turning Moment on Crankshaft.—For the position of the machine shown in Fig. 174, let P be the total pressure on the piston due to the gas or steam pressure; then the net turning moment acting on the crankshaft is

$$P \times OQ' - [m_c \times Q''O \times \omega^2 \times OQ' + m_b \times G''O \times \omega^2 \times h]$$

after allowance has been made for the inertia of the piston and connecting rod. This turning moment will produce an acceleration or retardation of the flywheel according as it exceeds or is less than the torque necessary to deliver the output.

All of these quantities have been determined for the complete revolution of a steam engine and the results are given and discussed at the end of the present chapter.

240. The Forces Acting at the Bearings.—The methods described enable the pressures acting on the bearings due to the inertia forces to be easily determined, and this problem is left for the reader to solve for himself.

In high-speed machinery the pressures on the bearings due to the inertia of the parts may become very great indeed and all care is taken by designers to decrease them. Thus, in automobile engines, some of which attain as high a speed as 3,000 revolutions per minute, or over, during test conditions, the rods are made as light as possible and the pistons are made of aluminum alloy in order to decrease their weight. In one of the recent automobile engines of 3-in. bore and 5-in. stroke the piston weighs 17 oz. and the force necessary to accelerate the piston at the end of the stroke and at a speed of 3,000 revolutions per minute is over 800 pds., corresponding to an average pressure of over 110 pds. per square inch on the piston and the effect of the connecting rod would increase this approximately 50 per cent; thus during the suction stroke the tension in the rod is over 1,200 pds. at the head-end dead center and the compressive stress in the rod is much less than that corresponding to the gas pressure. At the crank-end dead center the accelerating force is also high, though less than at the head end, and here also the rod is in compression due to the inertia forces. If the gas pressure alone were considered, the rod would be in compression in all but the suction stroke.

241. Computation on an Actual Steam Engine.—In order that the methods may be clearly understood an example is worked out here of an engine running at 525 revolutions per minute, and of the vertical, cross-compound type with cranks at 180°, and developing 125 hp. at full load. Both cylinders are 7 in. stroke and 11 in. and 15½ in. diameter for the high- and low-pressure sides respectively. The weight of each set of reciprocating parts including piston, piston rod and crosshead is 161 lb., while the connecting rod weighs 47 lb. has a length between centers of 18 in. and its radius of gyration about its cen-

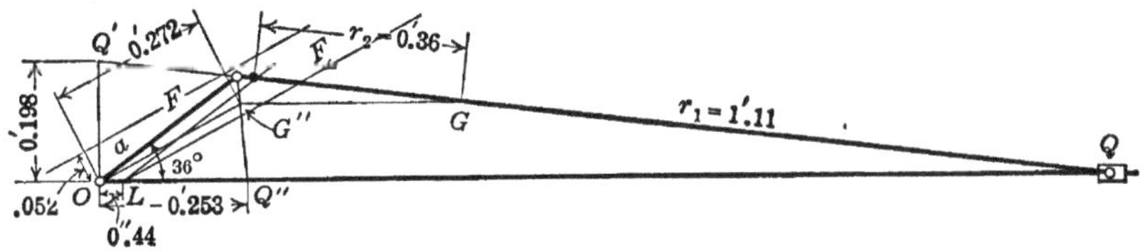

FIG. 175.

ter of gravity is 7.56 in., the latter point being located 13.3 in. from the center of the wristpin.

From the above data $\omega = 55$ radians per second,

$$m_c = \frac{161}{32.16} = 5, \quad m_b = \frac{47}{32.16} = 1.46 \text{ and } k_b = \frac{7.56}{12} = 0.63 \text{ ft.}$$

$$\text{Also } r_1 = \frac{13.3}{12} = 1.11 \text{ ft., } r_2 = \frac{0.63^2}{1.11} = 0.36 \text{ ft. and}$$

$$m_1 = 1.46 \times \frac{0.36}{1.11 + 0.36} = 0.35 \text{ while } m_2 = 1.11.$$

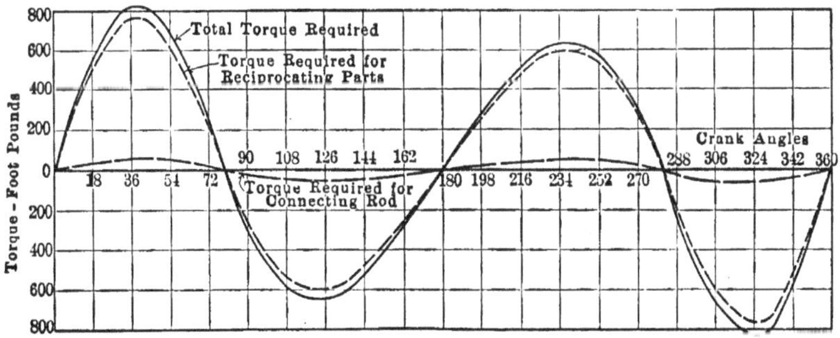

FIG. 176.—Effect of connecting rod and piston.

The construction for the crank angle 36° is shown in Fig. 175 with all dimensions marked on, and the complete results for the entire revolution for one side of the engine are set down on the accompanying table, all of the quantities being tabulated. The point L for this engine is located 0.44 in. from O and on the cylinder side of it. The table shows that at the head end a force of 5,262 lb. would be required to accelerate each piston which corresponds to a mean pressure for the high-pressure side of 55 lb. per square inch., in other words if the net steam pressure fell below 55 lb. at this point the high-pressure rod would be in tension instead of compression.

The disturbing effect of the connecting rod is much less marked as the table shows, but in accurate calculations cannot be neglected. The combined effect of the two as shown in the last column is quite decided.

In order that the results may be more clearly understood they have been plotted in Fig. 176, which shows the turning moment at the crankshaft required to move the piston and the crosshead separately, and also the combined effort required for both. The turning effort required for the rod is not quite one-twelfth that required for the piston.

Table Showing the Effect due to the Inertia of the Parts of an 11 by 7-in. Steam Engine Running at 525 Revolutions per Minute

Crank angle, θ	Piston, crosshead, etc.					Connecting rod					Total turning moment required at crank to move all parts, ft.-pds.	
	$Q''O$, ft.	Accel. of piston, ft.-sec.²	Force to accel. piston, pds.	$Q'O$, ft.	Turning moment on crank, ft.-pds.	$G''O$, ft.	Acceln. of G, ft.-sec.²	Force to accel., rod-pds.	h, ft.	Turning moment on crank, ft.-pds.	+	−
0	+0.348	+1,053	+5,262	0	0	0.349	1,056	1,542	0	0	0	
18	0.325	983	4,916	0.107	+526	0.297	898	1,311	0.033	+43	569	
36	0.253	765	3,827	0.198	758	0.272	823	1,201	0.052	62	820	
54	0.153	463	2,314	0.264	611	0.242	732	1,069	0.050	53	664	
72	+0.047	+142	+711	0.295	+210	0.217	656	958	0.023	+22	232	
90	−0.060	−181	−908	0.292	−265	0.215	650	950	0.023	−22	....	287
108	0.137	414	2,072	0.260	539	0.228	690	1,007	0.045	45	....	584
126	0.187	566	2,829	0.208	588	0.248	750	1,095	0.048	48	....	636
144	0.217	656	3,282	0.143	469	0.267	808	1,179	0.035	41	....	510
162	−0.232	−702	−3,509	0.073	−256	0.273	824	1,205	0.017	−20	....	276
180	0.235	711	3,555	0	0	0.235	711	1,038	0	0	....	0
198	+0.232	+702	+3,509	0.073	+256	0.273	824	1,205	0.017	+20	276	
216	0.217	656	3,282	0.143	469	0.267	808	1,179	0.035	41	510	
234	0.187	566	2,829	0.208	588	0.248	750	1,095	0.048	48	636	
252	0.137	414	2,072	0.260	539	0.228	690	1,007	0.045	45	584	
270	+0.060	+181	+908	0.292	+265	0.215	650	950	0.023	+22	287	
288	−0.047	−142	−711	0.295	−210	0.217	656	958	0.023	−22	....	232
306	0.153	463	2,314	0.264	611	0.242	732	1,069	0.050	53	....	664
324	0.253	765	3,827	0.198	758	0.272	823	1,201	0.052	62	....	820
342	0.325	983	4,916	0.107	−526	0.297	898	1,311	0.033	−43	....	569
360	−0.348	−1,053	−5,262	0	0	0.349	1,056	1,542	0	0	....	0

The relative effects of these turning moments is shown more clearly at Fig. 177 in which separate curves are drawn for the high- and low-pressure sides. The dotted curves in both cases show the torque **due to** the steam pressure found as in Chapter X, while the broken lines show the torque **required to** accelerate the parts and the curves in solid lines indicate the net resultant torque acting on the crankshaft. The reader will be at once

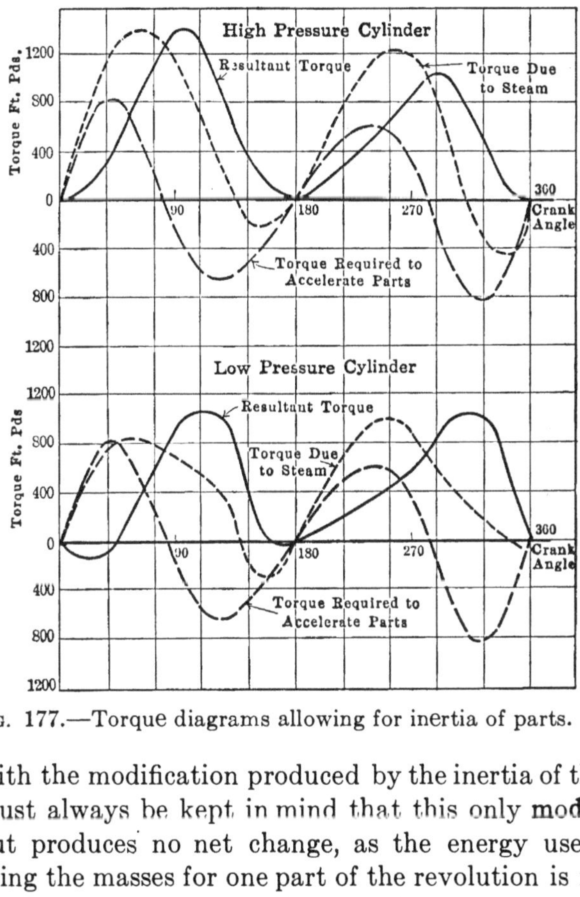

FIG. 177.—Torque diagrams allowing for inertia of parts.

struck with the modification produced by the inertia of the parts, but it must always be kept in mind that this only **modifies** the result but produces no net change, as the energy used up in accelerating the masses for one part of the revolution is returned when the masses are retarded later on in the cycle of the machine. That these forces must be reckoned with, especially in high-speed machinery, is very evident.

242. The Gnome Motor.—One further illustration of the principles stated here may be given in the Gnome motor, which has had much application in aeroplane work. The general form of the motor has already been shown in Fig. 12 in the early part of this book and it has been explained that the mechanism is exactly the same as in the ordinary reciprocating engine except that the crank is fixed and the connecting rod, cylinder and other parts make complete revolutions. The mechanism is shown in Fig. 178 in which a is the cylinder and parts secured to it, b is the connecting rod and c is the piston, and power is delivered from the rotating link a, which is assumed to turn at constant speed of ω radians per second.

At the wristpin two letters are placed, Q on the rod b and P on a directly below Q, and thus as the revolution proceeds P

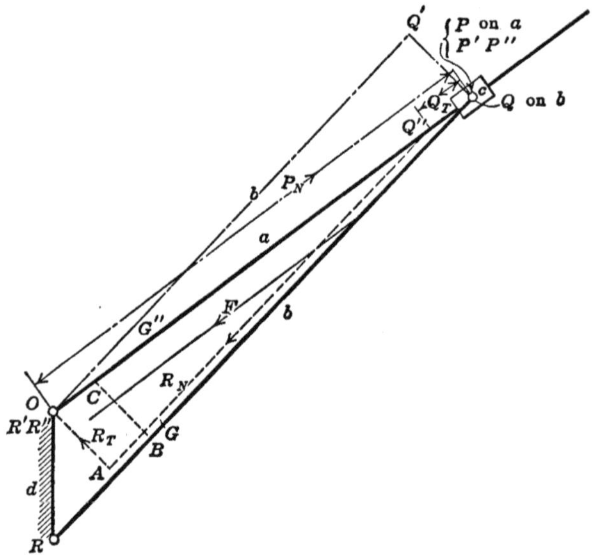

FIG. 178.—Gnome motor.

moves in and out along a; the motion of Q relative to P must, in the nature of the case, be one of sliding in the direction of a. The phorograph is obtained in a similar way to that for the Whitworth quick-return motion, Fig. 38, the only difference here being that the cylinder link a turns at uniform speed while in the former case the connecting rod did so. The image Q' lies on $P'Q'$ normal to a and on $R'Q'$ through O parallel to b.

To find the acceleration diagram the plan followed in Sec. 228 is employed. Thus, the acceleration of R relative to Q added to that of Q relative to P and that of P relative to O must be zero. Using the notation of Sec. 228 it follows that $R_T + R_N + Q_T + Q_N + P_T + P_N = 0$; and since the only acceleration which Q can have relative to P is tangential, it follows that $Q_N = 0$. Again, since a turns at a uniform speed, the value of P_T is also zero. Hence, the result is $R_T + R_N + Q_T + P_N = 0$.

Now, adopting the scale of $-\omega^2:1$, the acceleration $P_N = a\omega^2$ is represented by $OP'' = a$ and it has also been shown in Sec. 228 that $R_N = \frac{b'^2}{b} \times \omega^2$, so that $P''B$ is laid off along b and equal to $\frac{b'^2}{b}$, and thus $P''B$ will represent R_N. The vector diagram is closed by R_T and Q_T, the former perpendicular to b and the latter parallel to a, so that BC perpendicular to b represents R_T and hence $CR'' = Q_T$. It will make a more correct vector diagram to lay off $P''Q'' = CR''$ and make $Q''A$ and AR'' equal respectively to $P''B$ and BC. Then $OP'' = P_N$, $P''Q'' = Q_T$, $Q''A = R_N$, $AR'' = R_T$ and $Q''R''$ represents the rod b vectorially on the acceleration diagram, G'' corresponding to its center of gravity G. The acceleration of the center of gravity G of b is $G''O \times \omega^2$ and the angular acceleration of the rod is $R''A \times \frac{\omega^2}{b}$ as given in Sec. 229. The pull on b due to the centrifugal effect of the piston is $Q''O \times \omega^2 \times \frac{\text{weight of position}}{32.16}$ in the direction of a.

The resultant force F on the rod b may be found as in Sec. 230 and is in the direction $G''O$, that is, along a. Its position is shown on the figure and the pressure between the piston and cylinder due to this force is readily found knowing the value and position of F.

QUESTIONS ON CHAPTER XV

1. A weight of 10 lb. is attached by a rod 15 in. long to a shaft rotating at 100 revolutions per minute; find the acceleration of the weight and the tension in the rod.

2. If the shaft in question 1 increases in speed to 120 revolutions per minute in 40 sec., find the tangential acceleration of the weight and also its total acceleration.

3. A railroad train weighing 400 tons is brought to rest from 50 miles per hour in 1 mile. Find the average rate of retardation and the mean resistance used.

4. At each end of the stroke the velocity of a piston is zero; how is its acceleration a maximum?

5. Weigh and measure the parts of an automobile engine and compute the maximum acceleration of the parts and the piston pressure necessary to produce it.

6. Find the bending stresses in the connecting rod of the same engine, due to inertia, when the crank and rod are at right angles.

7. Divide the rod in question 6 up into its equivalent masses, locating one at the wristpin.

8. Make a complete determination for an automobile engine of the resulting torque diagram due to the indicator diagram and inertia of parts.

CHAPTER XVI

BALANCING OF MACHINERY

243. General Discussion on Balancing.—In all machines the parts have relative motion, as discussed in Chapter I. Some of the parts move at a uniform rate of speed, such as a crankshaft or belt-wheel or flywheel, while other parts, such as the piston, or shear blade or connecting rod, have variable motion. The motion of any of these parts may cause the machine to vibrate and to unduly shake its foundation or the building or vehicle in which it is used. It is also true that the annoyance caused by this vibration may be out of all proportion to the vibration itself, the results being so marked in some cases as to disturb buildings many blocks away from the place where the machine is. This disturbance is frequently of a very serious nature, sometimes forcing the abandonment of the faulty machine altogether; therefore the cause of vibration in machinery is worthy of careful examination.

It is not possible in the present treatise to discuss the general question of vibrations, as the matter is too extensive, but it may be stated that one of the most common causes is lack of balance in different parts of the machine and the present chapter is devoted entirely to the problem of balancing. Where any of the links in a machine undergo acceleration forces are set up in the machine tending to shake it, and unless these forces are balanced, vibrations of a more or less serious nature will occur, but balancing need only be applied where accelerations of the parts occur.

It must be borne in mind, however, that the accelerations are not confined solely to such parts as the piston or the connecting rod which have a variable motion, but the particles composing any mass which is rotating with uniform velocity about a fixed center also have acceleration[1] and may throw the machine out of balance, because, as explained in Sec. 226,

[1] In connection with this the first part of Chapter XV should be read over again.

a mass has acceleration along its path when its velocity is changing, and also acceleration normal to its curved path even when its velocity is constant. In discussing the subject it is most convenient to divide the problem up into two parts, dealing first with links which rotate about a fixed center and second with those which have a different motion, in all cases plane motion being assumed.

THE BALANCING OF ROTATING MASSES

244. Balancing a Single Mass.—Let a weight of w lb. which has a mass $m = \frac{w}{g}$ rotate about a shaft with a fixed center, at a fixed radius r ft., and let the radius have a uniform angular velocity of ω radians per second. Then, referring to Sec. 226 this mass will have no acceleration along its path since ω is assumed constant, but it will have an acceleration toward the axis of rotation of $r\omega^2$ ft. per second per second, and hence a radial force of amount $\frac{w}{g} r\omega^2 = mr\omega^2$ pds. must be applied to it to maintain it at the given radius r. This force must be applied by the shaft to which the weight is attached, and as the weight revolves there will be a pull on the shaft, always in the radial direction of the weight, and this pull will thus produce an unbalanced force on the shaft, which must be balanced if vibration is to be avoided.

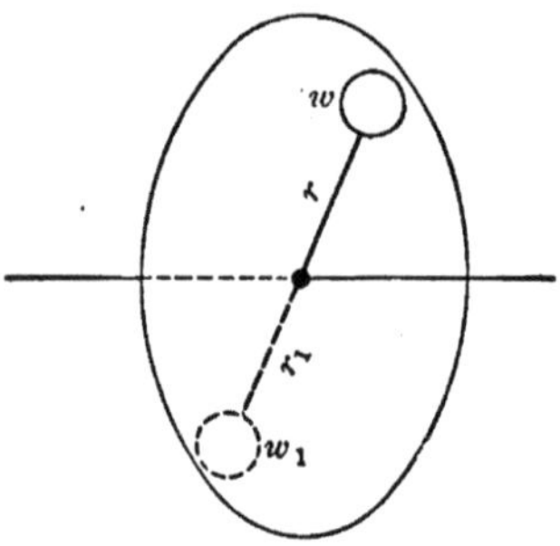

Fig. 179.

Let Fig. 179 represent the weight under consideration in one of its positions; then if vibration is to be prevented another weight w_1 must be attached to the same shaft so that its acceleration will be always in the same direction but in opposite sense to that of w, and this is possible only if w_1 is placed at some radius r_1 and diametrically opposite to w. Clearly, the relation between the two weights and radii is given by $\frac{w}{g} r\omega^2 = \frac{w_1}{g} r_1\omega^2$ or $rw = r_1w_1$, since $\frac{\omega^2}{g}$ is common to both sides, from which the product r_1w_1 is found, and having arbitrarily selected one of these quantities such as r_1, the value of w_1 is easily determined.

If the two weights are placed as explained there will be no resultant pull on the shaft during rotation, and hence no vibration; in other words the shaft with its weights is **balanced.**

It sometimes happens that the construction prevents the placing of the balancing mass directly opposite to the weight w, as for example in the case of the crankpin of an engine, and then the balancing weights must be divided between two planes which are usually on opposite sides of the disturbing mass, although they may be on the same side of it if desired. Let Fig. 180 represent the crankshaft of an engine, and let the crankpin correspond to an unbalanced weight w lb. at radius r. The planes A and B are those in which it is possible to place counterbalance weights and the magnitude and position of the weights are

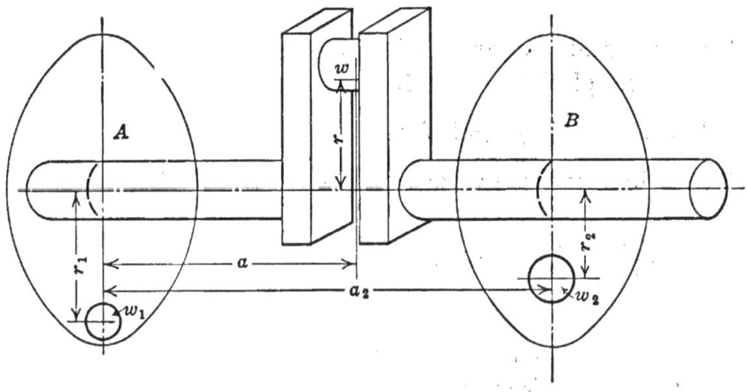

FIG. 180.—Crank-shaft balancing.

required. Let the weights be w_1 and w_2 lb. and their radii of rotation be r_1 and r_2 respectively; then clearly the vector sum $(w_1r_1 + w_2r_2 - wr)\,\frac{\omega^2}{g} = 0$, or $w_1r_1 + w_2r_2 = wr$. Let all the masses be in the plane containing the axis of the shaft and the radius r.

Now it is not sufficient to have the relation between the masses and radii determined by the formula $w_1r_1 + w_2r_2 = wr$ alone, because this condition only means that the shaft will be in static equilibrium, or will be balanced if the shaft is supported at rest on horizontal knife edges. When the shaft revolves, however, there may be a tendency for it to "tilt" in the plane containing its axis and the radii of the three weights, and this can only be avoided by making the sum of the moments of the quantities $r \times w \times \frac{\omega^2}{g}$ about an axis through the shaft normal to the last-mentioned plane, equal to zero.

For convenience, select the axis in the plane in which w_1 revolves, and let a and a_2 be the respective distances of the planes of rotation of w and w_2 from the axis; then the moment equation gives

$$(wra - w_2r_2a_2)\frac{\omega^2}{g} = 0 \text{ or } wra = w_2r_2a_2.$$

Combining this relation with the former one

$$w_1r_1 + w_2r_2 = wr$$

gives

$$w_1r_1 = wr\left(1 - \frac{a}{a_2}\right) \text{ and } w_2r_2 = wr\frac{a}{a_2}$$

so that w_1r_1 and w_2r_2 are readily determined.

As an example let $w = 10$ lb., $r = 2$ in., $a = 4$ in. and $a_2 = 10$ in.; then $w_2r_2 = 8/12$, and if r_2 be taken as 4 in. $w_2 = \frac{8/12}{r_2} = 8/4 = 2$ lb., since the radii are to be in feet. Further, the value of $w_1r_1 = wr\left(\frac{a_2 - a}{a_2}\right) = 10 \times \frac{2}{12}\left(\frac{\frac{10}{12} - \frac{4}{12}}{\frac{10}{12}}\right) = 1$, from which if w_1 be arbitrarily chosen as 4 lb., it will have to revolve at a radius of 1/4 ft. or 3 in. from the shaft center. In this way the two weights are found in the selected planes which will balance the crankpin.

245. Balancing Any Number of Rotating Masses Located on Different Planes Normal to a Shaft Revolving at Uniform Speed.—Let there be any number of masses, say four, of weights w_1, w_2, w_3 and w_4, rotating at respective radii r_1, r_2, r_3 and r_4 on a shaft with fixed axis and which is turning at ω radians per second, the whole being as shown at Fig. 181. It is required to balance the arrangement.

As before, this may be done by the use of two additional weights revolving with the shaft and located in two planes of revolution which may be arbitrarily selected; these are shown in the figure, the one containing the point O, and the other at A, and the quantities a_1, a_2, a_3, a_4 and a_5 represent the distances of the several planes of revolution from O.

It is convenient to use the left-hand plane, or that through O, as the plane of reference and, in fact, **the reference plane must always contain one of the unknown masses,** and it will be evident that if the masses are balanced the vector sum $\frac{w}{g} \times r \times \omega^2$ must be zero. Further, the vector sum of the tilting moments $w \times r \times a \times \frac{\omega^2}{g}$ of the various masses in planes containing the masses and the shaft must also be zero; otherwise, although the system may be in equilibrium when at rest, it will not be so while it is in motion. Now, since ω^2 and g are the same for all the

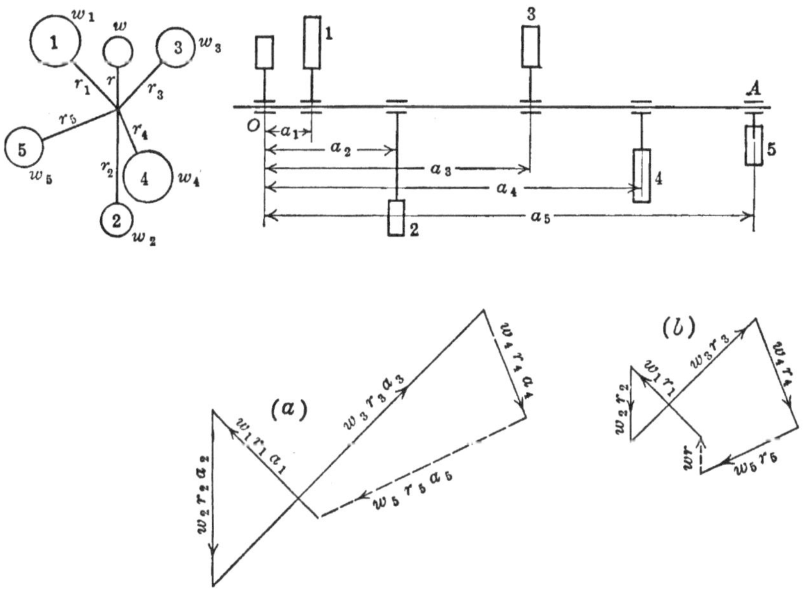

FIG. 181.—Balancing revolving masses.

masses, therefore, the above equations may be reduced to the form: (1) vector sum of the products $w \times r$ must be zero; and (2) vector sum of the products wra must be zero. Since the first of these is the condition to be observed if the shaft is stationary, it may be called the static condition, while the second is the dynamic condition coming into play only when the shaft is revolving.

Now the tilting moment $w \times r \times a$ has a tendency to tilt the shaft in the plane containing r and the shaft, and it will be most convenient to represent it by a vector parallel to the trace of this plane on the plane of revolution, or what is the same thing, by a vector parallel with the radius r itself, and a similar method will be used with other tilting moments. Two balancing weights will be required, w at an arbitrarily selected radius r in the normal plane through O, and w_5 at a selected radius r_5 in the normal plane through A.

Now from the static condition the vector sum

$$wr + w_1r_1 + w_2r_2 + w_3r_3 + w_4r_4 + w_5r_5 = 0$$

where w and w_5 are unknown, and these cannot yet be found because the directions of the radii r and r_5 are not known. Again, since the reference plane passes through O, tilting moments about O must balance, or

$$w_1r_1a_1 + w_2r_2a_2 + w_3r_3a_3 + w_4r_4a_4 + w_5r_5a_5 = 0$$

and here the only unknown is $w_5r_5a_5$ which may therefore be determined. The vector polygon for finding this quantity is shown at (*a*) in Fig. 181 and on dividing by a_5 the value of w_5r_5 is given. The force polygon shown at (*b*) may now be completed, and the only other unknown $w \times r$ found, and thus the magnitude and positions of the balancing weights w and w_5 may be found. The construction gives the value of the products wr and w_5r_5 so that either w or r may be selected as desired and the remaining factor is easily computed.

By a method similar to the above, therefore, any number of rotating masses in any positions may be balanced by two weights in arbitrarily selected planes. Many examples of this kind occur in practice, one of the most common being in locomotives (see Sec. 253), where the balancing weights must be placed in the driving wheels and yet the disturbing masses are in other planes.

246. Numerical Example on Balancing Revolving Masses.— Let there be any four masses of weights $w_1 = 10$ lb., $w_2 = 6$ lb. $w_3 = 8$ lb. and $w_4 = 12$ lb., rotating at radii $r_1 = 6$ in., $r_2 = 8$ in., $r_3 = 9$ in. and $r_4 = 4$ in. in planes located as shown on Fig. 12.8 It is required to balance the system by two weights in the planse through the points O and A respectively.

The data of the problem give $w_1r_1 = 10 \times \frac{6}{12} = 5$, $w_2r_2 = 6 \times \frac{8}{12} = 4$, $w_3r_3 = 8 \times \frac{9}{12} = 6$ and $w_4r_4 = 12 \times \frac{4}{12} = 4$, and further $w_1r_1a_1 = 5 \times \frac{6}{12} = 2.5$, $w_2r_2a_2 = 4 \times \frac{10}{12} = 3.33$, $w_3r_3a_3 = 6 \times \frac{15}{12} = 7.5$ and $w_4r_4a_4 = 4 \times \frac{18}{12} = 6$.

The first thing is to draw the tilting-couple vector polygon as shown on the left of Fig. 182 and the only unknown here is $w_5r_5a_5$ which may thus be found and scales off as 4.65. Dividing by $a_5 = \frac{12}{12} = 1$ ft. gives $w_5r_5 = 4.65$ and the direction of r_5 is also given as parallel to the vector $w_5r_5a_5$.

Next draw the vector diagram for the products $w.r$ as shown on the right of Fig. 182, the only unknown being the product wr

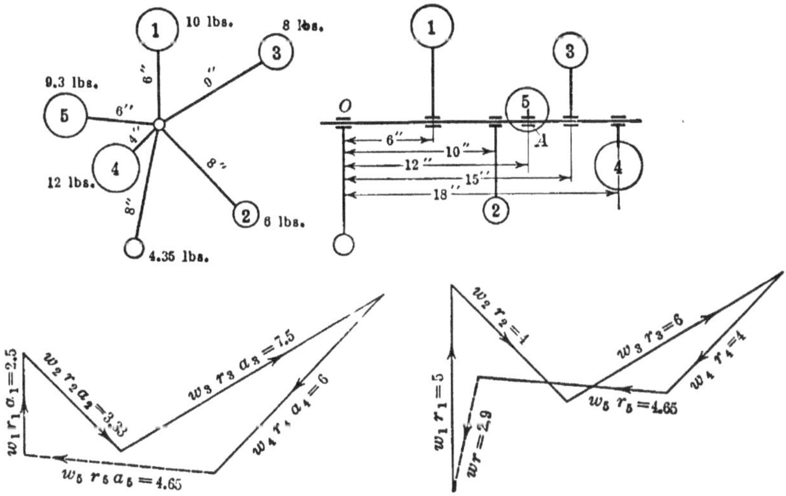

FIG. 182.

for the plane through O. From the polygon this scales off as 2.9 and the direction of r is parallel to the vector in the diagram.

In this way the products wr and w_5r_5 are known in magnitude and direction, and then, on assuming the radii, the weights are easily found. This has been done in the diagram. It is advisable to check the work by choosing a reference plane somewhere between O and A and making the calculations again.

BALANCING OF NON-ROTATING MASSES

247. The Balancing of Reciprocating and Swinging Masses.—The discussion in the preceding sections shows that it is always possible to balance any number of rotating masses by means of two properly placed weights in any two desired planes of revolution, and the method of determining these weights has been fully explained. The present and following sections deal with a much more difficult problem, that of balancing masses which do not revolve in a circle, but have either a motion of translation at variable speed, such as the piston of an engine or else a swinging motion such as that of a connecting rod or of the jaw of a rock crusher or other similar part. Such problems not only present much difficulty, but their exact solution is usually impossible and all that can generally be done is to partially balance the parts and so minimize the disturbing effects.

248. Balancing Reciprocating Parts Having Simple Harmonic Motion.—The first case considered is that of the machine shown

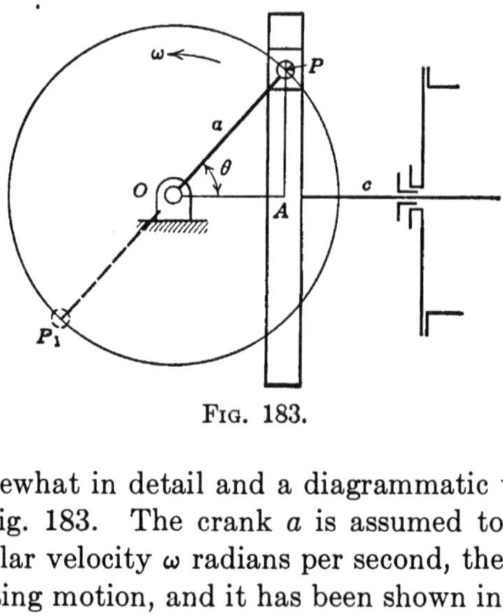

FIG. 183.

in Fig. 6 somewhat in detail and a diagrammatic view of which is given in Fig. 183. The crank a is assumed to revolve with uniform angular velocity ω radians per second, the piston c having reciprocating motion, and it has been shown in Sec. 234 that the acceleration of c is, at any instant, equal to the projection of a upon the direction of c multiplied by ω^2, or the acceleration of the piston is $OA \times \omega^2 = a \cos \theta \times \omega^2$. The force necessary to produce this acceleration of the piston then is $F = \frac{w}{g} \times a \cos \theta \times \omega^2$ where w is the weight of the piston, and it is this force F which must be applied to give a balance. A little consideration will show that this force F is constant in direction, always coinciding with the direction of motion of c, but it is variable in magnitude, since the latter depends on the crank angle θ.

Suppose now that at P is placed a weight w exactly equal to the weight of the reciprocating parts; then the centrifugal force acting radially is $\frac{w}{g}a\omega^2$ and the resolved part of this in the direction of motion of c is clearly $\frac{w}{g} \times a \cos\theta \times \omega^2$, that is to say, the horizontal resolved part of the force produced by the weight w at P is the same as that due to the motion of the piston. It therefore follows that if at P_1, located on PO produced so that $OP_1 = OP$, there is placed a concentrated weight of w lb., the horizontal component of the force produced by it will balance the reciprocating masses; the vertical component, however, of the force due to w at P_1 is still unbalanced and will cause vibrations vertically. Thus, the only effect produced by the weight w at P_1 is to change the horizontal shaking forces due to c into vertical forces, and the machine still has the unbalanced vertical forces.

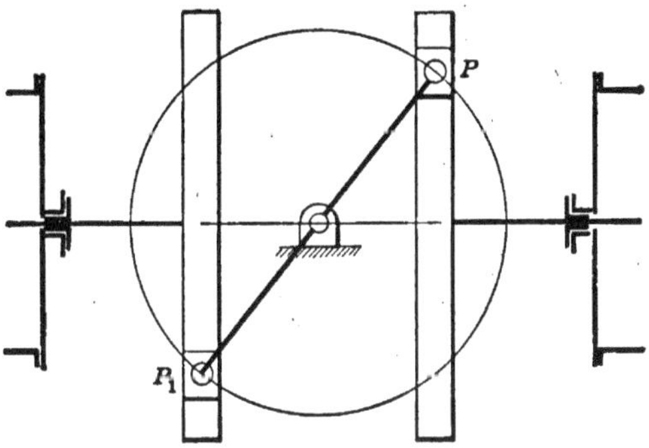

FIG. 184.

Frequently in machinery there is no real objection to this vertical disturbing force, because it may be taken up by the foundation of the machine, but in portable machines, such as locomotives or fire engines or automobiles, it may cause trouble also. It is seen, however, that complete balance is not obtained in this way, that is, a single revolving mass cannot be made to balance a reciprocating mass.

There is only one way in which such a mass can be completely balanced and that is by duplication of the machine. Thus, if it were possible to use P_1 as a crank and place a second piston, as shown in Fig. 184, the masses would be completely balanced. If the second machine cannot be placed in the same plane normal to the shaft as the first, then balance could be obtained by dividing it into two parts each having reciprocating weights

$\frac{w}{2}$ and moving in planes equidistant from the plane of the first machine.

When the reciprocating mass moves in such a way that its position may be represented by such a relation as $a \cos \theta$ it is said to have simple harmonic motion and its acceleration may always be represented by the formula $a \cos \theta \times \omega^2$. Balancing problems connected with this kind of motion are problems in primary balancing and are applicable to cases where the connecting rod is very long, giving approximate results in such cases, and exact results in cases where the rod is infinitely long, and in the case shown in Fig. 183, just discussed.

One method in which revolving weights may be used to produce exact balance in the case of a part having simple harmonic

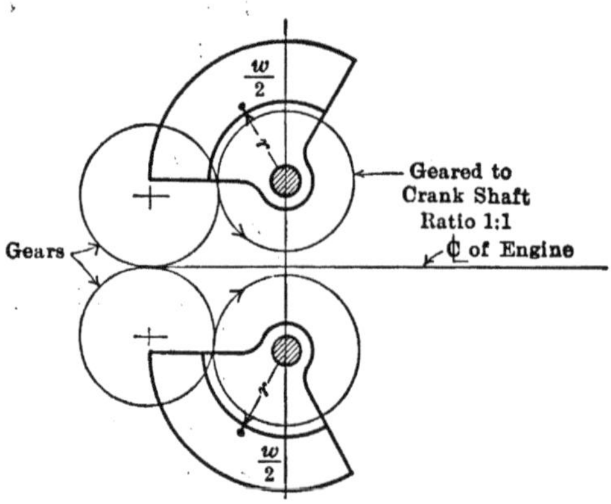

FIG. 185.—Engine balancing—primary balance.

motion is shown in Fig. 185, where the two weights $\frac{1}{2}w$ are equal and revolve at the speed of the crank and in opposite sense to one another, their combined weight being equal to the weight w of the reciprocating parts. Evidently here the vertical components of the two weights balance one another, leaving their horizontal components free to balance the reciprocating parts. Taking the combined effective weights as equal to that of the reciprocating parts, then they must rotate at a radius equal to that of the crank, and must be 180° from the latter when it is on the dead center.

249. Reciprocating Parts Operated by Short Connecting Rod.—The general construction adopted in practice for moving reciprocating parts differs from Fig. 183 in that the rod imparting the motion is not so long that the parts move with simple harmonic motion, and in the usual proportions adopted in engines the variation is quite marked, for the rods are never longer than six times the crank radius and are often as small as four and one-half times this radius.

The method to be adopted in such cases is to determine the acceleration of the reciprocating parts and to plot it for each of the crank angles as described in the preceding chapter. To illustrate this, suppose it is required to balance the reciprocating parts in the engine examined in Sec. 241; then the accelerations of these are found and set down as shown in the table belonging to this case. The accelerations shown in the third column have

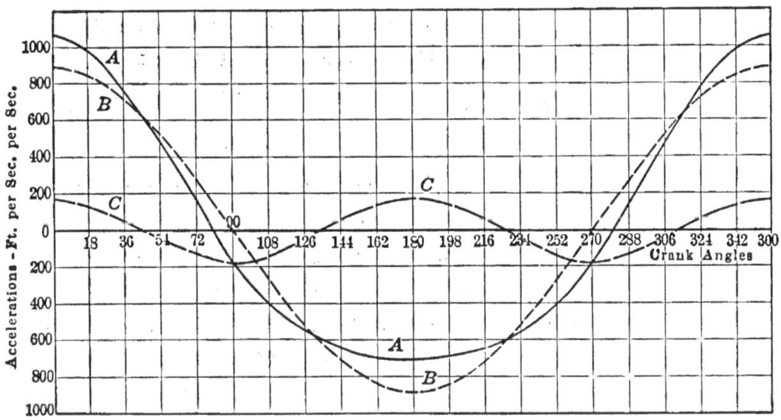

FIG. 186.

been plotted in the plain line A on Fig. 186. This curve must now be broken up into its corresponding harmonic components, and it is usual to assume that these are in phase at the inner dead center with the original curve. The dotted line B represents a simple harmonic or sine curve in phase with the plain curve and having maximum height of $\frac{1}{2}(1{,}053 + 711) = 882$, which is the mean value for the true curve heights at 0° and at 180° crank angles. The difference between these two curves has been plotted in the broken line curve C and will be found on examination to be almost a true sine curve, in fact, it differs so little from a sine curve that it would be impossible to distinguish between them on the scale of this drawing.[1]

It will be observed that the curve C is also in phase at the

[1] See Appendix A for mathematical proof of these statements.

inner dead center with the curve A but has twice the frequency and maximum height on the drawing of 171 ft. per second per second. It will also be found that 171 is $\frac{a}{b} \times 882$ or $\frac{3\frac{1}{2}}{18} \times 882 = 171$.

The reciprocating parts of this engine could also be balanced in the manner shown in Fig. 185, but it would require that instead of one pair of weights, two pairs should be used; one pair rotating at the speed of the crank and 180° from it at the dead centers, and another pair in phase at the inner dead center with the first but rotating at double the speed of the crank. The weight rotating at the speed of the crankshaft should be the same as that of the piston, namely 161 lb. ($m = 5$), if placed at at $3\frac{1}{2}$ in. radius, while the weight making twice the speed ($\omega = 110$) of the crank might also be placed at a radius of $3\frac{1}{2}$ in., in which case it would weigh $\dfrac{5 \times 171 \times 32.2}{\frac{3.5}{12} \times (110)^2} = 7.76$ lb. In order that these weights could rotate without interference they might have to be divided and separated axially, in which case the two halves of the same weight would have to be placed equidistant from the plane of motion of the connecting rod.

It is needless to say that the arrangement sketched above is too complicated to be used to any extent except in the most urgent cases, where some serious disturbance results. Counterweights attached directly to the crankshaft are sometimes used, but at best these can only balance the forces corresponding to the curve B and always produce a lifting effect on the engine. The reader must note that the above method takes no account of the weight of the connecting rod, which will be considered later.

If the method already described cannot be used, then the only other method is by duplication of the parts and this will be described at a later stage.

Where the acceleration of the reciprocating masses cannot be represented by a simple harmonic curve, but must have a second harmonic of twice the frequency, superposed on it, the problem is one of **secondary balancing,** so called because of the latter harmonic.

250. Balancing Masses Having Any General Form of Plane Motion.—It is impossible, in general, to balance masses moving in a more or less irregular way, such, for example, as the jaw of the rock crusher shown in Fig. 168, or the connecting rod of an engine. The general method, however, is to plot the curve of accelerations for the center of gravity of the mass, using crank angles as a base, and if the resulting curve at all approximates to a simple harmonic curve a weight may be attached to the crank, as already described, which will roughly balance the forces, or will at least reduce them very greatly. The magnitude of the weight and its position will be found by drawing a simple harmonic curve which approaches most nearly to the actual acceleration curve. As the accelerations are not all in the same direction, the most correct way is to plot two curves giving the resolved parts of the acceleration in two planes and balance each separately, but usually an approximate result is all that is desired, and, as the shaking forces are mainly in one plane, the resolved part in this plane alone is all that is usually balanced.

In engines, the method of balancing the connecting rod is somewhat different to that outlined above. The usual plan is to divide the rod up into two equivalent masses in the manner described in Sec. 238, one of the masses m_1 being assumed as located at the wristpin and the location of the other mass m_2 is found as described in the section referred to. In this way the one mass m_1 may be regarded simply as an addition to the weight of the reciprocating masses and balancing of it effected as described in the last or following sections. The other mass m_2, however, gives trouble and cannot, as a matter of fact, be exactly balanced at all, so that there is still an unbalanced mass.

Consideration of a number of practical cases shows that m_2 in many steam engines lies close to the center of the crankpin. The engine discussed in Sec. 241, for instance, has the mass m_1 concentrated at the wristpin and the mass m_2 will be only 0.36 in. away from the crankpin; for long-stroke engines, however, the mass m_2 may be some distance from the crankpin and in such cases the method described below will not give good results. In automobile engines the usual practice is to make the crank end of the rod very much heavier and the crankpin larger than the same quantities at the piston end and hence the first statement of this paragraph is not true. In one rod examined the length between centers was 12 in., and the center of gravity 3.03 in. from the crankpin center; the weight of the rod was 2.28 lb. and selecting the mass m_1 at the wristpin its weight would be 0.43 lb. and the remaining weight would be 1.85 lb. concentrated 0.92 in.

from the crankpin and on the wristpin side of it, so that considerable error might result by assuming the latter mass at the crankpin center.

The fact that the mass m_2 does not fall exactly at the crankpin has been already explained in Sec. 241, and in the engine there discussed the resultant force on the rod passes through L, slightly to the right of the crankshaft, instead of passing through this center, as it would do if the mass m_2 fell at the crankpin. If an approximation is to be used, and it appears to be the only thing to do under existing conditions, m_2 may be assumed to lie at the crankpin, and thus the rod is divided into two masses; one, m_1 concentrated at the wristpin and balanced along with the reciprocating masses, and the other, m_2, concentrated at the crankpin and balanced along with the rotating masses.

It should be pointed out in passing, that the method of dividing the rod according to the first of two equations of Sec. 238, that is, so that their combined center of gravity lies at the true center of gravity of the rod, to the neglect of the third equation, leads to errors in some rods. Much more reliable results are obtained by finding m_1 and m_2 according to the three equations in Sec. 238 and the examples of Sec. 241, except that m_2 is assumed to be at the crankpin center. Dividing the mass m so that its components m_1 and m_2 have their center of gravity coinciding with that of the actual rod will usually give fairly good results, if the diameters of the crankpin and wristpin do not differ unduly.

251. Balancing Reciprocating Masses by Duplication of Parts.—Owing to the complex construction involved when the reciprocating masses are balanced by rotating weights, such a plan is rarely used, the more common method being to balance the reciprocating masses by other reciprocating masses. The method may best be illustrated in its application to engines, and indeed this is where it finds most common use, automobile engines being a notable example.

For a single-cylinder engine the disturbing forces due to the reciprocating masses are proportional to the ordinates to such a curve as A, Fig. 186, or what is the same thing to the sum of the ordinates to the curves B and C. Suppose now a second engine, an exact duplicate of the first, was attached to the same shaft as the former engine and let the cranks be set 180° apart as at Fig. 187(a), then it is at once evident that there will be a tilting moment normal to the shaft in the plane passing through the axis of the shaft and containing the reciprocating masses, and further, a study of Fig. 186 will show that while the ordinates to the two curves B belonging to these machines neutralize, still the two curves C are additive and there is unbalancing due to the forces corresponding to curves C. In Fig. 187 are shown at (*b*) and (*c*) two other arrangements of two engines, both of which eliminate the tilting moments; in the arrangement (*b*) the cranks are at 180° and the unbalanced forces are completely eliminated, producing perfect balance, whereas at (*c*) the sum of the crank angles for the two opposing engines is 180° and the forces corresponding to curve B are balanced, while those corresponding to C are again unbalanced and additive, so that there is still an unbalanced force.[1] Since the disturbing forces are in the direction of motion of the pistons, nothing would be gained in this respect by making the cylinder directions different in the two cases.

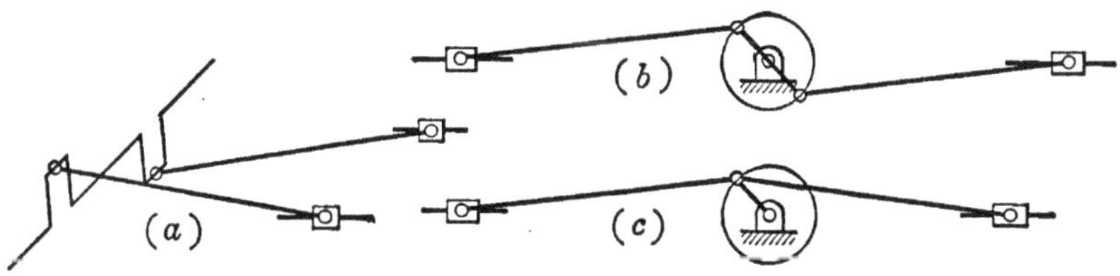

FIG. 187.—Different arrangements of engines.

If a three-cylinder engine is made, with the cylinders side by side and cranks set at 120° as in Fig. 188, an examination by the aid of Fig. 186 will show that the arrangement gives approximately complete balance, since the sum of ordinates to the curves B and C for the three will always be zero, but there is still a tilting moment normal to the axis of the crankshaft which is unavoidable. Four cylinders side by side on the same shaft, with the two outside cranks set together and the two inner ones also together and set 180° from the other, does away with the tilting moment of Fig. 187(*a*) but still leaves unbalanced forces proportional to the ordinates to the curves C. A six-cylinder arrangement with cylinders set side by side and made of two

[1] In order to get a clear grasp of these ideas the reader is advised to make several separate tracings of the curves B and C, Fig. 186, and to shift these along relatively to one another so as to see for himself that the statements made are correct.

parts exactly like Fig. 188, but with the two center cranks parallel gives complete balance with the approximations used here.

In what has been stated above the reader must be careful to remember that the rod has been divided into two equivalent masses and the discussion deals only with the balancing of the reciprocating part of the rod and the other reciprocating masses. The part acting with the rotating masses must also be balanced, usually by the use of a balancing weight or weights on the crankshaft according to the method already described in Sec. 245. It is to be further understood that certain approximations have been introduced with regard to the division of the connecting rod, and also with regard to the breaking up of the actual acceleration curve for the reciprocating masses into two simple harmonic curves, one having twice the frequency of the other. Such a division is a fairly close approximation, but is not exact.

FIG. 188.

The shape of the acceleration curves A, Fig. 186, and its components B and C, depend only upon the ratio of the crank radius to the connecting-rod length, and also upon the angular velocity ω. For the same value of $\frac{a}{b}$ the curves will have the same shape for all engines, and the acceleration scale can always be readily determined by remembering that at crank angle zero the acceleration is $\left(a + \frac{a^2}{b}\right) \omega^2$ ft. per second per second. These curves also represent the tilting moment to a certain scale since the moment is the accelerating force multiplied by the constant distance from the reference plane.

The chapter will be concluded by working out a few practical examples.

252. Determination of Crank Angles for Balancing a Four-Cylinder Engine.—An engine with four cylinders side by side and of equal stroke, is to have the reciprocating parts balanced by setting the crank angles and adjusting the weights of one of the pistons. It is required to find the proper setting and weight, motion of the piston being assumed simple harmonic. The dimensions of the engine and all the reciprocating parts but one set are given.

Let Fig. 189 represent the crankshaft and let w_2, w_3, w_4 represent the known weights of three of the pistons, etc., together with the part of the connecting rod taken to act with each of them as found in Sec. 250. It is required to find the remaining weight w_1 and the crank angles.

Choose the reference plane through w_1, and all values of r are the same; also the weights may be transferred to the respective crankpins, Sec. 248, as harmonic motion is assumed. Draw the wra triangle with sides of lengths w_2ra_2, w_3ra_3 and w_4ra_4 which gives the directions of the three cranks 2, 3, and 4. Next draw the wr polygon, from which w_1r is found, and thus w_1, and the

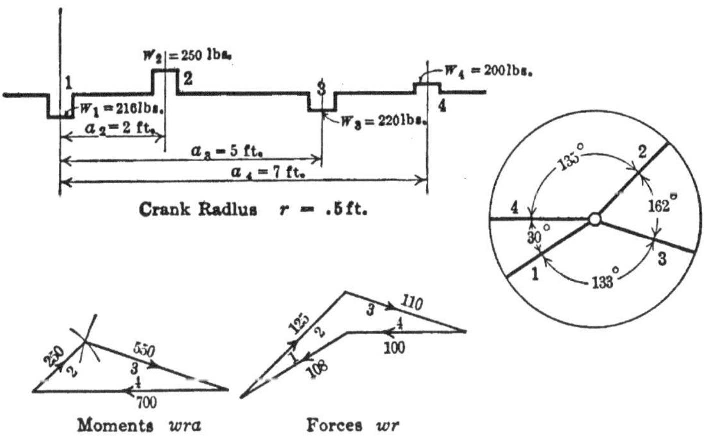

FIG. 189.—Balancing a four crank engine.

corresponding crank angle. The part of the rods acting at the respective crankpins, as well as the weight of the latter, must be balanced by weights determined as in Sec. 245. The four reciprocating weights operated by cranks set at the angles found will be balanced, however, if harmonic motion is assumed.

Example.—*Let* $w_2 = 250$ lb., $w_3 = 220$ lb. and $w_4 = 200$ lb., $r = 6$ in. and the distance between cylinders as shown. Then $w_2r_2 = 125$, $w_3r_3 = 110$, $w_4r_4 = 100$, $w_2r_2a_2 = 250$, $w_3r_3a_3 = 550$ and $w_4r_4a_4 = 700$. The solution is shown on Fig. 189 which gives the crank angles and weight $w_1 = 216$ lb. The rotating weights would have to be independently balanced.

253. Balancing of Locomotives.—In two-cylinder locomotives **the cranks are at 90°, and the** balance weights must be in the driving wheels. In order to avoid undue vertical forces it is usual to balance only a part of the reciprocating masses, usually about two-thirds, by means of weights in the driving wheels, and these balancing weights are also so placed as to compensate for the weights of the cranks. Treating the motion of the piston as simple harmonic, this problem gives no difficulty.

Example.—*Let* a locomotive be proportioned as shown on Fig. 190. The piston stroke is 2 ft. and the weight of the revolving masses is equivalent to 620 lb. attached to the crankpin. The reciprocating masses are assumed to have harmonic motion and to weigh 550 lb. and only 60 per cent. of these latter masses are to be balanced, so that weight at the crankpin corresponding to both of these will be $620 + 0.60 \times 550 \doteq 950$ lb. for each side.

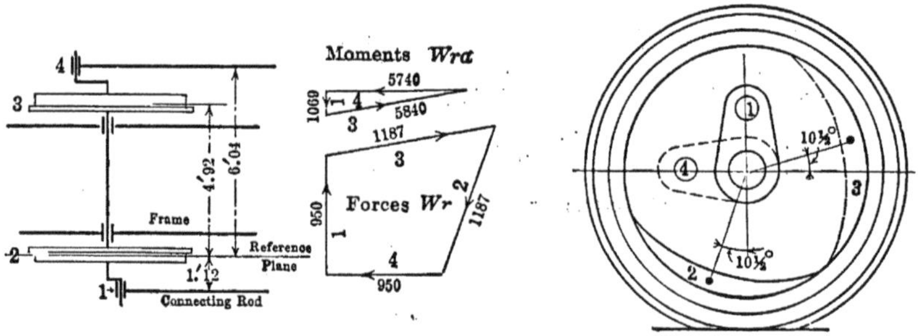

FIG. 190.—Locomotive balancing.

The reference plane for the tilting moments must always pass through one of the unknown masses, and the plane is here taken through the wheel 2. Note that the crank 1 being on opposite side of the reference plane to wheel 3 and crank 4, the sense of the moment vector must be opposite to what it would be if it were in the position 4. The crank 1 is thus drawn from the shaft in opposite sense to the vector $w_1r_1a_1$.

A diagrammatic plan of the locomotive in Fig. 190 gives the moment arms of the masses and the values of the corresponding moments are plotted on the right. Thus $w_1r_1a_1 = 950 \times 1 \times 1.12 = 1{,}069$ and $w_4r_4a_4 = 950 \times 1 \times 6.04 = 5{,}740$ and from the vector diagram $w_3r_3a_3$ scales off as 5,840. Since $a_3 = 4.92$ ft., $w_3r_3 = 1{,}187$. The force polygon may now be drawn with sides 950, 950 and 1,187 parallel to the moment vectors and then w_2r_2 is scaled off as 1,187. Selecting suitable radii r_2 and r_3 give the weights w_2 and w_3 and the end view of the wheels and axle on the right shows how the weights would be placed in accordance with the results.

The above treatment deals only with primary disturbing forces, only part of which are balanced, and further, it is to be noticed that there will be considerable variation in rail pressure, which might, with some designs, lift the wheel slightly from the tracks at each revolution, a very bad condition where it occurs.

254. Engines Used for Motor Cycles and Other Work.—In recent years engines have been constructed having more than one cylinder, with the axes of all the cylinders in one plane normal to the crankshaft. Frequently, in such engines, all the connecting rods are attached to a single crankpin, and any number of cylinders may be used, although with more than five, or seven cylinders at the outside, there is generally difficulty in making the actual construction. The example, shown in Fig.

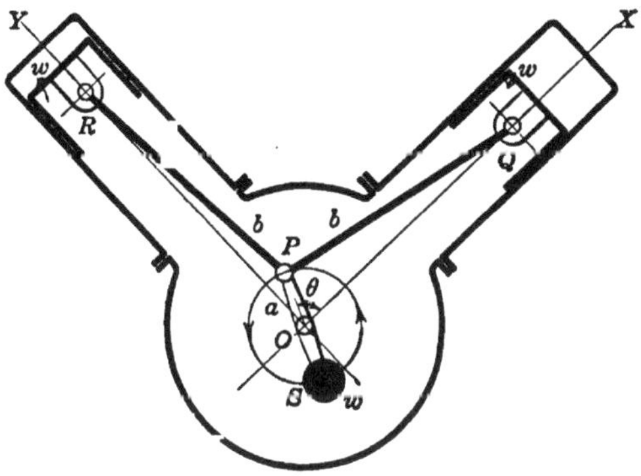

FIG. 191.—Motor cycle engine.

191, represents a two-cylinder engine with lines 90° apart and is a construction often used in motorcycles, in which case a vertical line passes upward through O, midway between the cylinders; in these motorcycle engines the angle between the cylinders is frequently less than 90°. The same setting has been in use for many years with steam engines of large size, in which case one of the cylinders is vertical, the other horizontal.

When a similar construction is used for more than two cylinders, the latter are usually evenly spaced; thus with three cylinders the angle between them is 120°.

These constructions introduce a number of difficult problems in balancing, which can only be touched on here, and the method of treatment discussed. The motorcycle engine of Fig. 191

with cylinders at 90° will alone be considered, and it will be assumed that both sets of moving parts are identical and that the weight of each piston together with the part of the rod that may be treated as a reciprocating mass is w lb. In the following discussion only the reciprocating masses are considered, and the part of the rods that may be treated as masses rotating with the crankpin, and also the crankpin and shaft are balanced independently; as the determination of the latter balance weights offers no difficulty the matter is not taken into account.

It has already been shown in Sec. 249 (and in Appendix A) that the acceleration of the piston may be represented by the sum of two harmonic curves, one of the frequency of the crank rotation, and another one of twice this frequency; these are shown in Fig. 186 in the curves B and C. It is further explained in Sec. 249 that the maximum ordinate to the curve C is $\frac{a}{b}$ times that to the curve B.

The discussion of Secs. 248 and 249 should also make it clear that if a weight w be secured at S, in Fig. 191, the component of the force produced by this weight in the direction OX will balance the primary component of the acceleration of the piston Q, that is, it will balance the accelerations corresponding to the ordinates to the curve B, Fig. 186. There is still unbalanced the accelerations corresponding to the curve C and also the vertical components of the force produced by the revolving weight w at S, these latter being in the direction OY. A very little consideration will show that the latter forces are exactly balanced by the reciprocating mass at R, or that the weight w at S produces complete primary balance.

The forces due to the ordinates to the curve C for the piston Q could be balanced by another weight, in phase with S when θ is zero, but rotating at double the angular speed of OS. If this weight is at the crank radius, its magnitude should be $\left(\frac{1}{2}\right)^2 \times \frac{a}{b}$ times w, since its angular speed is double that of OS, and also the maximum height of the curve C is $\frac{a}{b}$ times that of B, Sec. 249. The difficulty is that this latter weight has a component in the direction of OY which will not be balanced by the forces corresponding to the curve C for the piston R. It is not hard to see this latter point, for when θ becomes 90° the fast running weight should coincide with S to balance the forces of the piston R, whereas if it coincided with w when θ is zero it will be exactly opposite to it when $\theta = 90°$.

With such an arrangement as that shown there is, then, perfect primary balance but the secondary balance cannot be made at the same time. If the secondary balance weight coincides with w when θ is zero, then the unbalanced force in the direction OY will be a maximum, and if Q is exactly balanced the maximum unbalanced force in the direction OY is twice that corresponding to the ordinates to the curve C, or is $2 \times \frac{w}{g} \times \frac{a}{b} \times a\omega^2$. Owing to the difficult construction involved in putting in the secondary balance weight, the latter is not used, and then the maximum unbalanced force may readily be shown to be $\sqrt{2} \times \frac{w}{g} \times \frac{a}{b} \times a\omega^2$ pds.

The use of the curves like Fig. 186 will enable the reader to prove the correctness of the above results without difficulty.

In dealing with all engines of this type, no matter what the number or distribution of the cylinders, the primary and secondary revolving masses are always to be found, and by combining each of these separately for all the cylinders the primary and secondary disturbing forces may be found, and the former always balanced by a revolving weight on the crank, but the latter can be balanced only in some cases where it is possible to make the reciprocating masses balance one another. Thus, a six-cylinder engine of this type with cylinders at equal angles may be shown to be in perfect balance.

QUESTIONS ON CHAPTER XVI

1. What are the main causes of vibrations in rotating and moving parts of machinery? What is meant by balancing?

2. The chapter deals only with the balancing of forces due to the masses; why are not the fluid pressures considered?

3. If an engine ran at the same speed would there be any different arrangement for balancing whether the crank was rotated by a motor or operated by steam or gas? Why?

4. Two masses weighing 12 and 18 lb. at radii of 20 and 24 in. and inclined at 90° to one another revolve in the same plane. Find the position and size of the single balancing weight.

5. If the weights in question 4 revolve in planes 10 in. apart, find the weights in two other planes 15 and 12 in. outside the former weights, which will balance them.

6. Examine the case of V-type of engine similar to Fig. 191 with the cylinders at 60° and see if there are unbalanced forces and how much.

7. A gas engine 15 in. stroke has two flywheels with the crank between them, one being 18 in. and the other 24 in. from the crank. The equivalent rotating weight is 200 lb. at the crankpin, while the reciprocating weight is 250 lb. Find the weights required in the flywheels to balance all of these masses.

8. In a four-crank engine the cylinders are all equally spaced and the reciprocating weights for three of the engines are 300, 400 and 500 lb. Find the weight of the fourth set and the crank angles for balance.

APPENDIX A

Approximate Analytical Method of Computing the Acceleration of the Piston of an Engine

The graphical solution of the same problem is given in Sec. 236. Let Fig. 192 represent the engine and let θ be the crank angle reckoned from the head-end dead center, and further let x denote the displacement of the piston corresponding to the motion of the crank through angle θ. Taking ω to represent the angular velocity of the crank, and t the time required to pass through the angle θ, then $\theta = \omega t$.

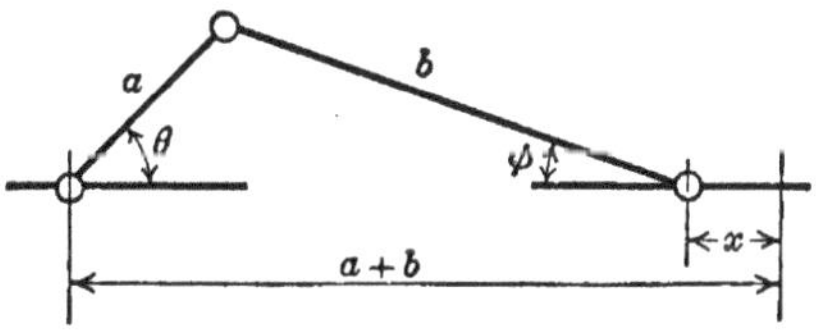

FIG. 192.

When the crank is in the position shown, the velocity of the piston is $\frac{dx}{dt}$ and its acceleration is $\frac{d^2x}{dt^2}$.

Examination of the figure shows that:

$$x = a + b - a \cos \theta - b \cos \phi$$

or

$$-x = a \cos \theta + b \cos \phi - (b + a).$$

Now $\cos \phi = \sqrt{1 - \sin^2 \phi} = \sqrt{1 - \frac{a^2}{b^2} \sin^2 \theta}$; since $\sin \phi = \frac{a}{b} \sin \theta$.

Further, since $\frac{a^2}{b^2} \sin^2 \theta$ is generally small compared with unity the value of $\sqrt{1 - \frac{a^2}{b^2} \sin^2 \theta}$ is equal $1 - \frac{a^2}{2b^2} \sin^2 \theta$ approximately. (It is in making this assumption that the approximation is introduced and for most cases the error is not serious.)

Thus

$$-x = a \cos \theta + b \left[1 - \frac{a^2}{2b^2} \sin^2 \theta\right] - (b + a)$$

$$\text{or} \qquad -x = a \cos \omega t - \frac{a^2}{2b} \sin^2 \omega t - a$$

$$\text{therefore} \quad -\frac{dx}{dt} = a\omega \sin \omega t - \frac{a^2}{b} \omega \sin \omega t \cos \omega t$$

$$= a\omega \sin \omega t - \frac{a^2}{2b} \omega \sin 2\omega t$$

$$\text{and} \qquad -\frac{d^2x}{dt^2} = a\omega^2 \cos \omega t - \frac{a^2}{b} \omega^2 \cos 2\omega t$$

$$= a\omega^2 \cos \theta - \frac{a^2}{b} \omega^2 \cos 2\theta.$$

Therefore the acceleration is

$$-f = a\omega^2 \cos \theta - \frac{a^2}{b} \omega^2 \cos 2\theta$$

$$= a\omega^2 \left[\cos \theta - \frac{a}{b} \cos 2\theta\right].$$

Since x is negative and the acceleration is also negative, the latter is toward the crankshaft, in the same sense as x.

The above expression will be found to be exactly correct at the two dead centers and nearly correct at other points. It shows that the acceleration curve for the piston is composed of two simple harmonic curves starting in phase, the latter of which has twice the frequency of the other, and an amplitude of $\frac{a}{b}$ times the former's value. This has been found to be the case in the curves plotted from the table at the end of Chapter XV and shown on Fig. 186, and the error due to the factor neglected is found very small in this case.

In the case shown in Fig. 186, ω is 55 radians per second and $a = \frac{3\frac{1}{2}}{12} = 0.292$ ft., so that the value of $a\omega^2 \cos \theta$ at crank angle $\theta = 0$ is $a\omega^2 = 0.292 \times (55)^2 = 882$ ft. per second per second, and this is the maximum height of the first curve. At the same angle $\theta = 0$ the value of $a\omega^2 \times \frac{a}{b} \cos 2\theta = 882 \times \frac{3\frac{1}{2}}{18} = 171$ ft. per second per second, which gives the maximum height of the curve of double frequency. These values are the same as those scaled from Fig. 186.

APPENDIX B

Experimental Method of Finding the Moment of Inertia of Any Body

For the convenience of those using this book the experimental method of finding the moment of inertia and radius of gyration of a body about its center of gravity is given herewith.

Suppose it is desired to find these quantities for the connecting rod shown in Fig. 193. Take the plane of the paper as the plane of motion of the rod. Balance the rod carefully across a knife edge placed parallel with the plane of motion of the rod, and the center of gravity G will be directly above the knife edge.

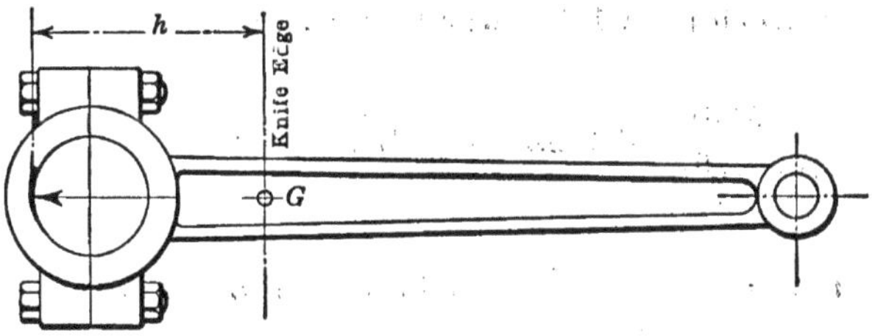

FIG. 193.—Inertia of rod.

Next secure a knife edge in a wall or other support so that its edge is exactly horizontal and hang the rod on it with the knife edge through one of the pin holes and let it swing freely like a pendulum. By means of a stop watch find exactly the time required to swing from one extreme position to the other; this can be most accurately found by taking the time required to do this say, 100 times. Let t sec. be the time for the complete swing.

Next measure the distance h feet from the knife edge to the center of gravity, and also weigh the rod and get its exact weight w lb.

Then it is shown in books on mechanics that

$$I = \frac{t^2}{\pi^2} \times wh - \frac{w}{32.2} \times h^2$$

in foot and pound units, gives the moment of inertia of the rod about its center of gravity.

As an example, an experiment was made on an automobile rod of 12-in. centers and weighing 2 lb. 4½ oz. or 2.281 lb. The crank and wristpins were respectively 2 in. and 11/16 in. diameter, and when placed sideways on a knife edge it was found to balance at a point 3.03 in. from the center of the crankpin. The rod was first hung on a knife edge projecting through the crankpin end, so that $h = 4.03$ in. or 0.336 ft., and it was found that it took 94⅗ sec. to make 200 swings; thus

$$t = \frac{94.6}{200} = 0.473 \text{ sec.}$$

$$\text{Then } I = \frac{(0.473)^2}{(3.1416)^2} \times 2.281 \times 0.336 - \frac{2.281}{32.2} \times (0.336)^2$$
$$= 0.00937 \text{ in foot and pound units.}$$

When suspended from the wristpin end, it is evident that $h =$ 9.315 in. or 0.776 ft., while t was found to be 0.539 sec. giving the value

$$I = \frac{(0.539)^2}{(3.1416)^2} \times 2.281 \times 0.776 - 0.0709 \times (0.776)^2$$
$$= 0.00940.$$

The average of these is 0.0094 which may be taken as the moment of inertia about the center of gravity. The square of the radius of gyration about the same point is

$$k^2 = \frac{I \times g}{w} = \frac{0.0094 \times 32.16}{2.281} = 0.132$$

or

$$k = 0.36 \text{ ft.}$$

INDEX

C

D

E

F

G

H

www.ingramcontent.com/pod-product-compliance
Ingram Content Group UK Ltd.
Pitfield, Milton Keynes, MK11 3LW, UK
UKHW032050260826
14628UKWH00008B/3288

9 789355 280619